Advances in Experimental Medicine and Biology

Volume 1136

More information about this series at http://www.springer.com/series/5584

Preface

I am pleased to introduce the book entitled *Hypoxia and Cancer Metastasis* in the Advances in Experimental Medicine and Biology series. The latest books on hypoxia and cancer, edited by Celeste Simon in 2010 and Giovanni Melillo in 2014, offered an excellent overview for scientists interested in the role of hypoxia in cancer biology. The current book aims to provide an update with a specific focus on hypoxia and metastasis. I am thankful that each contributor wholeheartedly welcomed the invitation to contribute chapters to provide a comprehensive (to-date) understanding of the role of hypoxia in metastasis. Each of the contributors has spent their career focused on understanding the role of hypoxia in human disease, in particular, cancer.

Metastasis is the leading cause of cancer death. Of all the processes involved in tumorigenesis, local invasion and the formation of metastases are the most clinically relevant because they result in deaths attributed to cancer. Intratumoral hypoxia is found in a majority of solid tumors and is associated with an increased risk of metastasis and cancer treatment failure. This has prompted the need for intensive investigation by cancer biologists, clinicians, and scientists, alike. Therefore, the purpose of this book is to provide a review of clinical and preclinical studies that explore the influence of hypoxia in the metastatic process. The first chapter provides an introduction and lays the groundwork for the book. Chapter 2 provides an up-to-date review on the clinical and preclinical methods used to quantify hypoxia in tumors. The chapters that follow examine the role of hypoxia in integral steps that are required for metastatic disease including the role of hypoxia in angiogenesis and lymphangiogenesis, secreted factors that influence homing and migration, and metabolic programming to enhance cell fitness and promote tumor heterogeneity. The closing chapters discuss the role of hypoxia in immune suppression and resistance to therapy. The final chapter discusses current clinical trials and emerging therapies aimed at targeting hypoxic cells.

The book is designed for researchers both new and heavily engaged in hypoxia research. The overall goal is to encourage the development of matched biomarkers and therapeutics for cancers in which hypoxia plays a detrimental role.

Baltimore, MD, USA Daniele M. Gilkes

Contents

1 Hypoxia Mediates Tumor Malignancy and Therapy Resistance

Weibo Luo and Yingfei Wang

Abstract

Hypoxia is a hallmark of the tumor microenvironment and contributes to tumor malignant phenotypes. Hypoxia-inducible factor (HIF) is a master regulator of intratumoral hypoxia and controls hypoxia-mediated pathological processes in tumors, including angiogenesis, metabolic reprogramming, epigenetic reprogramming, immune evasion, pH homeostasis, cell migration/invasion, stem cell pluripotency, and therapy resistance. In this book chapter, we reviewed the causes and types of intratumoral hypoxia, hypoxia detection methods, and the oncogenic role of HIF in tumorigenesis and chemo- and radio-therapy resistance.

Keywords

Angiogenesis · Cell motility · Epigenetics · HIF · Hypoxia · Hypoxia imaging · Metabolism · Stem cell · Therapy resistance · Tumorigenesis

Reduced O_2 availability, known as hypoxia, is a hallmark of the tumor microenvironment. This biological phenomenon was initially observed in lung carcinoma by Thomlinson and Gray in 1950s [146]. Accumulating studies have demonstrated that hypoxia exists in all solid tumors including breast, pancreas, brain, liver, lung, stomach, cervix, ovary, head-and-neck, prostate, bladder, kidney, skin, and colon tumors [6, 7, 14, 77, 118, 146, 151, 166], and that up to 50–60% of tumor regions exhibit much lower O_2 levels than their tissues of origin [152]. For example, the median partial pressure of O_2 (pO_2) is about 10 mmHg in breast tumors, which is significantly less than 65 mmHg in normal breast tissues [152]. While the presence of intratumoral hypoxia is independent of tumor size, grade, stage, or histology [106], hypoxia regulates many pathological processes to promote tumor malignancy and is significantly correlated with poor clinical outcome in cancer patients.

W. Luo (✉)
Department of Pathology, UT Southwestern Medical Center, Dallas, TX, USA

Department of Pharmacology, UT Southwestern Medical Center, Dallas, TX, USA
e-mail: Weibo.Luo@UTSouthwestern.edu

Y. Wang (✉)
Department of Pathology, UT Southwestern Medical Center, Dallas, TX, USA

Department of Neurology and Neurotherapeutics, UT Southwestern Medical Center, Dallas, TX, USA
e-mail: Yingfei.Wang@UTSouthwestern.edu

D. M. Gilkes (ed.), *Hypoxia and Cancer Metastasis*, Advances in Experimental Medicine and Biology 1136,
https://doi.org/10.1007/978-3-030-12734-3_1

1.1 Definition of Intratumoral Hypoxia

Intratumoral hypoxia is the result of an imbalance of reduced O_2 supply by abnormal tumor vessels and increased O_2 consumption by highly proliferative cancer cells. The tumor blood vessels are often occluded and leaky [19], resulting in transient perfusion-limited hypoxia within a tumor, also known as acute hypoxia. The time frame of acute hypoxia is variable in tumors, and it can last for several minutes to days [13]. Hypoxia also causes the significant morphological changes of endothelial cells in the tumor vessels, which functionally reduces O_2 diffusion to the surrounding tumor tissues [44]. As a result, the area within 70–100 μm away from the blood vessel is relatively oxygenated but oxygen diffusion is dramatically impaired beyond 100–150 μm distance to the blood vessel, leading to hypoxia or even anoxia in the necrotic regions. This type of intratumoral hypoxia has been termed 'diffusion-limited' hypoxia or chronic hypoxia. Chronic hypoxia exists from several hours to weeks in tumors [13]. Emerging studies showed that the pO_2 levels may fluctuate by cyclic hypoxia and reoxygenation within a tumor, which is defined as intermittent hypoxia [103]. Several studies have reported fluctuations of blood flow in human and murine tumors using laser Doppler and O_2 microelectrode approaches [61, 74, 115]. Intermittent hypoxia may robustly increase oxidative stress in tumors [94]. Martinive P et al. showed that intermittent hypoxia makes tumor cells and endothelial cells more resistant to apoptosis and radiotherapy [97]. Cancer cells exposed to intermittent hypoxia have been shown to be more aggressive and increase ability in tumor initiation and metastasis in xenograft mice [28, 123]. Overall, solid tumors suffer the temporal and spatial changes of pO_2, which lead to heterogeneous and gradient O_2 distribution (Fig. 1.1). Three types of hypoxia trigger the distinct intracellular signaling pathways in tumor cells by altering O_2 concentration and flux time duration and have a significant pathological role in tumorigenesis.

1.2 Detection of Intratumoral Hypoxia

As hypoxia drives tumorigenesis and also mediates resistance to chemo- and radio-therapy, the measurement of intratumoral hypoxia has great value for the basic cancer research and clinical prognosis/diagnosis in cancer patients. Many invasive and non-invasive approaches have been developed to detect and quantify the pO_2 levels within a tumor in experimental conditions as well

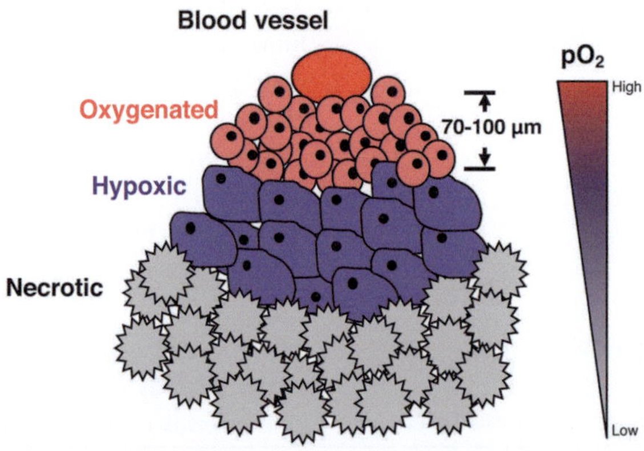

Fig. 1.1 **Heterogeneity of intratumoral hypoxia** Solid tumors display the gradient pO_2 levels. The tissues surrounding the blood vessel are relatively oxygenated, but O_2 diffusion is significantly impaired beyond 100 μm distance to the blood vessel, which leads to hypoxia or even anoxia in the necrotic regions

Table 1.1 Methods in detection of intratumoral hypoxia

	Application	References
Invasive methods		
Polarographic oxygen electrode	Direct method	[36]
Luminescence-based optical sensor	Direct method	[33]
Pimonidazole	Immunohistochemistry	[57]
HIF-1α	Immunohistochemistry	[125]
Carbonic anhydrase IX	Immunohistochemistry	[125]
18F-2-nitroimidazolpentafluoropropylacetamiede (EF5)	Immunohistochemistry	[57]
Non-invasive methods		
18F-fluoromisonidazole (18F-FMISO)	PET	[76]
2-deoxy-2-(18F)fluoro-D-glucose (18F-FDG)	PET	[138]
18F-fluoro-erythronitroimidazole (18F-FETNIM)	PET	[34]
18F-fluoro-azomycin-arabinoside (18F-FAZA)	PET	[114]
18F-2-nitroimidazolpentafluoropropylacetamiede (EF5)	PET	[168]
18F-EF3	PET	[92]
18F-FRP170	PET	[70]
18F-flortanidazole (HX4)	PET	[150]
[(68) Ga]-HP-DO3A-nitroimidazole	PET	[159]
Cu-ATSM	PET	[64]
^{131}I-IAZGP	PET	[120]
^{89}Zr-labeled cG250-F(ab')$_2$	PET	[63]
^{123}I-Iodoazomycin arabinoside	SPECT	[113]
^{99m}Tc-cyclam-2-nitroimidazole	Planar scintigraphy	[4]
^{68}Ga-metronidazole	MRI	[101]
Oxygen-enhanced MRI	MRI	[110]
Dynamic contrast-enhanced MRI	MRI	[43]
Blood oxygen level-dependent MRI	MRI	[110]
Tissue oxygen level dependent MRI	MRI	[8]

as the clinical settings (Table 1.1). The eppendorf polarographic needle electrode is the most direct method that utilizes the electrochemical reduction of O_2 at the cathode to measure pO_2 [36]. The electrode is sensitive and can rapidly detect pO_2 changes in tumors. However, this invasive approach measures local pO_2 levels surrounding the electrode, which cannot represent the entire tumor due to the heterogeneity of intratumoral hypoxia. The electrode itself consumes O_2 during measurement, leading to the overestimation of tumor hypoxia. Therefore, this method has limitations for clinical application.

Immunohistochemical methods have been also developed to detect intratumoral hypoxia in preclinical cancer models and biopsies from cancer patients (Table 1.1). Pimonidazole, which was originally developed as a hypoxic cell radiosensitizer for clinical use [108], binds to thiol groups from proteins, peptides, and amino acids specifically in viable hypoxic cancer cells. Pimonidazole-positive cells can be visualized by an antibody-based immunohistochemical assay under microscopy [57]. Likewise, hypoxia-inducible factor (HIF)-1α, a subunit of the hypoxia regulator HIF, and its target gene carbonic anhydrase IX are also frequently used as hypoxia markers to detect hypoxic regions in tumors by the immunohistochemical approach [125]. Therefore, immunohistochemical staining generates the high-resolution hypoxic regions at a single cell level within a tumor. Although immunohistochemical methods have been widely employed to quantify hypoxic regions in solid tumors, the disadvantage of these methods is that they display a static but not dynamic hypoxia status in tumors.

The non-invasive imaging methods including positron emission tomography (PET), single photon emission computed tomography

(SPECT), and magnetic resonance imaging (MRI) have been widely employed in clinical diagnosis. These approaches are considered as the practical, sensitive, and reproducible hypoxia imaging methods. A number of hypoxia PET radiotracers have been developed for the repetitive measurement of intratumoral hypoxia in the animal models and cancer patients (Table 1.1). The current hypoxia PET radiotracers are classified into two major families, 18F-labelled nitromidazole analogs and Copper(II)-labelled diacetyl-bis(N^4-methythiosemicarbzaone) (Cu-ATSM) analogs. 18F-fluoromisonidazole (18F-FMISO) is the first PET tracer for hypoxia detection in the clinical studies [76]. Like other 2-nitroimidazole analogs, 18F-FMISO is metabolized into reactive intermediates, and these intermediates are further reduced to the nitroradical anion that can bind to intracellular macromolecules in hypoxic cells. In contrast, 18F-FMISO intermediates are re-oxidized into the original compound under normoxia. As such, 18F-FMISO is irreversibly trapped and accumulated in viable hypoxic cells below 10 mmHg, which can be detected by a PET scanner for quantification of spatial and temporal pO_2 levels in many human cancers, including breast, brain, lung, head-and-neck, prostate, and kidney cancers [33, 65, 76, 119, 149]. However, 18F-FMISO has little clinical value in rectal and pancreatic cancers because of their non-specific accumulation in normoxic colon tissues or no accumulation in pancreatic tumors [122, 128].

Cu-ATSM is another class of hypoxia PET radiotracer that is used for clinical prognosis in cancer patients. Cu offers a series of radioisotopes (60Cu, 61Cu, 62Cu, 64Cu, and 67Cu) and 64Cu is often used for PET imaging due to its appropriate half-life (12.7 hr) and good imaging resolution [64]. 64Cu-ATSM is highly membrane permeable and requires less time to enter the cell, compared to 2-nitroimidazole analogs. After uptaken to the cell, Cu(II)-ATSM is reduced to Cu(I)-ATSM$^-$ by intracellular thiols. Reduced Cu(I)-ATSM$^-$ can be re-oxidized back to Cu(II)-ATSM by O_2 and then exits from the cell under normoxia. However, Cu(I)-ATSM$^-$ is dissociated under hypoxia and free Cu(I) is trapped within hypoxic cells, where the acidic environment facilitates destabilization of Cu(II)-ATSM and cellular Cu(I) trapping [40]. The application of the Cu-ATSM approach for prognostic evaluation has been studied in patients with head-and-neck cancer, cervical cancer and lung cancer [41, 51, 144].

18F-fluorodeoxyglucose (18F-FDG) is a D-glucose analog and has been extensively used as a hypoxia marker in preclinical and clinical cancer studies. The rationale of this radiotracer is that hypoxia increases the expression of the glucose transporters and glycolytic enzymes, and shifts glucose oxidation towards glycolysis in cancer cells [129]. Therefore, cancer cells rely on increased glucose uptake for their growth under hypoxia. 18F-FDG is accumulated in the cytosol after entering the cell through the glucose transporters because of lack of a 2-hydroxyl group and its phosphorylation by hexokinases, which blocks its degradation via the glycolytic pathway and prevents leave from cancer cells, respectively. 18F-FDG has a short half-life (110 min) and has been used for diagnosis in patients with colorectal cancer, breast cancer, head-and-neck cancer, kidney cancer, or lung cancer [52, 59, 60, 68, 107]. However, the cellular concentration of 18F-FDG is also very high in normal cells with high glucose consumption, such as normal cerebral cortex and basal ganglia in the brain, which generates a rather high background in the PET image [11].

MRI is another non-invasive imaging method that is increasingly being used to measure the temporal and spatial pO_2 levels in tumors because of its high spatial resolution and relatively low cost. It utilizes a large external magnetic field to image nuclei of atoms (such as ^1H, ^{19}F, or ^{13}C) in the body and pO_2 can be quantified in the presence of the contrast agent. Several functional MRI techniques, including blood oxygen level dependent (BOLD) MRI, tissue oxygen level dependent (TOLD) MRI, dynamic contrast enhanced (DCE) MRI, and oxygen enhanced (OE) MRI, have been developed to quantify pO_2 changes in tumors [8, 43, 110]. MRI has been applied to assess tumor hypoxia for prediction of outcome in the pre-clinical and clinical radiation studies [35]. Please see Chap. 2 by Dewhirst and colleagues for a review of the clinical and preclinical methods for quantifying the extent of hypoxia in human tumors.

1.3 HIF Is a Master Regulator in Response to Intratumoral Hypoxia

Numerous studies have demonstrated that intratumoral hypoxia promotes angiogenesis, glycolysis, cell invasiveness, cell survival, and immune evasion, leading to cancer progression and metastasis [131]. These cellular adaptive responses to intratumoral hypoxia are mainly mediated by a family of HIFs [130]. HIF is a heterodimeric transcription factor, consisting of α and β subunits [154]. So far, three HIF family members (HIF-1, HIF-2, and HIF-3) have been identified in mammals [53, 135, 147]. HIF-1α is ubiquitously expressed in most cell types, whereas HIF-2α expression is limited to certain cells, such as endothelial cells [147]. While a total of 19 distinct HIF-3α transcripts have been identified in the human genome database due to alternative mRNA splicing mechanisms, only 8 variants may encode HIF-3α protein [91]. HIF-1 and HIF-2 are the most well-studied members and mediate many common and unique biological responses to intratumoral hypoxia [132]. The functions of HIF-3 are diverse as different HIF-3α isoforms appear to be a gene activator or repressor through distinct mechanisms [99, 165].

HIF-α is regulated by the O_2-dependent ubiquitin-proteasome mechanism (Fig. 1.2) [91].

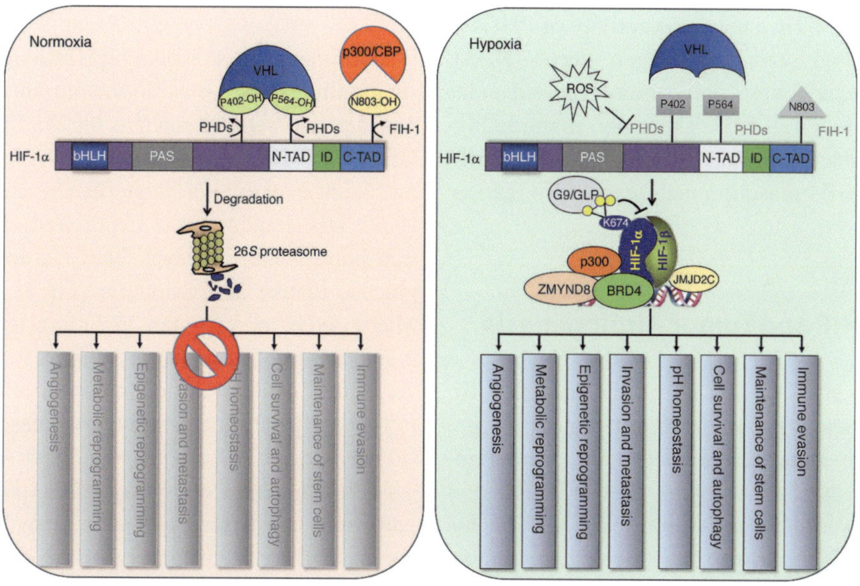

Fig. 1.2 Regulation of HIF transcriptional activity in cancer cells
HIF-α is prolyl hydroxylated by PHDs and targeted by the ubiquitin E3 ligase complex Cul2-ElonginB/C-VHL for proteasomal degradation under normoxia. HIF-α is also hydroxylated on the asparagine residue by FIH-1, which blocks the recruitment of the coactivators p300/CBP, leading to HIF inactivation. Under hypoxia, the enzymatic activity of PHDs and FIH-1 is impaired due to reduced O_2 and increased ROS. As a result, HIF-α escapes from proteasomal degradation and dimerizes with HIF-1β. The heterodimer binds to the hypoxia response element and cooperates with epigenetic regulators including p300, ZMYND8, BRD4, and JMJD2C to activate the target genes whose protein products mediate angiogenesis, metabolic reprogramming, epigenetic reprogramming, invasion/metastasis, pH homeostasis, cell survival, autophagy, maintenance of stem cells, and immune evasion. The methyltransferases G9a/GLP methylate lysine 674 on HIF-1α to inhibit HIF-1 transcriptional activity in glioblastoma cells

In well-oxygenated cells, HIF-α is hydroxylated on the proline (Pro) residues (e.g., Pro 402 and 564 on human HIF-1α) by a family of prolyl hydroxylases (PHDs) with O_2 as the substrate [66]. PHD2 is mainly responsible for prolyl hydroxylation of HIF-1α [16]. The Von Hippel-Lindau (VHL) protein recruits prolyl hydroxylated HIF-α to the ubiquitin E3 ligase complex Cullin2/Elongin-B/C, leading to HIF-α polyubiquitination and subsequent protein degradation in the 26*S* proteasome [98]. Under hypoxia, PHDs' enzymatic activity is robustly reduced and HIF-α fails to be prolyl hydroxylated, so that HIF-α protein escapes from proteasomal degradation and dimerizes with HIF-1β to enhance the transcription of hundreds of genes, whose protein products regulate angiogenesis, metabolism, epigenetics, pH homeostasis, stem cell pluripotency, cell survival, cell migration/invasion, immune evasion, and tumor growth and metastasis (Fig. 1.2). The enzymatic activity of PHDs is also inhibited by reactive oxygen species (ROS), whose production is significantly increased under hypoxia (Fig. 1.2) [23]. Therefore, multiple layers of molecular regulation have been shown to control HIF transcriptional activity in cancer cells.

1.4 HIF and Tumor Angiogenesis

Previous studies demonstrated that HIF-1α, HIF-2α, and HIF-1β knockout mice all are embryonic lethal due to defective vascularization, indicating an essential role of HIF-1 and HIF-2 in vascular remodeling [67, 78, 95]. This scenario has been also documented in tumors. HIF-1 and HIF-2 are the key regulators of hypoxia-induced angiogenesis in tumors [80]. It has been well-known that HIF-1 and HIF-2 directly induce the expression of many genes involved in multiple steps of angiogenesis, including *VEGFA*, *VEGFR2*, *PDGFB*, *bFGF*, *ANGPT1*, *ANGPT2*, *ANGPT3*, *ANGPT4*, *TEK*, *MMP-1*, *SDF-1*, and *CXCR4* [1, 2, 20, 47, 62, 71, 96, 109, 127, 137, 161]. Multiple cell types including cancer cells, endothelial cells, and regulatory T cells are involved in the hypoxia-induced release of angiogenic factors in tumors [47, 50, 96]. Thus, HIF-1 and HIF-2 actively participate in vascular permeability, endothelial cell proliferation, basement membrane degradation, sprouting, cell migration, and tube formation, leading to new blood vessel formation in tumors. New blood vessel formation is required to support rapidly proliferating cancer cells with O_2 and nutrients, thereby promoting tumor formation, development, and metastasis. Inhibition of angiogenesis by VEGF or VEGF receptor inhibitors impairs tumor growth in the experimental mouse models of human cancers as well as in cancer patients [124]. Please see Chap. 5 for an extensive review of hypoxic signalling in angiogenesis and lymphangiogenesis.

1.5 HIF and Metabolic Reprogramming in Tumors

Otto Warburg observed the high rate of glycolysis in tumor tissues even in the presence of oxygen, a phenomenon known as the Warburg effect [155]. HIF-1 represents one of the molecular mechanisms of the Warburg effect in renal cancer [129]. HIF-1 induces the expression of glucose transporters and glycolytic enzymes to enhance glucose uptake and oxidation (Fig. 1.3) [133]. Many glycolytic enzymes including hexokinase 2 and phosphofructokinase 1 appear to be oncogenes that mediate tumor initiation and growth [157, 163]. Pyruvate dehydrogenase kinase (PDK) 1 and PDK3 are also the direct HIF-1 target genes and induced by hypoxia to phosphorylate and inactivate pyruvate dehydrogenase complex, thereby shifting pyruvate metabolism towards lactate in cancer cells [72, 87]. Induction of lactate dehydrogenase A (LDHA) by HIF-1 further promotes glycolytic flux into lactate in cancer cells [133]. Lactate is secreted into the extracellular space by the monocarboxylate transporter (MCT) 4 [42]. High lactate levels in tumors are positively correlated with increased metastasis in head-and-neck cancer patients [18]. Interestingly, it was shown that extracellular lactate can be transported into oxygenated tumor cells by MCT1 and, as a fuel, supports the tricar-

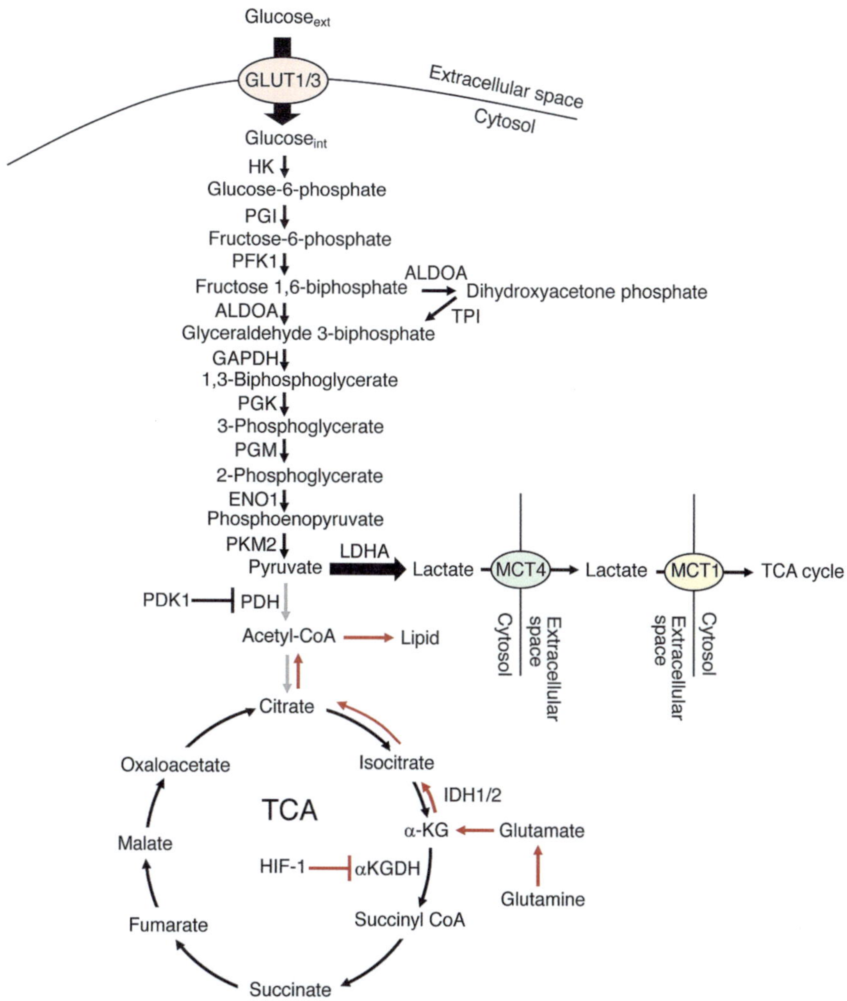

Fig. 1.3 **HIF-dependent metabolic reprogramming in cancer cells**
HIF directly induces the expression of the glucose transporters and glycolytic enzymes to upregulate glucose uptake and oxidation in cancer cells. PDK1 is also upregulated by HIF-1 in cancer cells and promotes the metabolic shift towards lactate production by inactivating PDH. The extracellular lactate is transported via MCT1 into the neighboring cancer cells as a fuel to support their survival. HIF also enhances reductive glutamine metabolism to sustain the TCA cycle and to increase lipogenesis in cancer cells

boxylic acid (TCA) cycle and respiration in these neighboring cancer cells [139]. These findings have been also observed in human patients with non-small-cell lung cancer [45]. Targeting MCT1 by the small molecule inhibitor blocks the growth of lung tumors in xenograft mice [116].

Recent studies uncovered the feedback role of the glycolytic enzymes in HIF-dependent glycolysis in multiple types of cancer cells. Pyruvate kinase M2 enhances the transcriptional activity of HIF-1 and HIF-2 to promote the expression of the glucose transporter GLUT1 and glycolytic enzymes LDHA and PDK1, thereby increasing glucose uptake and glycolysis in renal and cervical cancer cells [90]. Aldolase A (ALDOA) increases HIF-1α protein stability through lactate-mediated PHD inhibition to promote HIF transcriptional activity in lung cancer cells [27]. ALDOA activates HIF to increase MMP9 expression, thereby promoting lung cancer metastasis. In contrast, fructose-1,6-bisphosphatase 1 (FBP1) inhibits HIF-1 to decrease glycolysis in renal cancer cells [83]. The *FBP1* gene is lost in human clear cell renal cell carcinoma and loss of FBP1 promotes renal cancer progression by increasing HIF-dependent glycolysis. Therefore, the glycolytic enzymes regulate HIF to control glycolysis and tumor malignancy.

Glucose donates the carbon to generate acetyl-CoA, which enters the TCA cycle for the synthesis of ATP, intermediate metabolites, and amino acids. Activation of HIF leads to increased glycolysis but decreased glycolytic flux into the TCA cycle in cancer cells (Fig. 1.3) [72, 134]. Although glucose-derived carbon source is significantly reduced in hypoxic cells, HIF enhances glutaminolysis to sustain the TCA cycle and lipogenesis, which are necessary for cancer cell survival under hypoxia [102, 156]. Isocitrate dehydrogenase (IDH) 1/2 are responsible for the reductive carboxylation of glutamine-derived α-ketoglutarate for *de novo* lipogenesis under hypoxia, although it is unclear how hypoxia promotes the reductive activity of IDH1/2 in cancer cells. HIF-1 also increases the ubiquitin E3 ligase SIAH2-mediated ubiquitination and subsequent degradation of the E1 subunit of the α-ketoglutarate dehydrogenase complex, thereby facilitating reductive glutamine metabolism in hypoxic cells [142]. Therefore, activation of HIF reprograms the cellular metabolic pathways to supplement the intracellular metabolites that are necessary for cancer cell growth under hypoxia (Fig. 1.3). Chapter 5 contains further discussion of hypoxia and metabolic reprogramming in cancer.

1.6 HIF and Epigenetic Reprogramming in Tumors

Hypoxia mediates reprogramming of the chromatin landscape by inducing many HIF-dependent epigenetic regulators in cancer cells [91]. It has been reported that hypoxia induces the global increases in histone modifications including dimethyl lysine 4 of histone H3 (H3K4me2), trimethyl lysine 4 of histone H3 (H3K4me3), dimethyl lysine 9 of histone H3 (H3K9me2), dimethyl lysine 79 of histone H3 (H3K79me2), acetyl lysine 14 of histone H3 (H3K14ac), and dimethyl arginine 3 of histone H4 (H4R3me2) in hepatoma-derived Hepa 1-6 cells, but inhibits acetyl lysine 5 of histone H4 (H4K5ac), acetyl lysine 12 of histone H4 (H4K12ac), and acetyl lysine 5 of histone H2A (H2AK5ac) in breast cancer cells [69, 158]. H3K14ac and acetyl lysine 16 of histone H4 (H4K16ac) at the local hypoxia response element is also elevated by hypoxia in breast cancer cells [32]. Hypoxia upregulates many members of the histone demethylase Jumonji domain-containing protein (JMJD) family in cancer cells in a HIF-1- and/or HIF-2-dependent manner (Table 1.2). Although O_2 is required for the catalytic activity of the JMJD family members [136], several JMJD family members including JMJD1A, JMJD2B, and JMJD2C are reported to be active in hypoxic cells, which may be due to the compensatory mechanism by increased protein levels of these enzymes under hypoxia [49, 79, 89]. JMJD2C selectively demethylates trimethyl lysine 9 of histone H3 (H3K9me3) but not trimethyl lysine 36 of histone H3 (H3K36me3) at the hypoxia response element to increase HIF-1 binding and HIF-1 target gene expression in breast cancer cells (Fig. 1.2) [89]. JMJD2C is highly amplified in breast cancer and knockdown of JMJD2C inhibits triple-negative breast tumor growth and metastasis to lungs in mice [85, 89]. Likewise, JMJD1A (also named as KDM3A) is recruited to the promoter of the *SLC2A3* gene by HIF-1α and demethylates H3K9me2 to induce SLC2A3 expression in endothelial cells [104]. JMJD1A knockdown reduces colon cancer growth in mice [79]. It was also reported that JMJD2B demethylates H3K9me3 at the promoter of hypoxia-inducible genes *SLC2A1*, *UCA1*, *ELF3*, and *IFI6* to increase their expression in HCT116 cells [49]. In contrast, the demethylase activity of JMJD3 and JARID1A/B is impaired under hypoxia, leading to increased trimethyl K27 and trimethyl K4 of histone H3, respectively [26]. Histone deacetylase 3 is also induced by HIF-1 and decreases H3K4ac in breast cancer cells, which promotes H3K4me2 and H3K4me3 by recruiting the methyltransferase WDR5 [158]. Hypoxia induces the methyltransferase G9a at both transcriptional and post-translational levels in cancer cells, which is

Table 1.2 Hypoxia-induced JMJD family members in cancer cells

JMJD family members	Substrates	HIF regulation	References
KDM2A	H3K36me1/2	HIF-1	[5, 12, 160]
KDM2B	H3K36me1/2	HIF-1	[5, 12, 160]
JMJD1A	H3K9me1/2	HIF-1 and HIF-2	[5, 117, 162]
JMJD1B	H3K9me1/2	Indirect	[17, 117, 160]
JMJD1C	H3K9me1/2	Indirect	[73, 117, 160]
JMJD2A	H3K9me2/3; H3K36me2/3	Indirect	[5, 117]
JMJD2B	H3K9me2/3; H3K36me2/3	HIF-1 and HIF-2	[5, 162]
JMJD2C	H3K9me2/3; H3K36me2/3	HIF-1	[5, 117, 160]
JMJD2D	H3K9me2/3	Indirect	[5, 117]
UTX (KDM6A)	H3K27me2/3	ND	[5, 160]
JMJD3 (KDM6B)	H3K27me2/3	HIF-1 and HIF-2	[5, 81, 54]
JARID1A	H3K4me2/3	ND	[5, 160]
JARID1B	H3K4me1/2/3	HIF-1	[5, 160]
JARID1C	H3K4me2/3	ND	[5, 160]
JARID1D	H3K4me2/3	ND	[5, 160]
JARID2	No demethylase activity	ND	[84, 160]
JMJD6	H3R2me2 H4R3me2	HIF-1	[3, 24, 160]
PHF8	H3K4me3	HIF-1 and HIF-2	[93, 148, 160]

ND not determined

responsible for increased H3K9me2 under hypoxia [22, 29, 82]. Recent studies revealed that G9a and its paralog G9a-like protein directly methylate HIF-1α at lysine 674 to inhibit HIF-1 transcriptional activity and migration of glioblastoma cells (Fig. 1.2) [9]. The epigenetic reader ZMYND8 is upregulated in breast cancer cells by hypoxia in a HIF-1- and HIF-2-dependent manner [32]. ZMYND8 increases the recruitment of another epigenetic reader BRD4 to the hypoxia response element through its lysine acetylation by p300 and promotes the expression of the majority of HIF target genes in breast cancer (Fig. 1.2) [31, 32]. ZMYND8 mediates breast cancer progression and metastasis in mice, which requires HIF-1 and HIF-2 [32]. Together, mutual regulation of HIF and epigenetic regulators contributes to the malignant phenotype in human cancers.

1.7 Intratumoral Hypoxia and Therapy Resistance

Researchers in the early twentieth century uncovered a critical role of O_2 in radiosensitivity and subsequently discovered the existence of hypoxia in the tumor microenvironment [39]. Now it is well-known that hypoxia is a key factor that mediates resistance to radiotherapy in tumors [153]. O_2 is a potent radiosensitizer and oxidizes radiation-induced radicals to form more harmful molecules including ROS that can target DNA and enzymes to induce cell death. Radiosensitivity robustly decreases about threefold in cancer cells exposed to <10 mmHg compared to those under atmosphere conditions [121]. Many clinical trials have been conducted to increase O_2 levels in tumors by the use of normobaric or hyperbaric O_2 or the hypoxic cell radiosensitizer nitroimidazoles before or during irradiation to overcome radiotherapy resistance [100, 112]. However, O_2 modification has little impact on the clinical outcome in cancer patients [111].

Hypoxia confers resistance of cancer cells to radiation therapy by regulating angiogenesis/vasculogenesis, DNA damage response, energy metabolism, stem cell maintenance, and autophagy. Previous studies have shown that irradiation induces activation of HIF-1 by increasing ROS and HIF-1 is required for resistance to radiotherapy [56, 105, 164]. HIF-1-induced VEGFA and bFGF promote survival of endothelial cells to reduce radiosensitivity in tumors, and blockade of HIF-1 activity by its inhibitor YC-1 enhances irradiation-mediated antitumor efficacy in xenograft mice [105]. It was also shown that HIF-1 increases the recruitment of bone marrow-derived cells into glioblastoma in mice upon irradiation by inducing stromal cell-derived factor-1 and its receptor CXCR4, leading to vasculogenesis and tumor recurrence [75]. HIF-2α is highly expressed in $CD133^+$ lung cancer stem cells and may regulate radioresistance in lung cancer [141]. Several groups have reported that HIF-1-dependent autophagy contributes to hypoxia-induced resistance to radiation therapy in human osteosarcoma and breast cancer cells [46, 58].

HIF-independent mechanisms also contribute to hypoxia-mediated radioresistance. Severe hypoxia causes DNA replication stress, which leads to activation of ataxia telangiectasia and Rad3-related (ATR) and ataxia telangiectasia mutated (ATM) kinases and their downstream signaling cascades in cancer cells [15, 55]. Although it was reported that activation of the ATM downstream kinase Chk2 in hypoxic tumor regions increases sensitivity to irradiation in a mouse glioma model [140], another study showed that Chk1/2 are activated in $CD133^+$ human glioma stem cells upon irradiation and contribute to radioresistance [10]. Recent studies demonstrated that the ribonucleotide reductase subunit RRM2B is induced by severe hypoxia, which requires p53 but is independent of HIF-1, and maintains ongoing DNA replication to prevent accumulation of DNA damage in cancer cells under severe hypoxia [48]. RRM2B is a p53 target gene and also phosphorylated by ATM at serine 72 to increase its protein stability [25, 145]. Thus, RRM2B may regulate genome integrity leading to hypoxia-induced radioresistance.

Chemotherapy is currently one of the standard treatments for cancer patients. Hypoxia enhances chemotherapy resistance in human cancers. Diffusion of chemotherapeutic drugs in the hypoxic regions is significantly reduced due to abnormal blood vessels. Hypoxia induces the expression of the multidrug resistance proteins MDR1 and MRP1 in a HIF-1-dependent manner leading to decreased cellular uptake of the chemotherapeutic drugs and increased drug resistance of cancer cells [30, 37, 86, 143, 167]. On the other hand, chemotherapeutic drugs such as doxorubicin upregulate HIF-1α [21], and thus provide a feedback loop amplifying drug resistance responses in tumors. Previous studies also showed that chemotherapeutic drugs increase the enrichment of breast cancer stem cells, where HIF contributes to their survival [88, 126]. Pharmacological inhibition of HIF reduces the resistance of breast cancer cells to chemotherapeutic drugs paclitaxel or gemcitabine in vitro and in mice [126]. Graham and colleagues discuss the role of hypoxia in therapy resistance in Chap. 9.

1.8 Conclusions

Intratumoral hypoxia plays a crucial role in tumorigenesis and mediates resistance to chemo- and radio-therapy in both HIF-dependent and -independent manner. HIF mediates various pathological processes in response to hypoxia in tumors and promotes the initiation and progression of human cancers. Targeting HIF (also see Chap. 10) may reduce tumorigenesis and overcome therapy resistance. Indeed, a HIF-2 inhibitor, which blocks heterodimerization of HIF-2α and HIF-1β, is currently undergoing a clinical trial for the treatment of clear cell renal cell carcinoma [38]. Therefore, hypoxia/HIF is a valuable factor for prognosis and treatment of human cancers.

Acknowledgments Work in authors' laboratories was supported by grants from NIH (R00CA168746), CPRIT (RR140036), Susan G. Komen® (CCR16376227), Welch Foundation (I-1903-20160319), and American Cancer Society and UTSW Simmons Cancer Center (ACS-IRG-02-196) to W.L.; and NIH (R00NS078049, R35GM124693), Welch Foundation (I-1939-20170325), CPRIT (RP170671), Darrell K Royal Research Fund, TIBIR pilot grant, and UTSW startup funds to Y.W.. W.L. is a CPRIT Scholar in Cancer Research.

References

1. Abbott BD, Buckalew AR (2000) Placental defects in ARNT-knockout conceptus correlate with localized decreases in VEGF-R2, Ang-1, and Tie-2. Dev Dyn 219(4):526–538. https://doi.org/10.1002/1097-0177(2000)9999:9999<::AID-DVDY1080>3.0.CO;2-N
2. Abdulmalek K, Ashur F, Ezer N, Ye F, Magder S, Hussain SN (2001) Differential expression of Tie-2 receptors and angiopoietins in response to in vivo hypoxia in rats. Am J Physiol Lung Cell Mol Physiol 281(3):L582–L590. https://doi.org/10.1152/ajplung.2001.281.3.L582
3. Alahari S, Post M, Caniggia I (2015) Jumonji domain containing protein 6: a novel oxygen sensor in the human placenta. Endocrinology 156(8):3012–3025. https://doi.org/10.1210/en.2015-1262
4. Ali MS, Kong FL, Rollo A, Mendez R, Kohanim S, Smith DL, Yang DJ (2012) Development of (99m)Tc-N4-NIM for molecular imaging of tumor hypoxia. J Biomed Biotechnol 2012:828139. https://doi.org/10.1155/2012/828139
5. Allis CD, Berger SL, Cote J, Dent S, Jenuwien T, Kouzarides T, Pillus L, Reinberg D, Shi Y, Shiekhattar R, Shilatifard A, Workman J, Zhang Y (2007) New nomenclature for chromatin-modifying enzymes. Cell 131(4):633–636. https://doi.org/10.1016/j.cell.2007.10.039
6. Arteel GE, Thurman RG, Yates JM, Raleigh JA (1995) Evidence that hypoxia markers detect oxygen gradients in liver: pimonidazole and retrograde perfusion of rat liver. Br J Cancer 72(4):889–895
7. Awwad HK, El Merzabani MM, El Badawy S, Ezzat S, Akoush H, Abd El Moneim H, Saiid A, Soliman O, Khafagy M, Burgers MV (1980) Misonidazole in the preoperative and radical radiotherapy of bladder cancer. Cancer Clin Trials 3(3):275–280
8. Bane O, Besa C, Wagner M, Oesingmann N, Zhu H, Fiel MI, Taouli B (2016) Feasibility and reproducibility of BOLD and TOLD measurements in the liver with oxygen and carbogen gas challenge in healthy volunteers and patients with hepatocellular carcinoma. J Magn Reson Imaging 43(4):866–876. https://doi.org/10.1002/jmri.25051
9. Bao L, Chen Y, Lai HT, Wu SY, Wang JE, Hatanpaa KJ, Raisanen JM, Fontenot M, Lega B, Chiang CM, Semenza GL, Wang Y, Luo W (2018) Methylation of hypoxia-inducible factor (HIF)-1alpha by G9a/GLP inhibits HIF-1 transcriptional activity and cell migration. Nucleic Acids Res 46(13):6576–6591. https://doi.org/10.1093/nar/gky449
10. Bao S, Wu Q, McLendon RE, Hao Y, Shi Q, Hjelmeland AB, Dewhirst MW, Bigner DD, Rich JN (2006) Glioma stem cells promote radioresistance by preferential activation of the DNA damage response. Nature 444(7120):756–760. https://doi.org/10.1038/nature05236
11. Basu S, Alavi A (2009) Molecular imaging (PET) of brain tumors. Neuroimaging Clin N Am 19(4):625–646. https://doi.org/10.1016/j.nic.2009.08.012
12. Batie M, Druker J, D'Ignazio L, Rocha S (2017) KDM2 family members are regulated by HIF-1 in hypoxia. Cells 6(1):8. https://doi.org/10.3390/cells6010008
13. Bayer C, Vaupel P (2012) Acute versus chronic hypoxia in tumors: controversial data concerning time frames and biological consequences. Strahlenther Onkol 188(7):616–627. https://doi.org/10.1007/s00066-012-0085-4
14. Becker A, Hansgen G, Bloching M, Weigel C, Lautenschlager C, Dunst J (1998) Oxygenation of squamous cell carcinoma of the head and neck: comparison of primary tumors, neck node metastases, and normal tissue. Int J Radiat Oncol Biol Phys 42(1):35–41
15. Bencokova Z, Kaufmann MR, Pires IM, Lecane PS, Giaccia AJ, Hammond EM (2009) ATM activation and signaling under hypoxic conditions. Mol Cell Biol 29(2):526–537. https://doi.org/10.1128/MCB.01301-08
16. Berra E, Benizri E, Ginouves A, Volmat V, Roux D, Pouyssegur J (2003) HIF prolyl-hydroxylase 2 is the key oxygen sensor setting low steady-state levels of HIF-1alpha in normoxia. EMBO J 22(16):4082–4090
17. Brauchle M, Yao Z, Arora R, Thigale S, Clay I, Inverardi B, Fletcher J, Taslimi P, Acker MG, Gerrits B, Voshol J, Bauer A, Schubeler D, Bouwmeester T, Ruffner H (2013) Protein complex interactor analysis and differential activity of KDM3 subfamily members towards H3K9 methylation. PLoS One 8(4):e60549. https://doi.org/10.1371/journal.pone.0060549
18. Brizel DM, Schroeder T, Scher RL, Walenta S, Clough RW, Dewhirst MW, Mueller-Klieser W (2001) Elevated tumor lactate concentrations predict for an increased risk of metastases in head-and-neck cancer. Int J Radiat Oncol Biol Phys 51(2):349–353
19. Brown JM, Wilson WR (2004) Exploiting tumour hypoxia in cancer treatment. Nat Rev Cancer 4(6):437–447. https://doi.org/10.1038/nrc1367
20. Calvani M, Rapisarda A, Uranchimeg B, Shoemaker RH, Melillo G (2006) Hypoxic induction of an HIF-1alpha-dependent bFGF autocrine loop drives angiogenesis in human endothelial cells. Blood 107(7):2705–2712. https://doi.org/10.1182/blood-2005-09-3541
21. Cao Y, Eble JM, Moon E, Yuan H, Weitzel DH, Landon CD, Nien CY, Hanna G, Rich JN, Provenzale JM, Dewhirst MW (2013) Tumor cells upregulate normoxic HIF-1alpha in response to doxorubicin. Cancer Res 73(20):6230–6242. https://doi.org/10.1158/0008-5472.CAN-12-1345
22. Casciello F, Al-Ejeh F, Kelly G, Brennan DJ, Ngiow SF, Young A, Stoll T, Windloch K, Hill MM, Smyth

MJ, Gannon F, Lee JS (2017) G9a drives hypoxia-mediated gene repression for breast cancer cell survival and tumorigenesis. Proc Natl Acad Sci U S A 114(27):7077–7082. https://doi.org/10.1073/pnas.1618706114
23. Chandel NS, Maltepe E, Goldwasser E, Mathieu CE, Simon MC, Schumacker PT (1998) Mitochondrial reactive oxygen species trigger hypoxia-induced transcription. Proc Natl Acad Sci U S A 95(20):11715–11720
24. Chang B, Chen Y, Zhao Y, Bruick RK (2007) JMJD6 is a histone arginine demethylase. Science 318(5849):444–447. https://doi.org/10.1126/science.1145801
25. Chang L, Zhou B, Hu S, Guo R, Liu X, Jones SN, Yen Y (2008) ATM-mediated serine 72 phosphorylation stabilizes ribonucleotide reductase small subunit p53R2 protein against MDM2 to DNA damage. Proc Natl Acad Sci U S A 105(47):18519–18524. https://doi.org/10.1073/pnas.0803313105
26. Chang S, Park B, Choi K, Moon Y, Lee HY, Park H (2016) Hypoxic reprograming of H3K27me3 and H3K4me3 at the INK4A locus. FEBS Lett 590(19):3407–3415. https://doi.org/10.1002/1873-3468.12375
27. Chang YC, Chan YC, Chang WM, Lin YF, Yang CJ, Su CY, Huang MS, Wu ATH, Hsiao M (2017) Feedback regulation of ALDOA activates the HIF-1alpha/MMP9 axis to promote lung cancer progression. Cancer Lett 403:28–36. https://doi.org/10.1016/j.canlet.2017.06.001
28. Chen A, Sceneay J, Godde N, Kinwel T, Ham S, Thompson EW, Humbert PO, Moller A (2018) Intermittent hypoxia induces a metastatic phenotype in breast cancer. Oncogene 37(31):4214–4225. https://doi.org/10.1038/s41388-018-0259-3
29. Chen H, Yan Y, Davidson TL, Shinkai Y, Costa M (2006) Hypoxic stress induces dimethylated histone H3 lysine 9 through histone methyltransferase G9a in mammalian cells. Cancer Res 66(18):9009–9016. https://doi.org/10.1158/0008-5472.CAN-06-0101
30. Chen J, Ding Z, Peng Y, Pan F, Li J, Zou L, Zhang Y, Liang H (2014) HIF-1alpha inhibition reverses multidrug resistance in colon cancer cells via downregulation of MDR1/P-glycoprotein. PLoS One 9(6):e98882. https://doi.org/10.1371/journal.pone.0098882
31. Chen Y, Wang Y, Luo W (2018) ZMYND8 is a primary HIF coactivator that mediates breast cancer progression. Mol Cell Oncol 5(4):e1479619. https://doi.org/10.1080/23723556.2018.1479619
32. Chen Y, Zhang B, Bao L, Jin L, Yang M, Peng Y, Kumar A, Wang JE, Wang C, Zou X, Xing C, Wang Y, Luo W (2018) ZMYND8 acetylation mediates HIF-dependent breast cancer progression and metastasis. J Clin Invest 128(5):1937–1955. https://doi.org/10.1172/JCI95089
33. Cheng J, Lei L, Xu J, Sun Y, Zhang Y, Wang X, Pan L, Shao Z, Zhang Y, Liu G (2013) 18F-fluoromisonidazole PET/CT: a potential tool for predicting primary endocrine therapy resistance in breast cancer. J Nucl Med 54(3):333–340. https://doi.org/10.2967/jnumed.112.111963
34. Chung JK, Chang YS, Lee YJ, Kim YJ, Jeong JM, Lee DS, Jang JJ, Lee MC (1999) The effect of tumor size on F-18-labeled fluorodeoxyglucose and fluoroerythronitroimidazole uptake in a murine sarcoma model. Ann Nucl Med 13(5):303–308
35. Colliez F, Gallez B, Jordan BF (2017) Assessing tumor oxygenation for predicting outcome in radiation oncology: a review of studies correlating tumor hypoxic status and outcome in the preclinical and clinical settings. Front Oncol 7:10. https://doi.org/10.3389/fonc.2017.00010
36. Collingridge DR, Young WK, Vojnovic B, Wardman P, Lynch EM, Hill SA, Chaplin DJ (1997) Measurement of tumor oxygenation: a comparison between polarographic needle electrodes and a time-resolved luminescence-based optical sensor. Radiat Res 147(3):329–334
37. Comerford KM, Wallace TJ, Karhausen J, Louis NA, Montalto MC, Colgan SP (2002) Hypoxia-inducible factor-1-dependent regulation of the multidrug resistance (MDR1) gene. Cancer Res 62(12):3387–3394
38. Courtney KD, Infante JR, Lam ET, Figlin RA, Rini BI, Brugarolas J, Zojwalla NJ, Lowe AM, Wang K, Wallace EM, Josey JA, Choueiri TK (2018) Phase I Dose-Escalation Trial of PT2385, a first-in-class hypoxia-inducible factor-2alpha antagonist in patients with previously treated advanced clear cell renal cell carcinoma. J Clin Oncol 36(9):867–874. https://doi.org/10.1200/JCO.2017.74.2627
39. Crabtree HG, Cramer W (1933) The action of radium on cancer cells II. Some factors determining the susceptibility of cancer cells to radium. Proc R Soc B 113(782):238–250. https://doi.org/10.1098/rspb.1933.0044
40. Dearling JL, Packard AB (2010) Some thoughts on the mechanism of cellular trapping of Cu(II)-ATSM. Nucl Med Biol 37(3):237–243. https://doi.org/10.1016/j.nucmedbio.2009.11.004
41. Dehdashti F, Grigsby PW, Mintun MA, Lewis JS, Siegel BA, Welch MJ (2003) Assessing tumor hypoxia in cervical cancer by positron emission tomography with 60Cu-ATSM: relationship to therapeutic response-a preliminary report. Int J Radiat Oncol Biol Phys 55(5):1233–1238
42. Dimmer KS, Friedrich B, Lang F, Deitmer JW, Broer S (2000) The low-affinity monocarboxylate transporter MCT4 is adapted to the export of lactate in highly glycolytic cells. Biochem J 350(Pt 1):219–227
43. Drew PJ, Chatterjee S, Turnbull LW, Read J, Carleton PJ, Fox JN, Monson JR, Kerin MJ (1999) Dynamic contrast enhanced magnetic resonance imaging of the breast is superior to triple assessment for the preoperative detection of multifocal breast cancer. Ann Surg Oncol 6(6):599–603
44. Dudley AC (2012) Tumor endothelial cells. Cold Spring Harb Perspect Med 2(3):a006536. https://doi.org/10.1101/cshperspect.a006536

45. Faubert B, Li KY, Cai L, Hensley CT, Kim J, Zacharias LG, Yang C, Do QN, Doucette S, Burguete D, Li H, Huet G, Yuan Q, Wigal T, Butt Y, Ni M, Torrealba J, Oliver D, Lenkinski RE, Malloy CR, Wachsmann JW, Young JD, Kernstine K, DeBerardinis RJ (2017) Lactate metabolism in human lung tumors. Cell 171(2):358–371 e359. https://doi.org/10.1016/j.cell.2017.09.019
46. Feng H, Wang J, Chen W, Shan B, Guo Y, Xu J, Wang L, Guo P, Zhang Y (2016) Hypoxia-induced autophagy as an additional mechanism in human osteosarcoma radioresistance. J Bone Oncol 5(2):67–73. https://doi.org/10.1016/j.jbo.2016.03.001
47. Forsythe JA, Jiang BH, Iyer NV, Agani F, Leung SW, Koos RD, Semenza GL (1996) Activation of vascular endothelial growth factor gene transcription by hypoxia-inducible factor 1. Mol Cell Biol 16(9):4604–4613
48. Foskolou IP, Jorgensen C, Leszczynska KB, Olcina MM, Tarhonskaya H, Haisma B, D'Angiolella V, Myers WK, Domene C, Flashman E, Hammond EM (2017) Ribonucleotide reductase requires subunit switching in hypoxia to maintain DNA replication. Mol Cell 66(2):206–220 e209. https://doi.org/10.1016/j.molcel.2017.03.005
49. Fu L, Chen L, Yang J, Ye T, Chen Y, Fang J (2012) HIF-1alpha-induced histone demethylase JMJD2B contributes to the malignant phenotype of colorectal cancer cells via an epigenetic mechanism. Carcinogenesis 33(9):1664–1673. https://doi.org/10.1093/carcin/bgs217
50. Giatromanolaki A, Bates GJ, Koukourakis MI, Sivridis E, Gatter KC, Harris AL, Banham AH (2008) The presence of tumor-infiltrating FOXP3+ lymphocytes correlates with intratumoral angiogenesis in endometrial cancer. Gynecol Oncol 110(2):216–221. https://doi.org/10.1016/j.ygyno.2008.04.021
51. Grassi I, Nanni C, Cicoria G, Blasi C, Bunkheila F, Lopci E, Colletti PM, Rubello D, Fanti S (2014) Usefulness of 64Cu-ATSM in head and neck cancer: a preliminary prospective study. Clin Nucl Med 39(1):e59–e63. https://doi.org/10.1097/RLU.0b013e3182a756f0
52. Graves EE, Hicks RJ, Binns D, Bressel M, Le QT, Peters L, Young RJ, Rischin D (2016) Quantitative and qualitative analysis of [(18)F]FDG and [(18)F]FAZA positron emission tomography of head and neck cancers and associations with HPV status and treatment outcome. Eur J Nucl Med Mol Imaging 43(4):617–625. https://doi.org/10.1007/s00259-015-3247-7
53. Gu YZ, Moran SM, Hogenesch JB, Wartman L, Bradfield CA (1998) Molecular characterization and chromosomal localization of a third alpha-class hypoxia inducible factor subunit, HIF3alpha. Gene Expr 7(3):205–213
54. Guo X, Tian Z, Wang X, Pan S, Huang W, Shen Y, Gui Y, Duan X, Cai Z (2015) Regulation of histone demethylase KDM6B by hypoxia-inducible factor-2alpha. Acta Biochim Biophys Sin (Shanghai) 47(2):106–113. https://doi.org/10.1093/abbs/gmu122
55. Hammond EM, Denko NC, Dorie MJ, Abraham RT, Giaccia AJ (2002) Hypoxia links ATR and p53 through replication arrest. Mol Cell Biol 22(6):1834–1843
56. Harada H, Inoue M, Itasaka S, Hirota K, Morinibu A, Shinomiya K, Zeng L, Ou G, Zhu Y, Yoshimura M, McKenna WG, Muschel RJ, Hiraoka M (2012) Cancer cells that survive radiation therapy acquire HIF-1 activity and translocate towards tumour blood vessels. Nat Commun 3:783. https://doi.org/10.1038/ncomms1786
57. He F, Deng X, Wen B, Liu Y, Sun X, Xing L, Minami A, Huang Y, Chen Q, Zanzonico PB, Ling CC, Li GC (2008) Noninvasive molecular imaging of hypoxia in human xenografts: comparing hypoxia-induced gene expression with endogenous and exogenous hypoxia markers. Cancer Res 68(20):8597–8606. https://doi.org/10.1158/0008-5472.CAN-08-0677
58. He WS, Dai XF, Jin M, Liu CW, Rent JH (2012) Hypoxia-induced autophagy confers resistance of breast cancer cells to ionizing radiation. Oncol Res 20(5–6):251–258
59. Heiden BT, Chen G, Hermann M, Brown RKJ, Orringer MB, Lin J, Chang AC, Carrott PW, Lynch WR, Zhao L, Beer DG, Reddy RM (2018) 18F-FDG PET intensity correlates with a hypoxic gene signature and other oncogenic abnormalities in operable non-small cell lung cancer. PLoS One 13(7):e0199970. https://doi.org/10.1371/journal.pone.0199970
60. Heijmen L, de Geus-Oei LF, de Wilt JH, Visvikis D, Hatt M, Visser EP, Bussink J, Punt CJ, Oyen WJ, van Laarhoven HW (2012) Reproducibility of functional volume and activity concentration in 18F-FDG PET/CT of liver metastases in colorectal cancer. Eur J Nucl Med Mol Imaging 39(12):1858–1867. https://doi.org/10.1007/s00259-012-2233-6
61. Hill SA, Chaplin DJ (1996) Detection of microregional fluctuations in erythrocyte flow using laser Doppler microprobes. Adv Exp Med Biol 388:367–371
62. Hitchon C, Wong K, Ma G, Reed J, Lyttle D, El-Gabalawy H (2002) Hypoxia-induced production of stromal cell-derived factor 1 (CXCL12) and vascular endothelial growth factor by synovial fibroblasts. Arthritis Rheum 46(10):2587–2597. https://doi.org/10.1002/art.10520
63. Hoeben BA, Kaanders JH, Franssen GM, Troost EG, Rijken PF, Oosterwijk E, van Dongen GA, Oyen WJ, Boerman OC, Bussink J (2010) PET of hypoxia with 89Zr-labeled cG250-F(ab')2 in head and neck tumors. J Nucl Med 51(7):1076–1083. https://doi.org/10.2967/jnumed.109.073189
64. Holland JP, Lewis JS, Dehdashti F (2009) Assessing tumor hypoxia by positron emission tomography with Cu-ATSM. Q J Nucl Med Mol Imaging 53(2):193–200

65. Hugonnet F, Fournier L, Medioni J, Smadja C, Hindie E, Huchet V, Itti E, Cuenod CA, Chatellier G, Oudard S, Faraggi M, Hypoxia in Renal Cancer Multicenter G (2011) Metastatic renal cell carcinoma: relationship between initial metastasis hypoxia, change after 1 month's sunitinib, and therapeutic response: an 18F-fluoromisonidazole PET/CT study. J Nucl Med 52(7):1048–1055. https://doi.org/10.2967/jnumed.110.084517
66. Ivan M, Kondo K, Yang H, Kim W, Valiando J, Ohh M, Salic A, Asara JM, Lane WS, Kaelin WG Jr (2001) HIFalpha targeted for VHL-mediated destruction by proline hydroxylation: implications for O_2 sensing. Science 292(5516):464–468. https://doi.org/10.1126/science.1059817
67. Iyer NV, Kotch LE, Agani F, Leung SW, Laughner E, Wenger RH, Gassmann M, Gearhart JD, Lawler AM, Yu AY, Semenza GL (1998) Cellular and developmental control of O_2 homeostasis by hypoxia-inducible factor 1 alpha. Genes Dev 12(2):149–162
68. Jeong YJ, Jung JW, Cho YY, Park SH, Oh HK, Kang S (2017) Correlation of hypoxia inducible transcription factor in breast cancer and SUVmax of F-18 FDG PET/CT. Nucl Med Rev Cent East Eur 20(1):32–38. https://doi.org/10.5603/NMR.a2016.0043
69. Johnson AB, Denko N, Barton MC (2008) Hypoxia induces a novel signature of chromatin modifications and global repression of transcription. Mutat Res 640(1–2):174–179. https://doi.org/10.1016/j.mrfmmm.2008.01.001
70. Kaneta T, Takai Y, Iwata R, Hakamatsuka T, Yasuda H, Nakayama K, Ishikawa Y, Watanuki S, Furumoto S, Funaki Y, Nakata E, Jingu K, Tsujitani M, Ito M, Fukuda H, Takahashi S, Yamada S (2007) Initial evaluation of dynamic human imaging using 18F-FRP170 as a new PET tracer for imaging hypoxia. Ann Nucl Med 21(2):101–107
71. Kelly BD, Hackett SF, Hirota K, Oshima Y, Cai Z, Berg-Dixon S, Rowan A, Yan Z, Campochiaro PA, Semenza GL (2003) Cell type-specific regulation of angiogenic growth factor gene expression and induction of angiogenesis in nonischemic tissue by a constitutively active form of hypoxia-inducible factor 1. Circ Res 93(11):1074–1081. https://doi.org/10.1161/01.RES.0000102937.50486.1B
72. Kim JW, Tchernyshyov I, Semenza GL, Dang CV (2006) HIF-1-mediated expression of pyruvate dehydrogenase kinase: a metabolic switch required for cellular adaptation to hypoxia. Cell Metab 3(3):177–185. https://doi.org/10.1016/j.cmet.2006.02.002
73. Kim SM, Kim JY, Choe NW, Cho IH, Kim JR, Kim DW, Seol JE, Lee SE, Kook H, Nam KI, Kook H, Bhak YY, Seo SB (2010) Regulation of mouse steroidogenesis by WHISTLE and JMJD1C through histone methylation balance. Nucleic Acids Res 38(19):6389–6403. https://doi.org/10.1093/nar/gkq491
74. Kimura H, Braun RD, Ong ET, Hsu R, Secomb TW, Papahadjopoulos D, Hong K, Dewhirst MW (1996) Fluctuations in red cell flux in tumor microvessels can lead to transient hypoxia and reoxygenation in tumor parenchyma. Cancer Res 56(23):5522–5528
75. Kioi M, Vogel H, Schultz G, Hoffman RM, Harsh GR, Brown JM (2010) Inhibition of vasculogenesis, but not angiogenesis, prevents the recurrence of glioblastoma after irradiation in mice. J Clin Invest 120(3):694–705. https://doi.org/10.1172/JCI40283
76. Koh WJ, Bergman KS, Rasey JS, Peterson LM, Evans ML, Graham MM, Grierson JR, Lindsley KL, Lewellen TK, Krohn KA et al (1995) Evaluation of oxygenation status during fractionated radiotherapy in human nonsmall cell lung cancers using [F-18] fluoromisonidazole positron emission tomography. Int J Radiat Oncol Biol Phys 33(2):391–398. https://doi.org/10.1016/0360-3016(95)00170-4
77. Koong AC, Mehta VK, Le QT, Fisher GA, Terris DJ, Brown JM, Bastidas AJ, Vierra M (2000) Pancreatic tumors show high levels of hypoxia. Int J Radiat Oncol Biol Phys 48(4):919–922
78. Kotch LE, Iyer NV, Laughner E, Semenza GL (1999) Defective vascularization of HIF-1alpha-null embryos is not associated with VEGF deficiency but with mesenchymal cell death. Dev Biol 209(2):254–267. https://doi.org/10.1006/dbio.1999.9253
79. Krieg AJ, Rankin EB, Chan D, Razorenova O, Fernandez S, Giaccia AJ (2010) Regulation of the histone demethylase JMJD1A by hypoxia-inducible factor 1 alpha enhances hypoxic gene expression and tumor growth. Mol Cell Biol 30(1):344–353. https://doi.org/10.1128/MCB.00444-09
80. Krock BL, Skuli N, Simon MC (2011) Hypoxia-induced angiogenesis: good and evil. Genes Cancer 2(12):1117–1133. https://doi.org/10.1177/1947601911423654
81. Lee HY, Choi K, Oh H, Park YK, Park H (2014) HIF-1-dependent induction of Jumonji domain-containing protein (JMJD) 3 under hypoxic conditions. Mol Cells 37(1):43–50. https://doi.org/10.14348/molcells.2014.2250
82. Lee JS, Kim Y, Kim IS, Kim B, Choi HJ, Lee JM, Shin HJ, Kim JH, Kim JY, Seo SB, Lee H, Binda O, Gozani O, Semenza GL, Kim M, Kim KI, Hwang D, Baek SH (2010) Negative regulation of hypoxic responses via induced Reptin methylation. Mol Cell 39(1):71–85. https://doi.org/10.1016/j.molcel.2010.06.008
83. Li B, Qiu B, Lee DS, Walton ZE, Ochocki JD, Mathew LK, Mancuso A, Gade TP, Keith B, Nissim I, Simon MC (2014) Fructose-1,6-bisphosphatase opposes renal carcinoma progression. Nature 513(7517):251–255. https://doi.org/10.1038/nature13557
84. Li G, Margueron R, Ku M, Chambon P, Bernstein BE, Reinberg D (2010) Jarid2 and PRC2, partners in regulating gene expression. Genes Dev 24(4):368–380. https://doi.org/10.1101/gad.1886410
85. Liu G, Bollig-Fischer A, Kreike B, van de Vijver MJ, Abrams J, Ethier SP, Yang ZQ (2009) Genomic amplification and oncogenic properties of the

GASC1 histone demethylase gene in breast cancer. Oncogene 28(50):4491–4500
86. Liu L, Ning X, Sun L, Zhang H, Shi Y, Guo C, Han S, Liu J, Sun S, Han Z, Wu K, Fan D (2008) Hypoxia-inducible factor-1 alpha contributes to hypoxia-induced chemoresistance in gastric cancer. Cancer Sci 99(1):121–128. https://doi.org/10.1111/j.1349-7006.2007.00643.x
87. Lu CW, Lin SC, Chen KF, Lai YY, Tsai SJ (2008) Induction of pyruvate dehydrogenase kinase-3 by hypoxia-inducible factor-1 promotes metabolic switch and drug resistance. J Biol Chem 283(42):28106–28114. https://doi.org/10.1074/jbc.M803508200
88. Lu H, Chen I, Shimoda LA, Park Y, Zhang C, Tran L, Zhang H, Semenza GL (2017) Chemotherapy-induced Ca^{2+} release stimulates breast cancer stem cell enrichment. Cell Rep 18(8):1946–1957. https://doi.org/10.1016/j.celrep.2017.02.001
89. Luo W, Chang R, Zhong J, Pandey A, Semenza GL (2012) Histone demethylase JMJD2C is a coactivator for hypoxia-inducible factor 1 that is required for breast cancer progression. Proc Natl Acad Sci U S A 109(49):E3367–E3376. https://doi.org/10.1073/pnas.1217394109
90. Luo W, Hu H, Chang R, Zhong J, Knabel M, O'Meally R, Cole RN, Pandey A, Semenza GL (2011) Pyruvate kinase M2 is a PHD3-stimulated coactivator for hypoxia-inducible factor 1. Cell 145(5):732–744. https://doi.org/10.1016/j.cell.2011.03.054
91. Luo W, Wang Y (2018) Epigenetic regulators: multifunctional proteins modulating hypoxia-inducible factor-alpha protein stability and activity. Cell Mol Life Sci 75(6):1043–1056. https://doi.org/10.1007/s00018-017-2684-9
92. Mahy P, De Bast M, Gillart J, Labar D, Gregoire V (2006) Detection of tumour hypoxia: comparison between EF5 adducts and [18F]EF3 uptake on an individual mouse tumour basis. Eur J Nucl Med Mol Imaging 33(5):553–556. https://doi.org/10.1007/s00259-005-0049-3
93. Maina PK, Shao P, Jia X, Liu Q, Umesalma S, Marin M, Long D Jr, Concepcion-Roman S, Qi HH (2017) Histone demethylase PHF8 regulates hypoxia signaling through HIF1alpha and H3K4me3. Biochim Biophys Acta Gene Regul Mech 1860(9):1002–1012. https://doi.org/10.1016/j.bbagrm.2017.07.005
94. Malec V, Gottschald OR, Li S, Rose F, Seeger W, Hanze J (2010) HIF-1 alpha signaling is augmented during intermittent hypoxia by induction of the Nrf2 pathway in NOX1-expressing adenocarcinoma A549 cells. Free Radic Biol Med 48(12):1626–1635. https://doi.org/10.1016/j.freeradbiomed.2010.03.008
95. Maltepe E, Schmidt JV, Baunoch D, Bradfield CA, Simon MC (1997) Abnormal angiogenesis and responses to glucose and oxygen deprivation in mice lacking the protein ARNT. Nature 386(6623):403–407. https://doi.org/10.1038/386403a0
96. Manalo DJ, Rowan A, Lavoie T, Natarajan L, Kelly BD, Ye SQ, Garcia JG, Semenza GL (2005) Transcriptional regulation of vascular endothelial cell responses to hypoxia by HIF-1. Blood 105(2):659–669. https://doi.org/10.1182/blood-2004-07-2958
97. Martinive P, Defresne F, Quaghebeur E, Daneau G, Crokart N, Gregoire V, Gallez B, Dessy C, Feron O (2009) Impact of cyclic hypoxia on HIF-1alpha regulation in endothelial cells--new insights for anti-tumor treatments. FEBS J 276(2):509–518. https://doi.org/10.1111/j.1742-4658.2008.06798.x
98. Maxwell PH, Wiesener MS, Chang GW, Clifford SC, Vaux EC, Cockman ME, Wykoff CC, Pugh CW, Maher ER, Ratcliffe PJ (1999) The tumour suppressor protein VHL targets hypoxia-inducible factors for oxygen-dependent proteolysis. Nature 399(6733):271–275
99. Maynard MA, Evans AJ, Shi W, Kim WY, Liu FF, Ohh M (2007) Dominant-negative HIF-3 alpha 4 suppresses VHL-null renal cell carcinoma progression. Cell Cycle 6(22):2810–2816. https://doi.org/10.4161/cc.6.22.4947
100. Mendenhall WM, Morris CG, Amdur RJ, Mendenhall NP, Siemann DW (2005) Radiotherapy alone or combined with carbogen breathing for squamous cell carcinoma of the head and neck: a prospective, randomized trial. Cancer 104(2):332–337. https://doi.org/10.1002/cncr.21146
101. Mendichovszky I, Jackson A (2011) Imaging hypoxia in gliomas. Br J Radiol 84 Spec No 2(2):S145–S158. https://doi.org/10.1259/bjr/82292521
102. Metallo CM, Gameiro PA, Bell EL, Mattaini KR, Yang J, Hiller K, Jewell CM, Johnson ZR, Irvine DJ, Guarente L, Kelleher JK, Vander Heiden MG, Iliopoulos O, Stephanopoulos G (2011) Reductive glutamine metabolism by IDH1 mediates lipogenesis under hypoxia. Nature 481(7381):380–384. https://doi.org/10.1038/nature10602
103. Michiels C, Tellier C, Feron O (2016) Cycling hypoxia: a key feature of the tumor microenvironment. Biochim Biophys Acta 1866(1):76–86. https://doi.org/10.1016/j.bbcan.2016.06.004
104. Mimura I, Nangaku M, Kanki Y, Tsutsumi S, Inoue T, Kohro T, Yamamoto S, Fujita T, Shimamura T, Suehiro J, Taguchi A, Kobayashi M, Tanimura K, Inagaki T, Tanaka T, Hamakubo T, Sakai J, Aburatani H, Kodama T, Wada Y (2012) Dynamic change of chromatin conformation in response to hypoxia enhances the expression of GLUT3 (SLC2A3) by cooperative interaction of hypoxia-inducible factor 1 and KDM3A. Mol Cell Biol 32(15):3018–3032. https://doi.org/10.1128/MCB.06643-11
105. Moeller BJ, Cao Y, Li CY, Dewhirst MW (2004) Radiation activates HIF-1 to regulate vascular radiosensitivity in tumors: role of reoxygenation, free radicals, and stress granules. Cancer Cell 5(5):429–441
106. Moulder JE, Rockwell S (1984) Hypoxic fractions of solid tumors: experimental techniques, methods of analysis, and a survey of existing data. Int J Radiat Oncol Biol Phys 10(5):695–712

107. Nakaigawa N, Kondo K, Ueno D, Namura K, Makiyama K, Kobayashi K, Shioi K, Ikeda I, Kishida T, Kaneta T, Minamimoto R, Tateishi U, Inoue T, Yao M (2017) The acceleration of glucose accumulation in renal cell carcinoma assessed by FDG PET/CT demonstrated acquisition of resistance to tyrosine kinase inhibitor therapy. BMC Cancer 17(1):39. https://doi.org/10.1186/s12885-016-3044-0
108. Newman HF, Bleehen NM, Workman P (1986) A phase I study of the combination of two hypoxic cell radiosensitizers, Ro 03-8799 and SR-2508: toxicity and pharmacokinetics. Int J Radiat Oncol Biol Phys 12(7):1113–1116
109. Nilsson I, Shibuya M, Wennstrom S (2004) Differential activation of vascular genes by hypoxia in primary endothelial cells. Exp Cell Res 299(2):476–485. https://doi.org/10.1016/j.yexcr.2004.06.005
110. O'Connor J, Robinson S, Waterton J (2018) Imaging tumour hypoxia with oxygen-enhanced MRI and BOLD MRI. Br J Radiol 92:20180642. https://doi.org/10.1259/bjr.20180642
111. Overgaard J (2007) Hypoxic radiosensitization: adored and ignored. J Clin Oncol 25(26):4066–4074. https://doi.org/10.1200/JCO.2007.12.7878
112. Overgaard J, Hansen HS, Overgaard M, Bastholt L, Berthelsen A, Specht L, Lindelov B, Jorgensen K (1998) A randomized double-blind phase III study of nimorazole as a hypoxic radiosensitizer of primary radiotherapy in supraglottic larynx and pharynx carcinoma. Results of the Danish Head and Neck Cancer Study (DAHANCA) Protocol 5-85. Radiother Oncol 46(2):135–146
113. Parliament MB, Chapman JD, Urtasun RC, McEwan AJ, Golberg L, Mercer JR, Mannan RH, Wiebe LI (1992) Non-invasive assessment of human tumour hypoxia with 123I-iodoazomycin arabinoside: preliminary report of a clinical study. Br J Cancer 65(1):90–95
114. Piert M, Machulla HJ, Picchio M, Reischl G, Ziegler S, Kumar P, Wester HJ, Beck R, McEwan AJ, Wiebe LI, Schwaiger M (2005) Hypoxia-specific tumor imaging with 18F-fluoroazomycin arabinoside. J Nucl Med 46(1):106–113
115. Pigott KH, Hill SA, Chaplin DJ, Saunders MI (1996) Microregional fluctuations in perfusion within human tumours detected using laser Doppler flowmetry. Radiother Oncol 40(1):45–50
116. Polanski R, Hodgkinson CL, Fusi A, Nonaka D, Priest L, Kelly P, Trapani F, Bishop PW, White A, Critchlow SE, Smith PD, Blackhall F, Dive C, Morrow CJ (2014) Activity of the monocarboxylate transporter 1 inhibitor AZD3965 in small cell lung cancer. Clin Cancer Res 20(4):926–937. https://doi.org/10.1158/1078-0432.CCR-13-2270
117. Pollard PJ, Loenarz C, Mole DR, McDonough MA, Gleadle JM, Schofield CJ, Ratcliffe PJ (2008) Regulation of Jumonji-domain-containing histone demethylases by hypoxia-inducible factor (HIF)-1alpha. Biochem J 416(3):387–394. https://doi.org/10.1042/BJ20081238
118. Rampling R, Cruickshank G, Lewis AD, Fitzsimmons SA, Workman P (1994) Direct measurement of pO_2 distribution and bioreductive enzymes in human malignant brain tumors. Int J Radiat Oncol Biol Phys 29(3):427–431
119. Rasey JS, Koh WJ, Evans ML, Peterson LM, Lewellen TK, Graham MM, Krohn KA (1996) Quantifying regional hypoxia in human tumors with positron emission tomography of [18F]fluoromisonidazole: a pretherapy study of 37 patients. Int J Radiat Oncol Biol Phys 36(2):417–428
120. Riedl CC, Brader P, Zanzonico PB, Chun YS, Woo Y, Singh P, Carlin S, Wen B, Ling CC, Hricak H, Fong Y (2008) Imaging hypoxia in orthotopic rat liver tumors with iodine 124-labeled iodoazomycin galactopyranoside PET. Radiology 248(2):561–570. https://doi.org/10.1148/radiol.2482071421
121. Rockwell S, Dobrucki IT, Kim EY, Marrison ST, Vu VT (2009) Hypoxia and radiation therapy: past history, ongoing research, and future promise. Curr Mol Med 9(4):442–458
122. Roels S, Slagmolen P, Nuyts J, Lee JA, Loeckx D, Maes F, Stroobants S, Penninckx F, Haustermans K (2008) Biological image-guided radiotherapy in rectal cancer: is there a role for FMISO or FLT, next to FDG? Acta Oncol 47(7):1237–1248. https://doi.org/10.1080/02841860802256434
123. Rofstad EK, Gaustad JV, Egeland TA, Mathiesen B, Galappathi K (2010) Tumors exposed to acute cyclic hypoxic stress show enhanced angiogenesis, perfusion and metastatic dissemination. Int J Cancer 127(7):1535–1546. https://doi.org/10.1002/ijc.25176
124. Roskoski R Jr (2017) Vascular endothelial growth factor (VEGF) and VEGF receptor inhibitors in the treatment of renal cell carcinomas. Pharmacol Res 120:116–132. https://doi.org/10.1016/j.phrs.2017.03.010
125. Russell J, Carlin S, Burke SA, Wen B, Yang KM, Ling CC (2009) Immunohistochemical detection of changes in tumor hypoxia. Int J Radiat Oncol Biol Phys 73(4):1177–1186. https://doi.org/10.1016/j.ijrobp.2008.12.004
126. Samanta D, Gilkes DM, Chaturvedi P, Xiang L, Semenza GL (2014) Hypoxia-inducible factors are required for chemotherapy resistance of breast cancer stem cells. Proc Natl Acad Sci U S A 111(50):E5429–E5438. https://doi.org/10.1073/pnas.1421438111
127. Schito L, Rey S, Tafani M, Zhang H, Wong CC, Russo A, Russo MA, Semenza GL (2012) Hypoxia-inducible factor 1-dependent expression of platelet-derived growth factor B promotes lymphatic metastasis of hypoxic breast cancer cells. Proc Natl Acad Sci U S A 109(40):E2707–E2716. https://doi.org/10.1073/pnas.1214019109
128. Segard T, Robins PD, Yusoff IF, Ee H, Morandeau L, Campbell EM, Francis RJ (2013) Detection of

hypoxia with 18F-fluoromisonidazole (18F-FMISO) PET/CT in suspected or proven pancreatic cancer. Clin Nucl Med 38(1):1–6. https://doi.org/10.1097/RLU.0b013e3182708777
129. Semenza GL (2007) HIF-1 mediates the Warburg effect in clear cell renal carcinoma. J Bioenerg Biomembr 39(3):231–234
130. Semenza GL (2007) Life with oxygen. Science 318(5847):62–64
131. Semenza GL (2010) Defining the role of hypoxia-inducible factor 1 in cancer biology and therapeutics. Oncogene 29(5):625–634
132. Semenza GL (2012) Hypoxia-inducible factors in physiology and medicine. Cell 148(3):399–408
133. Semenza GL, Jiang BH, Leung SW, Passantino R, Concordet JP, Maire P, Giallongo A (1996) Hypoxia response elements in the aldolase A, enolase 1, and lactate dehydrogenase A gene promoters contain essential binding sites for hypoxia-inducible factor 1. J Biol Chem 271(51):32529–32537
134. Semenza GL, Roth PH, Fang HM, Wang GL (1994) Transcriptional regulation of genes encoding glycolytic enzymes by hypoxia-inducible factor 1. J Biol Chem 269(38):23757–23763
135. Semenza GL, Wang GL (1992) A nuclear factor induced by hypoxia via de novo protein synthesis binds to the human erythropoietin gene enhancer at a site required for transcriptional activation. Mol Cell Biol 12(12):5447–5454
136. Shi Y, Whetstine JR (2007) Dynamic regulation of histone lysine methylation by demethylases. Mol Cell 25(1):1–14
137. Shin DH, Dier U, Melendez JA, Hempel N (2015) Regulation of MMP-1 expression in response to hypoxia is dependent on the intracellular redox status of metastatic bladder cancer cells. Biochim Biophys Acta 1852(12):2593–2602. https://doi.org/10.1016/j.bbadis.2015.09.001
138. Som P, Atkins HL, Bandoypadhyay D, Fowler JS, MacGregor RR, Matsui K, Oster ZH, Sacker DF, Shiue CY, Turner H, Wan CN, Wolf AP, Zabinski SV (1980) A fluorinated glucose analog, 2-fluoro-2-deoxy-D-glucose (F-18): nontoxic tracer for rapid tumor detection. J Nucl Med 21(7):670–675
139. Sonveaux P, Vegran F, Schroeder T, Wergin MC, Verrax J, Rabbani ZN, De Saedeleer CJ, Kennedy KM, Diepart C, Jordan BF, Kelley MJ, Gallez B, Wahl ML, Feron O, Dewhirst MW (2008) Targeting lactate-fueled respiration selectively kills hypoxic tumor cells in mice. J Clin Invest 118(12):3930–3942. https://doi.org/10.1172/JCI36843
140. Squatrito M, Brennan CW, Helmy K, Huse JT, Petrini JH, Holland EC (2010) Loss of ATM/Chk2/p53 pathway components accelerates tumor development and contributes to radiation resistance in gliomas. Cancer Cell 18(6):619–629. https://doi.org/10.1016/j.ccr.2010.10.034
141. Sun JC, He F, Yi W, Wan MH, Li R, Wei X, Wu R, Niu DL (2015) High expression of HIF-2alpha and its anti-radiotherapy effect in lung cancer stem cells. Genet Mol Res 14(4):18110–18120. https://doi.org/10.4238/2015.December.22.37
142. Sun RC, Denko NC (2014) Hypoxic regulation of glutamine metabolism through HIF1 and SIAH2 supports lipid synthesis that is necessary for tumor growth. Cell Metab 19(2):285–292. https://doi.org/10.1016/j.cmet.2013.11.022
143. Sun Y, Guan Z, Liang L, Cheng Y, Zhou J, Li J, Xu Y (2016) HIF-1alpha/MDR1 pathway confers chemoresistance to cisplatin in bladder cancer. Oncol Rep 35(3):1549–1556. https://doi.org/10.3892/or.2015.4536
144. Takahashi N, Fujibayashi Y, Yonekura Y, Welch MJ, Waki A, Tsuchida T, Sadato N, Sugimoto K, Itoh H (2000) Evaluation of 62Cu labeled diacetyl-bis(N4-methylthiosemicarbazone) as a hypoxic tissue tracer in patients with lung cancer. Ann Nucl Med 14(5):323–328
145. Tanaka H, Arakawa H, Yamaguchi T, Shiraishi K, Fukuda S, Matsui K, Takei Y, Nakamura Y (2000) A ribonucleotide reductase gene involved in a p53-dependent cell-cycle checkpoint for DNA damage. Nature 404(6773):42–49. https://doi.org/10.1038/35003506
146. Thomlinson RH, Gray LH (1955) The histological structure of some human lung cancers and the possible implications for radiotherapy. Br J Cancer 9(4):539–549
147. Tian H, McKnight SL, Russell DW (1997) Endothelial PAS domain protein 1 (EPAS1), a transcription factor selectively expressed in endothelial cells. Genes Dev 11(1):72–82
148. Tong D, Liu Q, Liu G, Yuan W, Wang L, Guo Y, Lan W, Zhang D, Dong S, Wang Y, Xiao H, Mu J, Mao C, Wong J, Jiang J (2016) The HIF/PHF8/AR axis promotes prostate cancer progression. Oncogenesis 5(12):e283. https://doi.org/10.1038/oncsis.2016.74
149. Valk PE, Mathis CA, Prados MD, Gilbert JC, Budinger TF (1992) Hypoxia in human gliomas: demonstration by PET with fluorine-18-fluoromisonidazole. J Nucl Med 33(12):2133–2137
150. van Loon J, Janssen MH, Ollers M, Aerts HJ, Dubois L, Hochstenbag M, Dingemans AM, Lalisang R, Brans B, Windhorst B, van Dongen GA, Kolb H, Zhang J, De Ruysscher D, Lambin P (2010) PET imaging of hypoxia using [18F]HX4: a phase I trial. Eur J Nucl Med Mol Imaging 37(9):1663–1668. https://doi.org/10.1007/s00259-010-1437-x
151. Vaupel P, Briest S, Hockel M (2002) Hypoxia in breast cancer: pathogenesis, characterization and biological/therapeutic implications. Wien Med Wochenschr 152(13–14):334–342
152. Vaupel P, Mayer A, Höckel M (2004) Tumor hypoxia and malignant progression. Methods Enzymol 381:335–354
153. Vaupel P, Thews O, Hoeckel M (2001) Treatment resistance of solid tumors: role of hypoxia and anemia. Med Oncol 18(4):243–259. https://doi.org/10.1385/MO:18:4:243

154. Wang GL, Jiang BH, Rue EA, Semenza GL (1995) Hypoxia-inducible factor 1 is a basic-helix-loop-helix-PAS heterodimer regulated by cellular O_2 tension. Proc Natl Acad Sci U S A 92(12):5510–5514
155. Warburg O (1956) On the origin of cancer cells. Science 123(3191):309–314
156. Wise DR, Ward PS, Shay JE, Cross JR, Gruber JJ, Sachdeva UM, Platt JM, DeMatteo RG, Simon MC, Thompson CB (2011) Hypoxia promotes isocitrate dehydrogenase-dependent carboxylation of alpha-ketoglutarate to citrate to support cell growth and viability. Proc Natl Acad Sci U S A 108(49):19611–19616. https://doi.org/10.1073/pnas.1117773108
157. Wolf A, Agnihotri S, Micallef J, Mukherjee J, Sabha N, Cairns R, Hawkins C, Guha A (2011) Hexokinase 2 is a key mediator of aerobic glycolysis and promotes tumor growth in human glioblastoma multiforme. J Exp Med 208(2):313–326. https://doi.org/10.1084/jem.20101470
158. Wu MZ, Tsai YP, Yang MH, Huang CH, Chang SY, Chang CC, Teng SC, Wu KJ (2011) Interplay between HDAC3 and WDR5 is essential for hypoxia-induced epithelial-mesenchymal transition. Mol Cell 43(5):811–822. https://doi.org/10.1016/j.molcel.2011.07.012
159. Wu Y, Hao G, Ramezani S, Saha D, Zhao D, Sun X, Sherry AD (2015) [(68) Ga]-HP-DO3A-nitroimidazole: a promising agent for PET detection of tumor hypoxia. Contrast Media Mol Imaging 10(6):465–472. https://doi.org/10.1002/cmmi.1649
160. Xia X, Lemieux ME, Li W, Carroll JS, Brown M, Liu XS, Kung AL (2009) Integrative analysis of HIF binding and transactivation reveals its role in maintaining histone methylation homeostasis. Proc Natl Acad Sci U S A 106(11):4260–4265. https://doi.org/10.1073/pnas.0810067106
161. Yamakawa M, Liu LX, Date T, Belanger AJ, Vincent KA, Akita GY, Kuriyama T, Cheng SH, Gregory RJ, Jiang C (2003) Hypoxia-inducible factor-1 mediates activation of cultured vascular endothelial cells by inducing multiple angiogenic factors. Circ Res 93(7):664–673. https://doi.org/10.1161/01.RES.0000093984.48643.D7
162. Yang J, Ledaki I, Turley H, Gatter KC, Montero JC, Li JL, Harris AL (2009) Role of hypoxia-inducible factors in epigenetic regulation via histone demethylases. Ann N Y Acad Sci 1177:185–197. https://doi.org/10.1111/j.1749-6632.2009.05027.x
163. Yi W, Clark PM, Mason DE, Keenan MC, Hill C, Goddard WA 3rd, Peters EC, Driggers EM, Hsieh-Wilson LC (2012) Phosphofructokinase 1 glycosylation regulates cell growth and metabolism. Science 337(6097):975–980. https://doi.org/10.1126/science.1222278
164. Zhang M, Qiu Q, Li Z, Sachdeva M, Min H, Cardona DM, DeLaney TF, Han T, Ma Y, Luo L, Ilkayeva OR, Lui K, Nichols AG, Newgard CB, Kastan MB, Rathmell JC, Dewhirst MW, Kirsch DG (2015) HIF-1 alpha regulates the response of primary sarcomas to radiation therapy through a cell autonomous mechanism. Radiat Res 183(6):594–609. https://doi.org/10.1667/RR14016.1
165. Zhang P, Yao Q, Lu L, Li Y, Chen PJ, Duan C (2014) Hypoxia-inducible factor 3 is an oxygen-dependent transcription activator and regulates a distinct transcriptional response to hypoxia. Cell Rep 6(6):1110–1121. https://doi.org/10.1016/j.celrep.2014.02.011
166. Zhong H, De Marzo AM, Laughner E, Lim M, Hilton DA, Zagzag D, Buechler P, Isaacs WB, Semenza GL, Simons JW (1999) Overexpression of hypoxia-inducible factor 1alpha in common human cancers and their metastases. Cancer Res 59(22):5830–5835
167. Zhu H, Chen XP, Luo SF, Guan J, Zhang WG, Zhang BX (2005) Involvement of hypoxia-inducible factor-1-alpha in multidrug resistance induced by hypoxia in HepG2 cells. J Exp Clin Cancer Res 24(4):565–574
168. Ziemer LS, Evans SM, Kachur AV, Shuman AL, Cardi CA, Jenkins WT, Karp JS, Alavi A, Dolbier WR Jr, Koch CJ (2003) Noninvasive imaging of tumor hypoxia in rats using the 2-nitroimidazole 18F-EF5. Eur J Nucl Med Mol Imaging 30(2):259–266. https://doi.org/10.1007/s00259-002-1037-5

Clinical and Pre-clinical Methods for Quantifying Tumor Hypoxia

2

Ashlyn G. Rickard, Gregory M. Palmer, and Mark W. Dewhirst

Abstract

Hypoxia, a prevalent characteristic of most solid malignant tumors, contributes to diminished therapeutic responses and more aggressive phenotypes. The term hypoxia has two definitions. One definition would be a physiologic state where the oxygen partial pressure is below the normal physiologic range. For most normal tissues, the normal physiologic range is between 10 and 20 mmHg. Hypoxic regions develop when there is an imbalance between oxygen supply and demand. The impact of hypoxia on cancer therapeutics is significant: hypoxic tissue is 3× less radiosensitive than normoxic tissue, the impaired blood flow found in hypoxic tumor regions influences chemotherapy delivery, and the immune system is dependent on oxygen for functionality. Despite the clinical implications of hypoxia, there is not a universal, ideal method for quantifying hypoxia, particularly cycling hypoxia because of its complexity and heterogeneity across tumor types and individuals. Most standard imaging techniques can be modified and applied to measuring hypoxia and quantifying its effects; however, the benefits and challenges of each imaging modality makes imaging hypoxia case-dependent. In this chapter, a comprehensive overview of the preclinical and clinical methods for quantifying hypoxia is presented along with the advantages and disadvantages of each.

A. G. Rickard · G. M. Palmer · M. W. Dewhirst (✉)
Department of Radiation Oncology, Duke University, Durham, NC, USA
e-mail: mark.dewhirst@duke.edu

Keywords

Hypoxia · Quantification · Cycling hypoxia · MRI/EPR · Optical imaging · PET

2.1 Introduction

Hypoxia, a prevalent characteristic of most solid malignant tumors, contributes to diminished therapeutic responses and more aggressive phenotypes [1, 2]. The term hypoxia has two definitions. One definition would be a physiologic state where the oxygen partial pressure is below the normal physiologic range. For most normal tissues, the normal physiologic range is between 10 and 20 mmHg. However, portions of some normal tissues exist at oxygen pressure <10 mmHg, such as thymus [3], the kidney [4] and the pericentral vein tissues of the liver [5]. For these tissues, hypoxia would be an oxygen level that is much less than 10 mmHg. Hypoxic regions develop when there is an imbalance between oxygen supply and demand. There are a number of physiologic factors that contribute to restricted oxygen delivery, which have been reviewed in detail [6].

D. M. Gilkes (ed.), *Hypoxia and Cancer Metastasis*, Advances in Experimental Medicine and Biology 1136,
https://doi.org/10.1007/978-3-030-12734-3_2

Radiobiologically, hypoxia is defined an oxygen partial pressure of less than 10 mmHg in a local tissue volume [7]. Fully hypoxic cells are three times less sensitive to radiation than aerobic cells and hypoxia is very prevalent in nearly all solid tumors [1]. However, hypoxia contributes to a worse response to chemotherapy and even immunotherapy [8, 9]. Thus, hypoxia is predicted to be a nearly ubiquitous cause for treatment failure and worse overall outcome.

Several papers have been published showing that presence of hypoxia in human tumors predicts for worse response to radiotherapy and worse overall survival [1, 10–12]. Despite the widespread evidence for hypoxia being a source for treatment resistance and more aggressive behavior, hypoxia measurement has not been used in clinical trials testing the value of hypoxia modification, nor is it a part of routine clinical practice [13]. Multiple clinical trials failed for the hypoxic-targeting cytotoxin tirapazamine when combined with other therapies; while tirapazamine might not be effective despite convincing *in vitro* results, a significant factor is that none of these trials screened for patients with a high hypoxic tumor fraction nor did they monitor changes in hypoxia before, during or after treatment [14]. Measuring hypoxia could have widespread impacts, and there are number of different imaging methods that can measure hypoxia. The purpose of this review is to indicate which features of tumor hypoxia these methods measure, if they are quantifiable and to compare the pros and cons of each.

2.2 Causes of Hypoxia

To critically evaluate hypoxia imaging methods, it is first necessary to discuss the key physiologic features that lead to its development. At its most simple level, hypoxia at the cellular level is the result of imbalance between oxygen supply and demand. However, the deficiencies in oxygen delivery are complex and need to be discussed in some detail. Further, the influence of oxygen delivery vs demand on hypoxia are quite different.

2.2.1 The Relationship Between Hypoxia and Tumor Vasculature

Hypoxia is highly dependent on the pathophysiology of tumor vasculature. These features have been summarized as having seven components, and Fig. 2.1 demonstrates some of the physiology

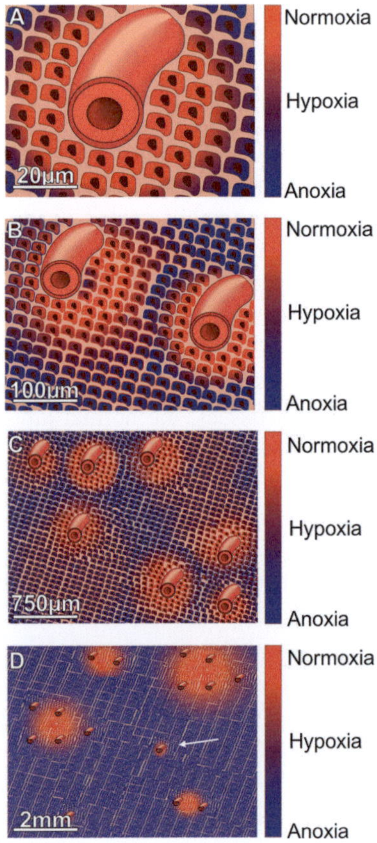

Fig. 2.1 (**a**) A blood vessel surrounded by tumor cells with a color bar representing the decrease in pO_2 (red is well-oxygenated and blue is hypoxic) as the distance from the blood vessel increases. At this scale, only optical imaging can quantify the gradient of pO_2. (**b**) At a larger scale than (**a**), optical imaging is still the only imaging modality that can measure the small hypoxic (blue) areas. (**c**) At this scale, MR methods in addition to optical imaging are able to quantify the hypoxic areas; however, some of the detail between vessels will likely be lost due to a lack of resolution. (**d**) At the largest scale, PET is able to quantify larger hypoxic and normoxic areas; however, the singular blood vessels (white arrow) will likely be lost due to partial volume artifacts, and a pixel size of 3 mm–5 mm will mask many hypoxic details

[6]. (1) Lack of sufficient arteriolar supply. When the arterial supply to a tumor is too scant, tumor microvessels become poor avenues for oxygen delivery as they become sparse and distant from arterial sources [15–17]. (2) Low vascular density. Tumor blood vessel density is inconsistent with inefficient organization of blood vessels leading to pockets of well-oxygenated areas and highly-hypoxic regions. The diffusion distance of oxygen is 180–200 μm, and in tumors, the distance between vessels can be much larger than the diffusion distance; regions with >180 μm between vessels will be hypoxic [18]. (3) Inefficient organization of vessels. Similarly, the chaotic organization of vasculature leads to extreme variations in the tissue pO_2. This leads to an inefficient oxygen delivery system and pockets of hypoxia [19, 20] (4) Variations in red blood cell flux. Vascular hypoxia occurs when there are large variations in the red blood cell (RBC) flux or when there are few or no RBCs in the vasculature [21]. (5) Hypoxic RBCs. Hypoxic red blood cells can affect this as they shrink, stiffen, and become sluggish, decreasing RBC flux and affecting the distribution of RBCs across vessel bifurcations [22]. Functional by-passes also affect RBC distribution in branching vessels. (6) Functional shunts. Large-diameter shunts between arteries and veins shuttle blood away from the tumor. These shunts are dynamic and problematic as they are constantly remodeling and disrupting consistent oxygen delivery [23]. (7) Oxygen supply and demand. Hypoxia results from the demand for oxygen being higher than the supply [24]. Microscopic hypoxic sub-regions are beyond the resolution of most measurement devices. As such, researchers are limited to a macroscopic fields and effects, which has hindered a complete understanding of the complex kinetics of hypoxia [25].

2.2.2 Chronic and Cyclical Hypoxia

Because of these dynamic changes in a tumor microenvironment, hypoxia is broken up into two categories: chronic and cyclical hypoxia (previously described as "acute" and "intermittent") [6, 26, 27]. Chronic hypoxia is defined by a tumor region that experiences pO_2 values of less than 10 mmHg for a long period of time [28, 29]. Cycling hypoxia, on the other hand, is dynamic. It has been established that this phenomenon is caused by instabilities in red cell flux in the microvessel network of tumors [30]. The instabilities of red cell flux lead to fluctuations in pO_2 that occur with multiple frequencies. The fastest frequency is in the range of 2–3 cycles per hour [31]. Slower fluctuations, on the order of hours have also been described [32]. Spatially, cyclical hypoxia is generally dependent on networks of microvessels instead of a large single vessel; this leads to larger cycling pockets of hypoxia across the tumor [27, 28]. This heterogeneous, temporal hypoxia is currently a subject of study, as it is challenging to measure. Preclinical models have provided much of the data on cycling hypoxia, including the verification of many of the probes used clinically. However, as will be described below, certain MRI methods have been able to capture this dynamic behavior in human head and neck cancers [33].

2.3 Effects of Hypoxia

2.3.1 The Impact of Hypoxia on Therapy

Tumor hypoxia has a detrimental effect on the efficacy of cancer therapy. Hypoxic tumor areas are three times less radiosensitive than well-oxygenated sub-volumes [34, 35]. Until recently, this knowledge was not applied in radiation therapy planning; however, with the implementation of intensity-modulated radiotherapy (IMRT), "dose-painting" specific, small structures in tumors is possible [36]. However, the presence of cycling hypoxia may cause a mismatch between the hypoxic subvolume and the IMRT Plan, unless the time interval between the imaging and treatment delivery is short (20–30 min). Clinicians are currently lacking a reliable imaging metric and subsequent model to measure and predict hypoxic volumes in tumors; thus, targeting these hypoxic areas are exceedingly difficult [27].

Chemotherapy is also negatively impacted by the presence of hypoxia. Well-oxygenated tissue is characteristic of efficient perfusion and vasculature, a feature that hypoxic tumors lack [21]. Nutrient and oxygen delivery are modulated by a set of complex signals that cause the formation and remodeling of vasculature [37, 38]. Chemotherapies depend on vasculature to travel and diffuse into cancerous tissue; therefore, hypoxic areas with poor perfusion and inefficient vasculature respond poorly to therapy [39]. Tromberg et al. demonstrated that hypoxia is associated with worse response to chemotherapy in breast cancer patients. Using diffuse optical spectroscopic imaging, they measured the baseline saturated oxygen and reported that an optimal tumor oxygenation threshold of $pO_2 = 76.7\%$ was associated with a complete pathological response [40].

2.4 Imaging Hypoxia

To image hypoxia, the modality must meet several requirements: (1) The system must be able to distinguish between hypoxia, normoxia and necrosis. A necrotic area of a tumor will invariably be hypoxic; however, it would not be a treatment target. Similarly, an aerobic tumor subregion would not benefit from a hypoxia modification strategy. (2) The system must be sensitive to hypoxic pO_2 values (0–15 mmHg) [7]. (3) The oxygen gradient between the microvessel and the surrounding tissue is very steep. Oxygenation can drop from normoxia to radiobiologic hypoxia within a few cell layers [28]. So, methods that measure primarily vascular pO_2 will underestimate the extent of hypoxia in tissue [28, 41]. Identifying oxygenated versus deoxygenated vasculature can be useful when modeling chaotic and dysfunctional vessels in the tumor microenvironment. However, cellular pO_2 depends on the diffusion distance from vasculature and measuring this gradient is the target [21]. Ideally, (4) the spatial resolution would be high enough to resolve the diffusion distance of oxygen (80–100 μm) [18]. Because this is generally unattainable, the limitations of volume averaging over a heterogeneous landscape of tumor hypoxia must be considered since the more severely hypoxic regions would contribute disproportionately to treatment resistance and recurrence and heterogeneous, diffusion limited hypoxia may not be accurately characterized with a large voxel size [13]. (5) To capture cycling hypoxia, the imaging method would need to obtain multiple measurements within an hour to capture the highest frequency of fluctuations which are in the range of 2–3 cycles per hour. (6) Hypoxia is known to occur in all solid tumors, so imaging methods that are not restricted by depth of tumor location would be the most broadly applicable. (7) Finally, the system should also be noninvasive and repeatable. Available imaging systems and their technical and practical considerations are summarized in Table 2.1.

The approach to imaging oxygen varies greatly, but can be placed in two categories: direct or indirect measurements. Direct measurements are calibrated to a pO_2 value while indirect measurements are either reflective of a surrogate measure like hemoglobin (Hb) saturation measurements that reflect relative differences.

Both direct and indirect methods can be quantified with absolute, comparable measurements for each subject; although, quantitative imaging is still a future goal for many systems. This is one of their main limitations as disease severity and treatment require standardized data instead of qualitative and subjective interpretations. For instance, the response evaluation criteria in solid tumors (RECIST) is the metric used to determine the clinical trial treatment outcome for solid tumors – apart from hepatocellular carcinomas that use mRECIST and follow a contrast-enhanced imaging protocol for assessing viable tumor tissue [42, 43]. The outcomes of RECIST depend heavily on anatomical changes in tumor burden, which do not account for endpoints that indicate functional or microscopic changes. FDG-PET is the only functional modality recommended; however, it is solely recommended for use in determining metastatic burden and identifying positive lymph nodes [42]. Further

Table 2.1 Description of preclinical and clinical imaging modalities for quantifying hypoxia

	Oxygen sensing probes	Phosphorescence lifetime imaging	Optical spectroscopy	PET	SPECT	BOLD MRI	OE-MRI	EPRI
Chemical probe	None	Phosphors (external)	None	Compound labeled with positron emitting isotope	Compound labeled with photon emitting isotope	None	None	Paramagnetic species
Direct/indirect	Direct	Direct	Indirect	Indirect	Indirect	Indirect	Indirect	Direct
Basis of oxygen measurement	Eppendorf: current generated from ionization of O_2; Oxylite: lifetime decay of fluorescent dye	Decay lifetime of phosphor upon interaction with O_2	Optical absorbance of hemoglobin	Accumulated positron signal of retained drug	Accumulated photon signal of retained drug	Magnetic state of deoxy-Hb in red blood cells	Hb saturation in blood plasma and interstitial fluid	Change in line width of probe upon interaction with O_2
Invasive/ non-invasive	Invasive	Minimally invasive	Non-invasive	Minimally invasive	Minimally Invasive	Non-Invasive	Non-Invasive	Minimally Invasive
Sensing depth	Several cm	mm-cm	mm-cm	No Limit	No Limit	No Limit	No Limit	mm-cm
Vascular/tissue pO_2	Both	Both	Vascular	Tissue	Tissue	Vascular	Both	Tissue
Temporal resolution	s	s	s	min-h	min-h	s	min	min-h
Range	Probe dependent: Eppendorfless sensitive at low pO_2; Oxylite highest at low pO_2	Probe dependent	0–100% Hb saturation	Marker-dependent; 0–10 mmHg ^{18}F-EF5 and ^{18}F-FMISO	0–10 mmHg	Change in T_2^* from air to hyperoxic gas breathing proportional to the starting level of Hbsat	Change in R_1 in response to hyperoxic gas challenge	0–100 mm Hg
Sensitivity	<1 mm Hg	<1 mm Hg	Hemoglobin saturation	Dependent on voxel spatial distribution of hypoxia	Dependent on voxel spatial distribution of hypoxia	Not applicable	>1 mmHg	<1 mmHg
Spatial resolution	Eppendorf: 20–30 μm; Oxylite: 250 μm	Sub mm-cm	mm-cm	Several mm	mm-cm	Machine dependent; sub mm-mm	mm	mm

(continued)

Table 2.1 (continued)

	Oxygen sensing probes	Phosphorescence lifetime imaging	Optical spectroscopy	PET	SPECT	BOLD MRI	OE-MRI	EPRI
Monitors changes in hypoxia	Continuous	min-min	Real time	>1 d	>1 d	s-s	s-min	min-min
Predominantly clinical or preclinical	Preclinical/clinical	Preclinical	Preclinical/ clinical	Clinical	Clinical	Clinical	Preclinical/ clinical	Preclinical (In clinical trials)
Cost: 1 = inexpensive, 5 = costly	2	3	1	5	4	5	5	3 (due to lower field strength)
Availability: 1 = poor, 5 = wide	4	2	2	1	2	3	2	1

Adapted from Ref. [7]

FDG-PET is not a surrogate for hypoxia, because avid glucose uptake by tumors can be driven by aerobic glycolysis (Warburg effect).

2.4.1 Standard Imaging Techniques in the Clinic

The six standard imaging methods utilized worldwide are x-ray imaging, computed tomography (CT), ultrasound imaging (US), optical imaging, nuclear imaging (including positron and single-photon emission tomography) and magnetic resonance imaging (MRI). Apart from nuclear imaging, these imaging modalities primarily focus on anatomical imaging with vastly different approaches; however, with the push for functional imaging capable of describing diseases and their severity via biomarkers, many clinics have adapted their existing systems to perform functional imaging. For instance, dynamic vascular imaging is made possible by the addition of a contrast agent such as iodine (for CT) or gadolinium (for MRI) [44, 45]. Ultrasound utilizes the Doppler effect to measure blood flow [46]. Positron emission tomography (PET) and single-photon emission tomography (SPECT) inherently image functional information by introducing radioactive compounds into the body that mimic biological processes. The most common radiotracer used in PET is fluorodeoxyglucose (FDG); a glucose-analog that used to measure tumor metabolism where an increased metabolism correlates with an active and aggressive cancer [47].

2.5 Positron Emission Tomography

PET is the modality most commonly used in both clinical trials and the clinic for measuring hypoxia; at the time of this review there are 49 active clinical trials using PET to image hypoxia in cancer (clinicaltrials.gov). PET exploits the beta-decay of radioactive compounds to form a 3D image of specific functional processes. Its signal-to-noise ratio is limited by the pharmacokinetics of the injected radioisotope. In PET, this noise can be significant: the radiotracer is meant to accumulate in specific areas; however, all tissue will generally collect some of the isotope. This can make it difficult to isolate areas of high signal. If the PET scanner does not have a high spatial and temporal resolution, the coincidence events become difficult to localize. This can be mitigated – in part – by increasing the scan duration so more photons can be accumulated.

Despite these challenges, PET-CT is a highly useful imaging system. For hypoxia, there are several radiotracers that have been featured in clinical trials (Fig. 2.2e). Isotopes containing oxygen-sensitive nitroimidazoles metabolize differently in aerobic vs anaerobic conditions [48]. The nitro group is reduced in a single-electron process that is oxidized back to the native drug in a normoxic environment. Under hypoxic conditions and in the presence of the protein nitroreductases, nitroimidazoles accumulate in the tissue and undergoes a 6-electron reduction. During this reduction, highly reactive species interact with biomolecules, forming covalent bonds and accumulating in areas of hypoxia. Interestingly, because metabolism is required for this reaction, apoptotic and necrotic cells do not accumulate these drugs. Thus, nitroimidazoles are an ideal candidate for revealing areas of viable hypoxia [49, 50].

2.5.1 [^{18}F]F-FMISO

^{18}F-labeled fluoromisonidazole ([^{18}F]F-FMISO) is the most widely studied PET hypoxia radiotracer and has been the most commonly used hypoxic marker in clinical trials [51, 52]. After an accumulation time of approximately 2–3 h, [^{18}F]F-FMISO is no longer affected by perfusion, making it specific to hypoxic areas, not just the well-vascularized areas of the tumor [53]. [^{18}F]F-FMISO is sensitive to pO_2 between 2 and 10 mmHg [50] in viable cells, with a low radiation dose of 3.7mBq/kg [50, 54, 55]. This radiotracer has been used in 49 clinical trials at the time of writing, and has been studied in an extensive number of cancer types: head and neck [56, 57], non-small cell lung cancer [58], soft tissue

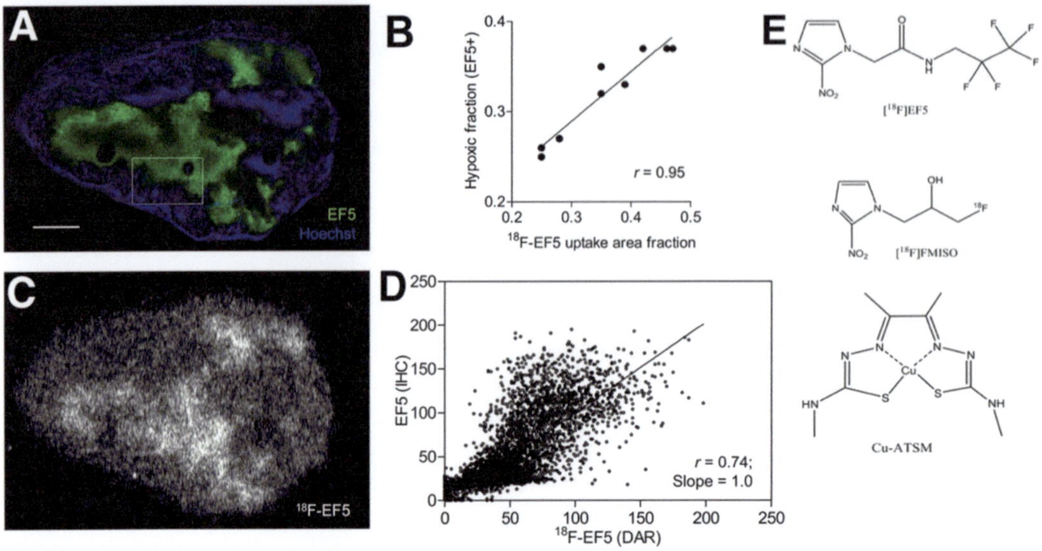

Fig. 2.2 Examples of PET imaging of hypoxia along with structural formulae of common hypoxic radiotracers. (**a**) Comparison of *18*F-EF5 uptake with EF5 binding and hypoxia in PC3 tumor model. Chitneni et al. studied various xenograft tumor models. EF5 is a gold-standard immunohistochemical hypoxia marker that can be used to measure hypoxic tumor areas in an *ex vivo* tumor section. To translate this technology to be used with PET imaging, the radioactive PET tracer *18*F was conjugated with EF5. In (**a**), the appearance of the immunohistochemical EF5 marker in a tumor section is shown in green. Perfusion distribution was observed by injecting Hoechst 33342 intravenously which stains cells around perfused vessels (shown in blue). Scale bar = 2 mm. (**b**) A strong correlation of r = 0.95 between *18*F-EF5 and positive EF5 signal (averaged over 3 animals) shows a linear relationship between the two. This correlation validates the PET modality as a valid hypoxic-measuring technique. (**c**) The PET image with the *18*F-EF5 radiotracer shows uptake that agrees with EF5 uptake in (**a**). (**d**) Spatial correlation analysis showing close correlation between EF5 binding with immunohistochemistry and *18*F-EF5 uptake in tumor sections, as measured using autoradiography **a** and **c** (r = 0.74). Each point in scatterplot represents 126-μm pixel and corresponding marker values on co-registered image (Reprinted with permission by [69]). (**e**) The structural formula of *18*F-EF5, [18]FFMISO and Cu-ATSM, the most common PET radiotracers for monitoring hypoxia in the clinic

sarcoma [59], renal tumors [60] and brain tumors [61].

Currently, investigators are focusing [^{18}F] F-FMISO research on two major areas: 1) The feasibility of IMRT planning based on hypoxic subvolumes in tumors [56, 57] and 2) prognostic imaging to predict the therapeutic outcome based on the hypoxic fractions [62–64]. Hendrickson et al. studied how RT boosts to hypoxic subvolumes in head and neck cancer patients could improve the local tumor control. Using [^{18}F] F-FMISO to determine hypoxic sub-volumes for RT-boost planning, they created example IMRT plans for 10 patients with a simultaneous integrated boost to the hypoxic sub-volumes. They predict an average improvement of 17% to the tumor control probability with a modest boost in radiation to hypoxic areas and no increased complications; however, they did not actually use these plans clinically [57]. Rischin et al. investigated the association between tumor hypoxia, treatment regimen, and locoregional failure (LRF) in patients with squamous cell carcinoma of the head and neck. They found that for 8 out of 13 patients with a high hypoxic fraction had a higher risk of LRF ($p = 0.038$ and hazard ratio of 7.1). Patients were placed in two groups: those treated with chemotherapy and with or without tirapazamine [65].

^{18}F-EF5 is another nitroimidazole radiotracer, based on the efficacy of the standard immunohistochemistry hypoxia marker EF5. EF5, when injected prior to tumor removal, is inversely – but directly – related to pO_2 through intracellular metabolism [66]. When EF5 is labeled with ^{18}F, it can be used in PET imaging. Most of the research on this radiotracer has been limited to pre-clinical studies; however, researchers have been successful in providing early treatment responses and outcome predictions in chemoradiation therapy [67] and radiation therapy [68]. Figure 2.2a–d shows data from a study by Chitneni et al. where they verified that ^{18}F-EF5 correlated with EF5 in several murine tumors [69]. Figure 2.2a shows a PC3 tumor with EF5 in green. They accounted for perfusion effects by injecting Hoechst 33342 intravenously which stains cells around perfused vessels (shown in blue). Figure 2.2b shows the PET image, and its signal appears to correlate well with Fig. 2.2a. Figure 2.2c, d shows the actual correlation between average EF5 and ^{18}F-EF5 signals is linear and high ($r = 0.95$) (2C) and in the tumor sections 2A and 2C. This study validated ^{18}F-EF5 as a reasonable marker for hypoxia by showing that it correlates well with EF5 [69]. In a clinical study by Qian et al. ^{18}F-EF5 PET was found to be capable of imaging hypoxia in 40% of patients with non-small cell lung cancer and that hypoxic tumors correlated with an increase in local tumor recurrence after conformal radiotherapy. Thirty percent of patients with imageable hypoxic tumors suffered a local recurrence in 12 months versus 0% recurrence for patients without hypoxic tumors ($p < 0.01$) [70].

2.5.2 Cu-ATSM

Cu-ATSM is a metal-chelated compound instead of a nitroimidazole-derivative like [^{18}F]F-FMISO and ^{18}F-EF5. A small molecule with high permeability to the cell membrane, copper has four positron-emitting radionuclides (^{60}Cu, ^{61}Cu, ^{62}Cu, and ^{64}Cu) suitable for PET imaging [71]. Cu-ATSM redox chemistry results in it being trapped in viable hypoxic cells at approximately 10 mmHg [51]. With a faster clearing time than both [^{18}F]F-FMISO and ^{18}F-EF5, Cu-ATSM requires a much shorter imaging time, mitigating cost and discomfort for patients. There is some controversy and conflicting data in Cu-ATSM that shows that inexplicably, some tumor types do not show hypoxia selectivity; Lapi et al. summarize the data in their review [72]. A clinical study on patients with lung cancer report that ^{60}Cu-ATSM uptake predicted treatment response [73]. This same group did a small study on patients with advanced squamous cell carcinoma of the cervix and found that tumor uptake of ^{60}Cu-ATSM was inversely related to progression-free (log-rank test $p = 0.0005$) and overall survival (log-rank test $p = 0.015$) [74]. Overall, ^{60}Cu-ATSM could be a powerful hypoxia marker, but still requires assays to determine the exact method of tumor retention for each cancer type particularly because they may not identify all areas of hypoxia. A possible solution that Zhang et al. proposed is combining both the hypoxia marker ^{60}Cu-ATSM with a perfusion marker ^{62}Cu-PTSM. In a pilot study with 10 patients, they found that normalizing ^{60}Cu-ATSM-PET images with ^{62}Cu-PTSM perfusion could provide a feasible method for obtaining more hypoxic information in a single scan [75].

PET imaging is a powerful and well-studied technique for imaging hypoxia. It has been proven to be highly selective and sensitive to hypoxic tissue, quantitative, and useful in predicting therapeutic outcomes in patients. It is not without some challenges, including high cost, long scan times, patient exposure to radiation, and low tumor/blood or tumor/muscle ratios. The latter is a significant problem in obtaining enough signal to conclusively define hypoxic regions. Furthermore, PET imaging is unable to monitor cyclical hypoxia effectively. [^{18}F]F-FMISO and ^{18}F-EF5 both have slow blood clearing times – up to 3 h – which delays hypoxic-sensitive imaging. This significantly hinders the ability to monitor the high-frequency hypoxia changes and impacts IMRT planning based on hypoxia subvolumes. Moreover, PET imaging has a low spatial resolution with large voxels, causing issue with partial volume effects. Therefore, a small hypoxic region in the middle of voxel may not be detected [13]. Despite this, PET is still an excellent candidate for hypoxic imaging; however, depending on the clinical indication, another modality might be more useful or practical.

2.6 Magnetic Resonance Imaging

Magnetic resonance imaging (MRI) is another nuclear imaging technique; however, unlike PET, its source of contrast is not radiation but rather the proton. It is non-invasive, with a high spatial resolution (<1 mm^2) and high soft-tissue contrast, making it a highly competitive modality for PET. Briefly, MRI harnesses the energy emitted from hydrogen nuclei excited under the presence of an oscillating magnetic field. The contrast depends on the rate that the excited nuclei relax back into their equilibrium state. A reconstructed MRI image is a 3D map of where protons (i.e.: hydrogen) are localized in the body. However, different sequences of the magnetic pulses and gradient fields have varying effects on the image with contrast dependent on high water content, high fat content, areas of high diffusion, etc. In functional MRI (fMRI), other compounds are excited by their specific resonant frequencies. While standard MRI images inherently have high SNR due to the abundance of hydrogen in the body, fMRI may require clever pulse sequences and reconstruction techniques to obtain a high enough signal for compounds other than hydrogen.

Hypoxia imaging via MRI has focused on several key techniques: blood oxygen level dependent (BOLD) MRI (also referred to intrinsic-susceptibility (IS)-MRI), oximetry, and oxygen-enhanced (OE) MRI. These are indirect oxygen imaging techniques and they mainly correlate with vascular or interstitial pO_2, respectively. The inference of tissue pO_2 based on vascular values is complex and not always straightforward, and an additional MRI scan using dynamic contrast-enhanced techniques is necessary to eliminate necrotic regions from analysis.

2.6.1 BOLD Imaging

BOLD imaging is a powerful imaging technique that uses clinical MRI scanners, and most MR-based hypoxia imaging has used this method. At the time of writing, BOLD imaging is part of 14 active clinical trials focusing on hypoxia imaging in prostate cancer, sarcoma, glioblastoma and head and neck squamous cell carcinoma (clinicaltrials.gov). A functional MRI technique, BOLD imaging exploits the intrinsic magnetic properties of hemoglobin. Oxygenated hemoglobin is weakly diamagnetic without unpaired electrons; conversely, deoxygenated hemoglobin, with 4 unpaired electrons exposing the iron center, is strongly paramagnetic [76]. The BOLD technique is sensitive to the local magnetic field distortions that deoxygenated hemoglobin causes in the tissue proximate to the blood vessels [77]. In a BOLD MR image, oxygenated and deoxygenated hemoglobin result in areas of high signal and low signal, respectively. These images can be quantified by measuring the T_2* relaxation time. The T_2*-relaxation parameter is the source of image contrast for the gradient pulse echo technique. It separates the diffusion effects of hemoglobin from static signal such as iron content in the muscle or fibrotic structures, and provides the basis for the linear relationship between R_2* and hemoglobin saturation (note: $1/T_2^* = R_2^*$) [55, 76, 78]. BOLD imaging and subsequent R_2*-based hemoglobin saturation measurements require knowledge of the blood volume and hematocrit for quantification [55]. This modality cannot provide absolute, direct pO_2 values, though the relative changes in R_2* can be quantified.

Several studies have linked direct pO_2 measurements with R_2* values [79–81]. Chopra et al. compared the BOLD R_2*signal with polarographic electrodes for nine patients with prostate cancer. They observed a trend to negative correlation between R_2* and pO_2 ($r = -0.66$, $p = 0.07$) and a positive correlation between R_2* and a hypoxic fraction (defined as regions <5 mmHg; $r = 0.76$, $p = 0.02$) [80]. This suggests that hypoxic tumors tend to exhibit higher R_2* values. Panek et al. conclusively demonstrated for the first time that cycling hypoxia occurs in human tumors used the BOLD MRI method. In patients with head-and-neck cancer, serial BOLD-MRI acquisitions monitored temporal oscillations in levels of paramagnetic deoxyhemoglobins

(Fig. 2.3b–d) [33]. Spontaneous R_2^* fluctuations with a median periodicity of 15 min were recorded which could be correlated with areas of impaired tumor vasculature, decreased perfusion and intermittent blood flow. This study illustrates the utility of BOLD-MRI to expand our understanding of cycling hypoxia which previously had not been confirmed conclusively in human tumors. In another human study, 36 patients with varying tumor types reported a BOLD intensity change in 56% of patients when breathing carbogen (95% O_2, 5% CO_2) gas during image acquisition [82]. This study also confirms that BOLD can detect hypoxic changes over time. Jian et al. tested the BOLD method on nine breast cancer patients before, during, and after chemotherapy. They found that the patients with complete pathological response had a significantly higher BOLD signal, corresponding to a well-oxygenated tumor [83].

These positive results have encouraged researchers to continue to study and employ this imaging technique. However, the intrinsic limitations of BOLD imaging must be considered. As previously mentioned, BOLD measurements reflect vascular pO_2 and the tissue immediately surrounding a vessel, which is not a direct tissue pO_2 measurement. When these R_2^* changes are recorded under a hyperoxic breathing challenge, it is assumed that the perfusion and hematocrit do not change when the subject goes from air to carbogen or oxygen breathing. However, Neeman et al. showed that in response to hyperoxia, the hematocrit decreased while plasma flow increased [84]. Dunn et al. also described vasoconstriction as an effect of hyperoxic and hypercarbic gas on tumor blood flow [85]. These findings suggest that hyperoxic breathing challenges may not be effective in changing large solid tumor oxygenation. Another limitation is the restricted ability to map the entire tumor at the required temporal resolution, which confines cycling hypoxia measurements to a single region of the tumor [33]. Echo-planar imaging (EP-MRI), which has been previously combined with BOLD-MRI, could improve this constraint [86]. Finally, although BOLD-MRI signal fluctuations indicating cycling hypoxia correlate with areas of impaired perfusion, they do not correlate well with areas of chronic hypoxia [33]. Combining multiple MRI methods, including a breathing challenge, could provide both chronic and cycling hypoxia information. Nonetheless, BOLD continues to be an option for noninvasively imaging patients and determining – to some extent – their hypoxic tumor burden.

2.6.2 Magnetic Resonance Spectroscopy

Magnetic resonance spectroscopy (MRS) is used to measure the presence of chemical compounds and their concentration in a region of interest. For hypoxic measurements, MRS utilizes the addition of compounds to increase the signal. While there are many compounds being investigated, most MRS oxygen measurements utilize ^{19}F-labeled compounds for two reasons: (1) there is a lack of endogenous ^{19}F in the body; therefore, when it is introduced in high enough quantities, it accumulates in tissue and causes bright "hot-spots" and (2) in NMR, ^{19}F has 83% the sensitivity of ^{1}H allowing for high sensitivity [87]. In the PET imaging section, we introduced nitroimidazole-based hypoxic imaging. For MRS, instead of being labeled with radioactive ^{18}F, nitroimidazole compounds are labeled with ^{19}F. Despite the success in PET-hypoxic imaging, due to the heterogeneity of tumor vasculature, the difficulty in quantifying accumulation based on pO_2 outweighs the benefits. Therefore, ^{19}F MRS is more often achieved using perfluorocarbon-based compounds (PFCs).

PFC probes provide absolute pO_2 measurements [88]. MRS measures the spin-lattice relaxation rate (R_1), which is linear with dissolved oxygen concentration [88]. PFC are also advantageous in that they are generally not affected by blood, pH, common proteins or other compounds found in the microenvironment. Several probes have been developed, including perfluoro-15-crown-5-ether and hexafluorobenzene. The latter is inexpensive, sensitive to oxygen and widely available, making it the PFC of choice [71]. The main challenge with these compounds is their

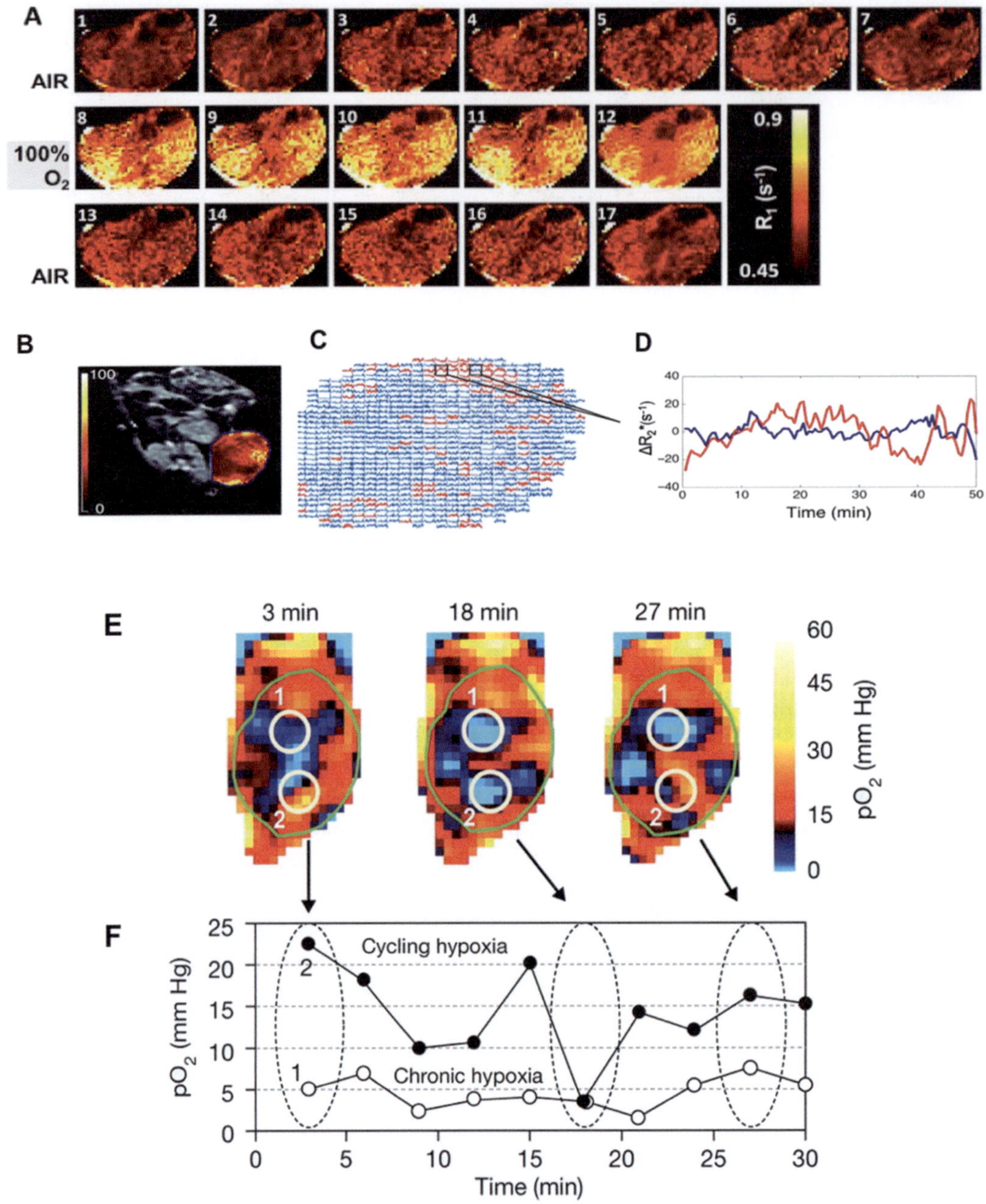

Fig. 2.3 Quantifying hypoxia via magnetic resonance imaging (MRI) and electron paramagnetic imaging (EPR)I. (**a**) Oxygen-enhanced MRI correlates the longitudinal relaxation coefficient R_1 to oxygenation. Shown is the R_1 map of a murine 786-0-R xenograft while the mouse breathed air (top row), 100% oxygen (middle row) and back to air (bottom row). As R_1 increases (yellow), oxygenation across the tumor increases, as expected. This validates oxygen-enhanced MRI as a method for detecting and quantifying change in tumor oxygenation via changes in R_1. This method has been validated with other standard methods (Reprinted with permission by [94]). (**b**) In blood-oxygen level dependent (BOLD) MRI, R_2* values correlate with the levels of O_2 saturation in hemoglobin. In the image, a BOLD MR image of a patient's head and neck tumor with the R_2* map. Note, the tumor is delineated in blue. (**c**) and (**d**) The R_2* time series showing non-random fluctuations in R2* values within the tumor over time. This demonstrates that hypoxia can be quantified on a pixel-by-pixel basis, and information about cycling hypoxia can be revealed over time (Reprinted with permission by [33]). (**e**) Electron paramagnetic resonance requires the injection of a spin trap to obtain a high

extreme hydrophobicity. Therefore, *in vivo* studies generally inject the PFC directly into the tissue of interest [88–91]. In addition to absolute pO_2quantification, ^{19}F-labeled PFC have been used in longitudinal monitoring of hypoxia. In a preclinical study, PFC-loaded alginate capsules were co-injected with tumor cells into the flanks of rats [91]. These biocompatible capsules shielded the PFC during tumor growth, minimizing migration and immune system response, and allowing for repeated measurements [91]. A significant limitation in this method is that the signal is restricted to the region of the probe. Though still predominantly a preclinical probe, ^{19}F-PFC MRS is a promising modality for directly measuring and quantifying hypoxia *in vivo.*

2.6.3 Oxygen-Enhanced MRI

Oxygen-Enhanced (OE) MRI is an emerging technique to measure hypoxia in tumors. It measures the longitudinal relaxation rate (R_1), which is sensitive to the oxygen concentration in interstitial fluid and blood plasma [92, 93]. O'Connor et al. were the first to validate OE-MRI and to quantify spatial variations in hypoxia. Under a breathing challenge, well-perfused areas obtain higher levels of dissolved oxygen in the interstitial tissue, and blood plasma remains hyperoxic, both of which increase the R_1 value (Fig. 2.3a) [94]. This change in R_1 (ΔR_1) is theoretically proportional to the change in dissolved oxygen in a given voxel. However, several preclinical studies observe the opposite in tumor subregions: because the hemoglobin has a low oxygen saturation, incoming O_2 molecules bind preferentially to them. This does not alter the plasma pO_2 or the ΔR_1; therefore, hypoxic areas are refractory and can be identified and quantified as areas with a $\Delta R_1 \leq 0$ [95–98]. O'Connor's group tested this hypothesis by localizing the ΔR_1 values against the hypoxia marker pimonizole in three murine, xenografted tumor cell lines. They describe a linear correlation between pimonizole binding and ΔR_1, and they used dynamic-contrast enhanced (DCE)-MRI to correct for perfusion in tumors with extensive necrosis [94]. Finally, by treating the animals with hydralazine, a vasoactive drug that is known to decrease perfusion and increase hypoxia [13], they reported a significant increase ($p = 0.045$) in the hypoxic fraction ($\Delta R_1 = 0$) 17.3% to 30.5% compared to untreated mice [94]. In this way, OE-MRI was demonstrated to be a feasible method for mapping hypoxia in tumors.

One advantage of this system is its ability to repeat measurements in the same area without having altered the microenvironment or introduced exogenous agents. Repeated, longitudinal measurements enable assessment of cycling hypoxia and hypoxia-modifying therapies [13]. Moreover, the spatial resolution in MRI is high compared to PET where partial volume artifacts can obscure smaller – though still important – hypoxic regions. The inclusion of DCE-MRI in standard OE-MRI acquisitions is time-consuming but doable in a single session, and, compared to BOLD-MRI, OE-MRI is not influenced by hemoglobin saturation and the vasoactive effects of hyperoxic gas breathing refractory to a change in blood oxygenation [84, 85]. Finally, the concept of the refractory change in hemoglobin satruation in hypoxic tumor regions and its physiological basis has been discussed in-depth in a commentary by Dewhirst and Birer [13]. They conclude that the study described above by O'Connor et al. [94] and others [99–103] confirm the effects of longitudinal oxygen gradients [13]. As blood traverses from the lungs through the body, arteries are constantly losing oxygen; 80% of oxygen is lost before entering vascular beds. However, due

Fig. 2.3 (continued) signal. This EPR image of a murine SCCVII tumor xeno graft shows areas of normoxia (yellow) and hypoxia or anoxia (blue). This oxygen map was obtained every 3 min during a total acquisition time of 30 min. Images shown are at 3 min, 18 min, and 27 min. Cycling hypoxia, normally challenging to image due to timing constraints, is demonstrated by changing values in the ROIs and placement of hypoxic areas. (**f**) pO_2 assessed and quantified in ROIs compare open circles (chronic hypoxia) and closed circles (cycling hypoxia) over time. This demonstrates the high resolution (spatial and temporal) that allows for quantification of both chronic and cycling hypoxia in mice. (Reprinted with permission by [121])

to redundancy in arterioles and microvessels, normal tissue does not experience any lack of oxygen. Tumors, conversely, often experience a lack of sufficient oxygen due to this effect and inadequate arteriole supply. Therefore, tumor tissue distant from arterioles are both hypoxic and refractory, resisting a change in R_1 after a breathing challenge, and the OE-MRI method could provide a much-needed method for clinical measurements of hypoxia in tumors [13, 103].

2.7 Electron Paramagnetic Resonance Imaging

Electron paramagnetic resonance (EPR) spectroscopy and imaging is unique in its ability to sustain real-time assessment of hypoxia over the course of several hours [104]. It is an emerging modality in its application to clinical hypoxia; however, it is making quick progress in expanding from EPR spectroscopy to EPR imaging and combining with clinical MRI scanners [105]. EPR oximetry is sensitive to unpaired electrons, which are not abundant in the body. Therefore, the addition of a paramagnetic spin probe (like a contrast agent) is necessary to obtain a high signal [106, 107]. Once these probes have been injected, they can remain in the site for months, enabling repeated measurements with some probes reporting an oxygen-sensitivity of <0.2 mmHg [108]. The interactions between unpaired electrons and the spin probe result in changes to the T_2 relaxation time; an increase in oxygenation results in a (usually linear) increase in the spin-spin relaxation rate R_2 ($R_2 = 1/T_2$) [109]. In EPR spectroscopy, this R_2 signal produces a spectrum, which results in a change in the width of the spectral peaks: an increase in oxygen concentration produces a broader peak [109].

EPR spectroscopy cannot provide anatomical information nor can it measure past a depth of 10 mm [106]. Despite these limitations, EPR oximetry has been used extensively in preclinical animal studies [110–112] including tumor reoxygenation in murine tumors after RT [113, 114], vascular changes inducing tumor oxygenation changes [115, 116], and the dependency of post radiotherapy cure rates on hypoxic fractions [117, 118].

Extending EPR oximetry to practical EPR imaging with reasonable acquisition times is a challenge; however, high-resolution, quantitative pO_2 maps have been produced with the introduction of triarylmethyl (TAM) radical probes [119]. Elas et al. performed a preclinical study comparing sequences of tumor pO_2 values from EPR oxygen images with sequences of oxygen measurements made along a track with an Oxylite oxygen probe. They report a high spatial resolution (~1 mm) and high hypoxic sensitivity of 3 mmHg in FSa fibrosarcoma xenografts in murine models [120]. The correlation between calibrated EPR images and pO_2 measurements by the Oxylite showed a good correlation in both pO_2 magnitude and spatial distribution patterns, validating EPRI as a reasonable method for mapping tumor oxygenation. Cycling hypoxia has also been demonstrated (Fig. 2.3e, f) [121]. In a preclinical murine SCCVII tumor xenograft, subjects were imaged every 3 min for 30 min. Figure 2.3e shows the change in hypoxia (blue) over time at 3 min, 18 min and 27 min. Figure 2.3f shows a quantification of specific ROIs delineated in 2E – open circles are areas of chronic hypoxia and closed circles indicated cycling hypoxia. Over time, the hypoxic areas were tracked and showed that EPR can obtain the spatial and temporal resolution needed to monitor hypoxia [121].

EPRI has also shown utility in tracking longitudinal oxygen measurements and localizing oxygen maps to anatomy via MRI. A study by Matsumoto et al. used pulsed EPRI combined with MRI to obtain quantitative, 3D maps of pO_2 in mice [122]. In an air/carbogen (95% O_2 plus 5% CO_2) breathing cycle, they studied the relationship between tumor blood perfusion and pO_2. They found that significant hypoxia existed even in regions that exhibited blood flow, which provides *in vivo* confirmation of hypoxia characteristics [122]. Matsumoto et al. treated mice with the multi-tyrosine kinase inhibitor sunitinib and by monitoring pO_2 and vascular density over time via EPR and MR imaging, they demonstrated vascular normalization of tumor [123]. They further report that, during a window of enhanced tumor oxygenation, radiation and sunitinib combination therapy resulted in a synergistic delay in tumor; moreover, during this time, cycling hypoxia was also suppressed [123]. This study demonstrates the capabilities of EPR to track temporal changes in hypoxia and potentially identify a target time when radiation therapy can be most effective. At the time of writing, there are three, recruiting clinical trials utilizing EPRI or EPR oximetry (clinicaltrials.gov).

2.8 Optical Methods

Optical imaging encompasses a wide array of methods capable of measuring hypoxia. Wavelengths between 100 nm and 1 mm describe the range of the optical spectrum, and researchers have developed many methods to harness oxygen detection in this range including oxygen-sensing electrodes, fluorescent lifetime sensors, diffuse reflectance spectroscopy, hyperspectral imaging, diffuse optical topography and tomography, photoacoustics, structured illumination, etc. [71]. These technologies share a common method: they obtain oxygen measurements by quantifying intrinsic sources of optical absorption, scattering, and fluorescence in tissue. Optical imaging or spectroscopy are advantageous in their high spatial and temporal resolution. Limitations in optical methods lie primarily in penetration-depth restrictions due to relatively high tissue absorption and scattering.

2.8.1 Near-Infrared Spectroscopy

Near-infrared spectroscopy (NIRS) measures the absorption and scattering events for light between 780 and 2500 nm. It has vast applications in many fields; however, in medicine, it provides information on hemoglobin oxygen saturation. Optical imaging does not always require an exogenous compound to measure oxygenation. Hemoglobin saturation has different absorption properties depending on if it is oxygenated (HbO_2) or deoxygenated (Hb) [124]. Hemoglobin is the foremost absorber in tissue, and in any given vessel, there is some mix of HbO_2 and Hb. The fraction of HbO_2 can be determined by measuring the absorption properties of the tissue across various wavelengths, and the vascular pO_2 can be determined through a hemoglobin disassociation curve (Fig. 2.4g) [125, 126]. NIRS is one method capable of differentiating oxygenated from deoxygenated hemoglobin.

While NIRS is not strictly an imaging method as it only provides precise, quantitative point-measurements in tissue, it can be expanded to obtain 3D information that can map an entire area. Diffuse optical tomography (DOT) is the 3D imaging version and has been studied in gliomas [127] and breast cancer [128–131]. In a clinical study, Gunther et al. monitored 40 breast-cancer patients receiving neoadjuvant chemotherapy [130]. Two weeks after treatment began, dynamic DOT images were acquired. After a 5-month treatment follow-up, researchers report that DOT identified patients with a pathological complete response with a positive predictive value of 70.6% (12 of 17), a negative predictive value of 94.1% (16 of 17), a sensitivity of 92.3% (12 of 13), and a specificity of 76.2% (16 of 21) based on their Hb saturation and their water and lipid content [130]. This study highlights the feasibility of predicting treatment outcome based on these biomarkers using DOT. The main advantages of DOT and NIRS, in general, is its quantitative, precise measurements. It has high temporal and reasonable spatial resolution, making it one of the few systems to feasibly measure cycling hypoxia. While NIRS experiences less attenuation effects of tissue than lower optical wavelengths, it is still limited to measuring superficial tumors. Moreover, Hb saturation measurements are an indirect method of measuring vascular hypoxia and can potentially miss important hypoxic information in tumor cells.

2.8.2 Phosphorescence-Lifetime Imaging

Phosphorescence-lifetime imaging (PLIM) involves the injection into the tissue or vasculature

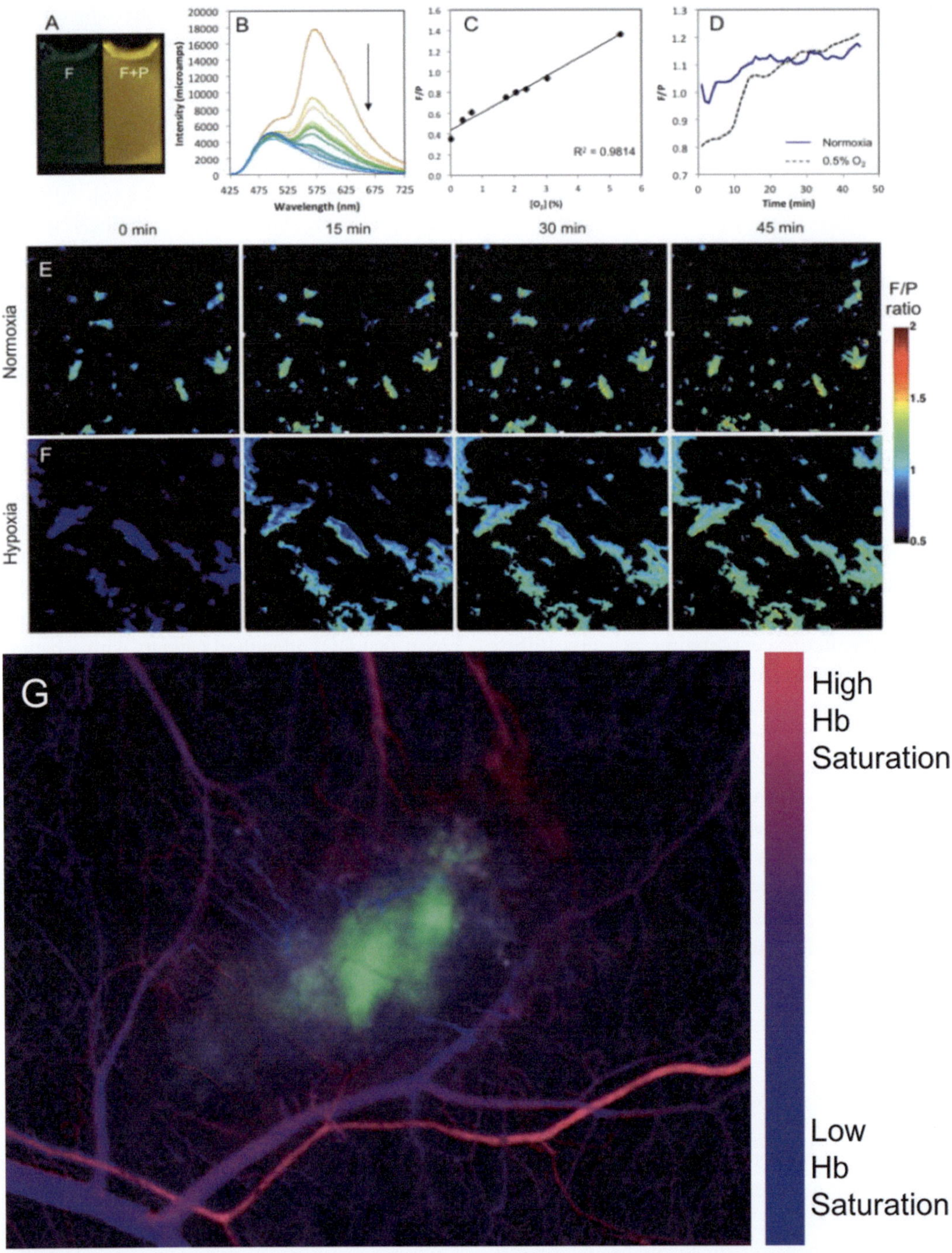

Fig. 2.4 Direct and indirect methods of quantifying hypoxia using optical imaging methods. (**a**) Image of boron nanoparticles (BNP) (BF2dnm(I)PLA 9b (NPI)) in air (fluorescence) and N_2 (fluorescence + phosphorescence) at λ_{ex} = 354 nm. (**b**) Spectra of BNPs at varying oxygen concentrations (0–21%). The arrow indicates decreasing phosphorescence with increasing $\%O_2$. Note that fluorescence does not change, allowing for ratiometric (F/P) imaging. (**c**) The fluorescence to phosphorescence (F/P) ratio with a linear fit in the 0–5.3% O_2 range. The BNP are most sensitive and linearly and directly correlate to oxygen values in hypoxic values. (**d**) A filter system attached to the microscope was used to split the wavelengths and determine the F/P ratio. At two different

and recording the phosphorescence. These injected phosphors generally use nanoparticle technology to encapsulate phosphorescent dye, preventing degradation in the harsh microenvironment. This phosphorescence emission can be recorded by many modalities and emit in all optical wavelengths, though preferentially near IR [71]. Oxygen-sensing phosphors rely on molecular oxygen quenching the excited phosphor in its triplet state [132]. This principle is competitive to phosphorescence intensity, providing a straightforward and direct calibration between increasing O_2 and decreasing phosphorescent intensity.

Oxygen-sensitive boron nanoparticles (BNP) are dual-emissive with both fluorescence and oxygen-sensitive phosphorescence; the ratio of fluorescence to phosphorescence (F/P) normalizes the measurements and corrects for intersubject differences, variations in detector sensitivity and fluctuations in light intensity [133]. Once calibrated to a F/P versus oxygen tension curve, these BNPs provide a direct, absolute tissue oxygen measurement. These ratiometric sensors BNPs have been used in *in vivo* preclinical studies in murine dorsal-skin-fold window chambers to study the relationship between hemoglobin saturation and tissue oxygenation [134]. Palmer et al. describe how changes in hemoglobin saturation in 4T1 tumors drive fluctuations in tissue oxygenation with a reportedly high correlation (r = 0.77) (Fig. 2.4a–f) [134]. This study highlights BNP's unique ability to report tissue oxygenation instead of Hb saturation measurements. While BNPs are not used clinically, its data is invaluable in the pursuit to understand the complexities of hypoxia in the tumor microenvironment.

Sergei Vinogradov's group has investigated lifetime fluorescent imaging using an Oxyphor probe, and used it in a study involving R3230Ac tumors growing in dorsal flap window chambers in rats [135]. The Oxyphor R2 probe has an interesting capability due to dual phosphorescent peaks (419 nm and 524 nm) with different lifetimes. The 419 nm excitation measures pO_2 in a layer less than 50um from the surface whereas the 524 nm excitation can image at deeper layers. This allows "optical sectioning" of tumor oxygenation. Using this technique, they determined that the tumor was much more hypoxic than the surrounding tissue and confirmed previous results that the growing edge of the tumors exhibit a lower pO_2 than at the central core of the tumor [135]. In a similar study Cardenas-Navia et al. sought to measure the spatial and temporal heterogeneity in rat fibrosarcomas, 9 L gliomas, and R3230 mammary adenocarcinomas grown in dorsal skin-fold window chambers [136]. Using phosphorescence-lifetime imaging, they quantified fluctuations in vascular pO_2, imaging every 2.5 min for a duration of 60–90 min. Their results confirmed the variability of oxygen tumor delivery and, more importantly, demonstrated that perfusion-limited or diffusion-limited hypoxia are extreme cases of O_2 delivery or metabolism dominant areas in tumors [136].

2.9 Conclusion

Hypoxia is one of the most important factors in clinical outcomes for cancer therapies. Functional imaging of hypoxia, however, has not been incorporated into clinical practice. Further, there are

Fig. 2.4 (continued) oxygen concentrations, it shows different trends in mouse tumor microenvironments over 45 min. (**e**) Filtered microscope images of 4T1 cells under normoxic conditions over 45 min. Color bar indicates F/P ratio, which correlates to oxygenation. Note that over time, the values stay in the same range. (**f**) Filtered microscope images of 4T1 cells as they transition from 0.5% O_2 to normoxia over 45 min. The microenvironment starts hypoxic, slowly increasing in oxygenation back to baseline. This suggest that the BNPs are sensitive enough to measure cycling hypoxia and directly quantify oxygen from the F/P ratio (Reprinted with permission by [133]). (**g**) Hemoglobin saturation, an indirect measurement of tissue oxygenation, is shown with high hemoglobin oxygen saturation in red and low hemoglobin oxygen saturation is shown in blue. A 4T1 tumor, colored white, is centered in the image. Transcriptional activity of hypoxia-inducible factor 1 alpha (HIF1A), an indicator of hypoxia, is shown in green. Note the longitudinal gradient of hemoglobin saturation along the vasculature, with lower saturations around the upper right pole of the tumor, which also shows increased HIF-1 expression. Refer to [138] for methods regarding hemoglobin saturation imaging and the murine tumor model

no widely approved methods to reduce hypoxia. In a way, this is a chicken and egg dilemma. Without a validated and proven method to measure hypoxia, it is a challenge to prove that reduction of hypoxia has any therapeutic value. This is aptly seen in the extensive review by Overgaard and Horsman. In reviewing the clinical literature, they identified over 70 randomized trials, conducted since the 1960s in which methods to reduce the influence of hypoxia on radiotherapy response were tested. Hypoxia imaging was not used as a selection tool in any of these prior trials and with the exception of head and neck cancer, no tumors demonstrated a benefit from modification of tumor hypoxia [137].

A significant challenge to using hypoxia imaging is the cost of implementing a new system, both in money and in time. PET and MR imaging – two powerful modalities for quantitatively imaging hypoxia – are also the most expensive systems and cost thousands of dollars per imaging session. Incorporating functional imaging with longer sessions, adding technology to clinical machines, and changing the workflow and analysis all deliver significant hindrances. However, these problems have motivated the drive for more inexpensive and available systems, particularly in optics. Although many of these optical systems are still being validated preclinically, the field has an opportunity to provide direct, quantitative, inexpensive and non-invasive systems for measuring hypoxia for tumors where there is an optical window. As more data is being produced describing the benefits of characterizing tumor hypoxia and potential therapeutics, the need for clinical hypoxia imaging will eclipse the cost and continue to move the technological research forward.

Another challenge in hypoxia imaging affects the outcomes of clinical trials: patients are not pre-screened for tumors with a high hypoxic fraction prior to being included in a hypoxia-modification clinical trial. Executing well-planned clinical trials with proper patient selection will maximize the possibility of achieving a positive result from future clinical trials with hypoxia modification treatments. Positive trials will subsequently provide rationale for including hypoxia imaging as part of clinical staging.

Overall, quantitatively imaging hypoxia and understanding its effects can significantly benefit patients. Hypoxia influences disease progression, therapeutic efficacy and prognosis. This has garnered the attention of researchers across the world who are investigating molecular effects of hypoxia in preclinical models and monitoring hypoxia in the clinic.

References

1. Vaupel P, Mayer A (2007) Hypoxia in cancer: significance and impact on clinical outcome. Cancer Metastasis Rev 26(2):225–239
2. Moon EJ, Brizel DM, Chi JT, Dewhirst MW (2007) The potential role of intrinsic hypoxia markers as prognostic variables in cancer. Antioxid Redox Signal 9(8):1237–1294
3. Braun RD, Lanzen JL, Snyder SA, Dewhirst MW (2001) Comparison of tumor and normal tissue oxygen tension measurements using OxyLite or microelectrodes in rodents. Am J Physiol Heart Circ Physiol 280(6):H2533–H2544
4. Samoszuk MK, Walter J, Mechetner E (2004) Improved immunohistochemical method for detecting hypoxia gradients in mouse tissues and tumors. J Histochem Cytochem 52(6):837–839
5. Arteel GE, Thurman RG, Yates JM, Raleigh JA (1995) Evidence that hypoxia markers detect oxygen gradients in liver: pimonidazole and retrograde perfusion of rat liver. Br J Cancer 72(4):889–895
6. Dewhirst MW, Cao Y, Moeller B (2008) Cycling hypoxia and free radicals regulate angiogenesis and radiotherapy response. Nat Rev Cancer 8(6):425–437
7. Chitneni SK, Palmer GM, Zalutsky MR, Dewhirst MW (2011) Molecular imaging of hypoxia. J Nucl Med 52(2):165–168
8. Brown JM, Wilson WR (2004) Exploiting tumour hypoxia in cancer treatment. Nat Rev Cancer 4(6):437–447
9. Sitkovsky MV, Hatfield S, Abbott R, Belikoff B, Lukashev D, Ohta A (2014) Hostile, hypoxia-A2-adenosinergic tumor biology as the next barrier to overcome for tumor immunologists. Cancer Immunol Res 2(7):598–605
10. Ruan K, Song G, Ouyang G (2009) Role of hypoxia in the hallmarks of human cancer. J Cell Biochem 107(6):1053–1062
11. Hatfield SM, Kjaergaard J, Lukashev D, Schreiber TH, Belikoff B, Abbott R et al (2015) Immunological mechanisms of the antitumor effects of supplemental oxygenation. Sci Transl Med 7(277):277ra30

12. Sethumadhavan S, Silva M, Philbrook P, Nguyen T, Hatfield SM, Ohta A et al (2017) Hypoxia and hypoxia-inducible factor (HIF) downregulate antigen-presenting MHC class I molecules limiting tumor cell recognition by T cells. PLoS One 12(11):e0187314
13. Dewhirst MW, Birer SR (2016) Oxygen-enhanced MRI is a major advance in tumor hypoxia imaging. Cancer Res 76(4):769–772
14. Marcu L, Olver I (2006) Tirapazamine: from bench to clinical trials. Curr Clin Pharmacol 1(1):71–79
15. Dewhirst MW, Ong ET, Braun RD, Smith B, Klitzman B, Evans SM et al (1999) Quantification of longitudinal tissue pO2 gradients in window chamber tumours: impact on tumour hypoxia. Br J Cancer 79(11–12):1717–1722
16. Dewhirst MW, Ong ET, Klitzman B, Secomb TW, Vinuya RZ, Dodge R et al (1992) Perivascular oxygen tensions in a transplantable mammary tumor growing in a dorsal flap window chamber. Radiat Res 130(2):171–182
17. Sorg BS, Hardee ME, Agarwal N, Moeller BJ, Dewhirst MW (2008) Spectral imaging facilitates visualization and measurements of unstable and abnormal microvascular oxygen transport in tumors. J Biomed Opt 13(1):014026
18. Tomlinson RH, Gray LH (1955) The histological structure of some human lung cancers and the possible implications for radiotherapy. Br J Cancer 9(4):539–549
19. Secomb TW, Hsu R, Dewhirst MW, Klitzman B, Gross JF (1993) Analysis of oxygen transport to tumor tissue by microvascular networks. Int J Radiat Oncol Biol Phys 25(3):481–489
20. Secomb TW, Hsu R, Park EY, Dewhirst MW (2004) Green's function methods for analysis of oxygen delivery to tissue by microvascular networks. Ann Biomed Eng 32(11):1519–1529
21. Dewhirst MW, Kimura H, Rehmus SW, Braun RD, Papahadjopoulos D, Hong K et al (1996) Microvascular studies on the origins of perfusion-limited hypoxia. Br J Cancer Suppl 27:S247–S251
22. Kavanagh BD, Coffey BE, Needham D, Hochmuth RM, Dewhirst MW (1993) The effect of flunarizine on erythrocyte suspension viscosity under conditions of extreme hypoxia, low pH, and lactate treatment. Br J Cancer 67(4):734–741
23. Eddy HA, Casarett GW (1973) Development of the vascular system in the hamster malignant neurilemmoma. Microvasc Res 6(1):63–82
24. Secomb TW, Hsu R, Ong ET, Gross JF, Dewhirst MW (1995) Analysis of the effects of oxygen supply and demand on hypoxic fraction in tumors. Acta Oncol 34(3):313–316
25. Grimes DR, Warren DR, Warren S (2017) Hypoxia imaging and radiotherapy: bridging the resolution gap. Br J Radiol 90(1076):20160939
26. Bristow RG, Hill RP (2008) Hypoxia and metabolism. hypoxia, DNA repair and genetic instability. Nat Rev Cancer 8(3):180–192
27. Dewhirst MW (2009) Relationships between cycling hypoxia, HIF-1, angiogenesis and oxidative stress. Radiat Res 172(6):653–665
28. Lanzen J, Braun RD, Klitzman B, Brizel D, Secomb TW, Dewhirst MW (2006) Direct demonstration of instabilities in oxygen concentrations within the extravascular compartment of an experimental tumor. Cancer Res 66(4):2219–2223
29. Cardenas-Navia LI, Yu D, Braun RD, Brizel DM, Secomb TW, Dewhirst MW (2004) Tumor-dependent kinetics of partial pressure of oxygen fluctuations during air and oxygen breathing. Cancer Res 64(17):6010–6017
30. Kimura H, Braun RD, Ong ET, Hsu R, Secomb TW, Papahadjopoulos D et al (1996) Fluctuations in red cell flux in tumor microvessels can lead to transient hypoxia and reoxygenation in tumor parenchyma. Cancer Res 56(23):5522–5528
31. Braun RD, Lanzen JL, Dewhirst MW (1999) Fourier analysis of fluctuations of oxygen tension and blood flow in R3230Ac tumors and muscle in rats. Am J Phys 277(2 Pt 2):H551–H568
32. Skala MC, Fontanella A, Lan L, Izatt JA, Dewhirst MW (2010) Longitudinal optical imaging of tumor metabolism and hemodynamics. J Biomed Opt 15(1):011112
33. Panek R, Welsh L, Baker LCJ, Schmidt MA, Wong KH, Riddell AM et al (2017) Noninvasive imaging of cycling hypoxia in head and neck cancer using intrinsic susceptibility MRI. Clin Cancer Res 23(15):4233–4241
34. Moeller BJ, Cao Y, Li CY, Dewhirst MW (2004) Radiation activates HIF-1 to regulate vascular radiosensitivity in tumors: role of reoxygenation, free radicals, and stress granules. Cancer Cell 5(5):429–441
35. Chaplin DJ, Durand RE, Olive PL (1986) Acute hypoxia in tumors: implications for modifiers of radiation effects. Int J Radiat Oncol Biol Phys 12(8):1279–1282
36. Thorwarth D, Eschmann SM, Paulsen F, Alber M (2007) Hypoxia dose painting by numbers: a planning study. Int J Radiat Oncol Biol Phys 68(1):291–300
37. Carmeliet P, Dor Y, Herbert JM, Fukumura D, Brusselmans K, Dewerchin M et al (1998) Role of HIF-1alpha in hypoxia-mediated apoptosis, cell proliferation and tumour angiogenesis. Nature 394(6692):485–490
38. Nam SY, Ko YS, Jung J, Yoon J, Kim YH, Choi YJ et al (2011) A hypoxia-dependent upregulation of hypoxia-inducible factor-1 by nuclear factor-kappaB promotes gastric tumour growth and angiogenesis. Br J Cancer 104(1):166–174
39. Cosse JP, Michiels C (2008) Tumour hypoxia affects the responsiveness of cancer cells to chemotherapy and promotes cancer progression. Anti Cancer Agents Med Chem 8(7):790–797
40. Ueda S, Roblyer D, Cerussi A, Durkin A, Leproux A, Santoro Y et al (2012) Baseline tumor oxygen saturation correlates with a pathologic complete

response in breast cancer patients undergoing neoadjuvant chemotherapy. Cancer Res 72(17):4318–4328
41. Dasu A, Toma-Dasu I (2008) Vascular oxygen content and the tissue oxygenation – a theoretical analysis. Med Phys 35(2):539–545
42. Eisenhauer EA, Therasse P, Bogaerts J, Schwartz LH, Sargent D, Ford R et al (2009) New response evaluation criteria in solid tumours: revised RECIST guideline (version 1.1). Eur J Cancer 45(2):228–247
43. Fournier L, Ammari S, Thiam R, Cuenod CA (2014) Imaging criteria for assessing tumour response: RECIST, mRECIST, Cheson. Diagn Interv Imaging 95(7–8):689–703
44. Martens MH, Lambregts DMJ, Kluza E, Beets-Tan RGH (2014) Tumor response to treatment: prediction and assessment. Curr Radiol Rep 2(9):62
45. Humbert O, Berriolo-Riedinger A, Cochet A, Gauthier M, Charon-Barra C, Guiu S et al (2014) Prognostic relevance at 5 years of the early monitoring of neoadjuvant chemotherapy using (18)F-FDG PET in luminal HER2-negative breast cancer. Eur J Nucl Med Mol Imaging 41(3):416–427
46. Di Renzo GC, Luzi G, Cucchia GC, Caserta G, Fusaro P, Perdikaris A et al (1992) The role of Doppler technology in the evaluation of fetal hypoxia. Early Hum Dev 29(1–3):259–267
47. Peeters SG, Zegers CM, Lieuwes NG, van Elmpt W, Eriksson J, van Dongen GA et al (2015) A comparative study of the hypoxia PET tracers [(1)(8)F]HX4, [(1)(8)F]FAZA, and [(1)(8)F]FMISO in a preclinical tumor model. Int J Radiat Oncol Biol Phys 91(2):351–359
48. Mirabello V, Cortezon-Tamarit F, Pascu SI (2018) Oxygen sensing, hypoxia tracing and in vivo imaging with functional metalloprobes for the early detection of non-communicable diseases. Front Chem 6:27
49. Chapman JD, Franko AJ, Sharplin J (1981) A marker for hypoxic cells in tumours with potential clinical applicability. Br J Cancer 43(4):546–550
50. Rasey JS, Grunbaum Z, Magee S, Nelson NJ, Olive PL, Durand RE et al (1987) Characterization of radiolabeled fluoromisonidazole as a probe for hypoxic cells. Radiat Res 111(2):292–304
51. Jerabek PA, Patrick TB, Kilbourn MR, Dischino DD, Welch MJ (1986) Synthesis and biodistribution of 18F-labeled fluoronitroimidazoles: potential in vivo markers of hypoxic tissue. Int J Rad Appl Instrum A 37(7):599–605
52. Ljungkvist AS, Bussink J, Kaanders JH, van der Kogel AJ (2007) Dynamics of tumor hypoxia measured with bioreductive hypoxic cell markers. Radiat Res 167(2):127–145
53. Rajendran JG, Krohn KA (2005) Imaging hypoxia and angiogenesis in tumors. Radiol Clin N Am 43(1):169–187
54. Gagel B, Reinartz P, Dimartino E, Zimny M, Pinkawa M, Maneschi P et al (2004) pO(2) Polarography versus positron emission tomography ([(18)F] fluoromisonidazole, [(18)F]-2-fluoro-2′-deoxyglucose). An appraisal of radiotherapeutically relevant hypoxia. Strahlenther Onkol 180(10):616–622
55. Padhani AR, Krohn KA, Lewis JS, Alber M (2007) Imaging oxygenation of human tumours. Eur Radiol 17(4):861–872
56. Lee NY, Mechalakos JG, Nehmeh S, Lin Z, Squire OD, Cai S et al (2008) Fluorine-18-labeled fluoromisonidazole positron emission and computed tomography-guided intensity-modulated radiotherapy for head and neck cancer: a feasibility study. Int J Radiat Oncol Biol Phys 70(1):2–13
57. Hendrickson K, Phillips M, Smith W, Peterson L, Krohn K, Rajendran J (2011) Hypoxia imaging with [F-18] FMISO-PET in head and neck cancer: potential for guiding intensity modulated radiation therapy in overcoming hypoxia-induced treatment resistance. Radiother Oncol 101(3):369–375
58. Eschmann SM, Paulsen F, Reimold M, Dittmann H, Welz S, Reischl G et al (2005) Prognostic impact of hypoxia imaging with 18F-misonidazole PET in non-small cell lung cancer and head and neck cancer before radiotherapy. J Nucl Med 46(2):253–260
59. Rajendran JG, Mankoff DA, O'Sullivan F, Peterson LM, Schwartz DL, Conrad EU et al (2004) Hypoxia and glucose metabolism in malignant tumors: evaluation by [18F]fluoromisonidazole and [18F] fluorodeoxyglucose positron emission tomography imaging. Clin Cancer Res 10(7):2245–2252
60. Lawrentschuk N, Poon AM, Foo SS, Putra LG, Murone C, Davis ID et al (2005) Assessing regional hypoxia in human renal tumours using 18F-fluoromisonidazole positron emission tomography. BJU Int 96(4):540–546
61. Bruehlmeier M, Roelcke U, Schubiger PA, Ametamey SM (2004) Assessment of hypoxia and perfusion in human brain tumors using PET with 18F-fluoromisonidazole and 15O-H2O. J Nucl Med 45(11):1851–1859
62. Spence AM, Muzi M, Swanson KR, O'Sullivan F, Rockhill JK, Rajendran JG et al (2008) Regional hypoxia in glioblastoma multiforme quantified with [18F]fluoromisonidazole positron emission tomography before radiotherapy: correlation with time to progression and survival. Clin Cancer Res 14(9):2623–2630
63. Hugonnet F, Fournier L, Medioni J, Smadja C, Hindie E, Huchet V et al (2011) Metastatic renal cell carcinoma: relationship between initial metastasis hypoxia, change after 1 month's sunitinib, and therapeutic response: an 18F-fluoromisonidazole PET/CT study. J Nucl Med 52(7):1048–1055
64. Grkovski M, Lee NY, Schoder H, Carlin SD, Beattie BJ, Riaz N et al (2017) Monitoring early response to chemoradiotherapy with (18)F-FMISO dynamic PET in head and neck cancer. Eur J Nucl Med Mol Imaging 44(10):1682–1691
65. Rischin D, Hicks RJ, Fisher R, Binns D, Corry J, Porceddu S et al (2006) Prognostic significance of [18F]-misonidazole positron emission tomography-detected tumor hypoxia in patients with advanced

head and neck cancer randomly assigned to chemoradiation with or without tirapazamine: a substudy of Trans-Tasman Radiation Oncology Group Study 98.02. J Clin Oncol 24(13):2098–2104
66. Dolbier WR Jr, Li AR, Koch CJ, Shiue CY, Kachur AV (2001) [18F]-EF5, a marker for PET detection of hypoxia: synthesis of precursor and a new fluorination procedure. Appl Radiat Isot 54(1):73–80
67. Chitneni SK, Bida GT, Yuan H, Palmer GM, Hay MP, Melcher T et al (2013) 18F-EF5 PET imaging as an early response biomarker for the hypoxia-activated prodrug SN30000 combined with radiation treatment in a non-small cell lung cancer xenograft model. J Nucl Med 54(8):1339–1346
68. Ali R, Apte S, Vilalta M, Subbarayan M, Miao Z, Chin FT et al (2015) 18F-EF5 PET is predictive of response to fractionated radiotherapy in preclinical tumor models. PLoS One 10(10):e0139425
69. Chitneni SK, Bida GT, Zalutsky MR, Dewhirst MW (2014) Comparison of the hypoxia PET tracer (18) F-EF5 to immunohistochemical marker EF5 in 3 different human tumor xenograft models. J Nucl Med 55(7):1192–1197
70. Qian Y, Von Eyben R, Liu Y, Chin FT, Miao Z, Apte S et al (2018) (18)F-EF5 PET-based imageable hypoxia predicts local recurrence in tumors treated with highly conformal radiation therapy. Int J Radiat Oncol Biol Phys 102:1183–1192
71. Manzoor AA, Yuan H, Palmer GM, Vigilanti BL, Dewhirst MW (2010) Imaging hypoxia. In: Molecular imaging: principles and practice. People's Medical Publishing House, Shelton, pp 756–780
72. Lapi SE, Lewis JS, Dehdashti F (2015) Evaluation of hypoxia with copper-labeled diacetyl-bis(N-methylthiosemicarbazone). Semin Nucl Med 45(2):177–185
73. Dehdashti F, Mintun MA, Lewis JS, Bradley J, Govindan R, Laforest R et al (2003) In vivo assessment of tumor hypoxia in lung cancer with 60Cu-ATSM. Eur J Nucl Med Mol Imaging 30(6):844–850
74. Dehdashti F, Grigsby PW, Mintun MA, Lewis JS, Siegel BA, Welch MJ (2003) Assessing tumor hypoxia in cervical cancer by positron emission tomography with 60Cu-ATSM: relationship to therapeutic response-a preliminary report. Int J Radiat Oncol Biol Phys 55(5):1233–1238
75. Zhang T, Das SK, Fels DR, Hansen KS, Wong TZ, Dewhirst MW et al (2013) PET with 62Cu-ATSM and 62Cu-PTSM is a useful imaging tool for hypoxia and perfusion in pulmonary lesions. AJR Am J Roentgenol 201(5):W698–W706
76. Thulborn KR, Waterton JC, Matthews PM, Radda GK (1982) Oxygenation dependence of the transverse relaxation time of water protons in whole blood at high field. Biochim Biophys Acta 714(2):265–270
77. Howe FA, Robinson SP, McIntyre DJ, Stubbs M, Griffiths JR (2001) Issues in flow and oxygenation dependent contrast (FLOOD) imaging of tumours. NMR Biomed 14(7–8):497–506
78. Ogawa S, Lee TM, Kay AR, Tank DW (1990) Brain magnetic resonance imaging with contrast dependent on blood oxygenation. Proc Natl Acad Sci U S A 87(24):9868–9872
79. Baudelet C, Gallez B (2002) How does blood oxygen level-dependent (BOLD) contrast correlate with oxygen partial pressure (pO2) inside tumors? Magn Reson Med 48(6):980–986
80. Chopra S, Foltz WD, Milosevic MF, Toi A, Bristow RG, Menard C et al (2009) Comparing oxygen-sensitive MRI (BOLD R2*) with oxygen electrode measurements: a pilot study in men with prostate cancer. Int J Radiat Biol 85(9):805–813
81. Hoskin PJ, Carnell DM, Taylor NJ, Smith RE, Stirling JJ, Daley FM et al (2007) Hypoxia in prostate cancer: correlation of BOLD-MRI with pimonidazole immunohistochemistry-initial observations. Int J Radiat Oncol Biol Phys 68(4):1065–1071
82. Taylor NJ, Baddeley H, Goodchild KA, Powell ME, Thoumine M, Culver LA et al (2001) BOLD MRI of human tumor oxygenation during carbogen breathing. J Magn Reson Imaging 14(2):156–163
83. Jiang L, Weatherall PT, McColl RW, Tripathy D, Mason RP (2013) Blood oxygenation level-dependent (BOLD) contrast magnetic resonance imaging (MRI) for prediction of breast cancer chemotherapy response: a pilot study. J Magn Reson Imaging 37(5):1083–1092
84. Neeman M, Dafni H, Bukhari O, Braun RD, Dewhirst MW (2001) In vivo BOLD contrast MRI mapping of subcutaneous vascular function and maturation: validation by intravital microscopy. Magn Reson Med 45(5):887–898
85. Dunn TJ, Braun RD, Rhemus WE, Rosner GL, Secomb TW, Tozer GM et al (1999) The effects of hyperoxic and hypercarbic gases on tumour blood flow. Br J Cancer 80(1–2):117–126
86. Mark CI, Mazerolle EL, Chen JJ (2015) Metabolic and vascular origins of the BOLD effect: implications for imaging pathology and resting-state brain function. J Magn Reson Imaging 42(2):231–246
87. Ruiz-Cabello J, Barnett BP, Bottomley PA, Bulte JW (2011) Fluorine (19F) MRS and MRI in biomedicine. NMR Biomed 24(2):114–129
88. Yu JX, Kodibagkar VD, Cui W, Mason RP (2005) 19F: a versatile reporter for non-invasive physiology and pharmacology using magnetic resonance. Curr Med Chem 12(7):819–848
89. Hunjan S, Zhao D, Constantinescu A, Hahn EW, Antich PP, Mason RP (2001) Tumor oximetry: demonstration of an enhanced dynamic mapping procedure using fluorine-19 echo planar magnetic resonance imaging in the Dunning prostate R3327-AT1 rat tumor. Int J Radiat Oncol Biol Phys 49(4):1097–1108
90. Mason RP, Antich PP, Babcock EE, Constantinescu A, Peschke P, Hahn EW (1994) Non-invasive determination of tumor oxygen tension and local variation with growth. Int J Radiat Oncol Biol Phys 29(1):95–103

91. Noth U, Rodrigues LM, Robinson SP, Jork A, Zimmermann U, Newell B et al (2004) In vivo determination of tumor oxygenation during growth and in response to carbogen breathing using 15C5-loaded alginate capsules as fluorine-19 magnetic resonance imaging oxygen sensors. Int J Radiat Oncol Biol Phys 60(3):909–919
92. Young IR, Clarke GJ, Bailes DR, Pennock JM, Doyle FH, Bydder GM (1981) Enhancement of relaxation rate with paramagnetic contrast agents in NMR imaging. J Comput Tomogr 5(6):543–547
93. Berkowitz BA (1997) Role of dissolved plasma oxygen in hyperoxia-induced contrast. Magn Reson Imaging 15(1):123–126
94. O'Connor JP, Boult JK, Jamin Y, Babur M, Finegan KG, Williams KJ et al (2016) Oxygen-enhanced MRI accurately identifies, quantifies, and maps tumor hypoxia in preclinical cancer models. Cancer Res 76(4):787–795
95. Matsumoto K, Bernardo M, Subramanian S, Choyke P, Mitchell JB, Krishna MC et al (2006) MR assessment of changes of tumor in response to hyperbaric oxygen treatment. Magn Reson Med 56(2):240–246
96. Hallac RR, Zhou H, Pidikiti R, Song K, Stojadinovic S, Zhao D et al (2014) Correlations of noninvasive BOLD and TOLD MRI with pO_2 and relevance to tumor radiation response. Magn Reson Med 71(5):1863–1873
97. Linnik IV, Scott ML, Holliday KF, Woodhouse N, Waterton JC, O'Connor JP et al (2014) Noninvasive tumor hypoxia measurement using magnetic resonance imaging in murine U87 glioma xenografts and in patients with glioblastoma. Magn Reson Med 71(5):1854–1862
98. Zhao D, Pacheco-Torres J, Hallac RR, White D, Peschke P, Cerdan S et al (2015) Dynamic oxygen challenge evaluated by NMR T1 and T2*--insights into tumor oxygenation. NMR Biomed 28(8):937–947
99. Falk SJ, Ward R, Bleehen NM (1992) The influence of carbogen breathing on tumour tissue oxygenation in man evaluated by computerised p02 histography. Br J Cancer 66(5):919–924
100. Le D, Mason RP, Hunjan S, Constantinescu A, Barker BR, Antich PP (1997) Regional tumor oxygen dynamics: 19F PBSR EPI of hexafluorobenzene. Magn Reson Imaging 15(8):971–981
101. Duling BR, Berne RM (1970) Longitudinal gradients in periarteriolar oxygen tension. A possible mechanism for the participation of oxygen in local regulation of blood flow. Circ Res 27(5):669–678
102. Dewhirst MW, Ong ET, Rosner GL, Rehmus SW, Shan S, Braun RD et al (1996) Arteriolar oxygenation in tumour and subcutaneous arterioles: effects of inspired air oxygen content. Br J Cancer Suppl 27:S241–S246
103. Erickson K, Braun RD, Yu D, Lanzen J, Wilson D, Brizel DM et al (2003) Effect of longitudinal oxygen gradients on effectiveness of manipulation of tumor oxygenation. Cancer Res 63(15):4705–4712
104. Colliez F, Gallez B, Jordan BF (2017) Assessing tumor oxygenation for predicting outcome in radiation oncology: a review of studies correlating tumor hypoxic status and outcome in the preclinical and clinical settings. Front Oncol 7:10
105. Krishna MC, Matsumoto S, Yasui H, Saito K, Devasahayam N, Subramanian S et al (2012) Electron paramagnetic resonance imaging of tumor pO(2). Radiat Res 177(4):376–386
106. Gallez B, Baudelet C, Jordan BF (2004) Assessment of tumor oxygenation by electron paramagnetic resonance: principles and applications. NMR Biomed 17(5):240–262
107. Subramanian S, Matsumoto K, Mitchell JB, Krishna MC (2004) Radio frequency continuous-wave and time-domain EPR imaging and overhauser-enhanced magnetic resonance imaging of small animals: instrumental developments and comparison of relative merits for functional imaging. NMR Biomed 17(5):263–294
108. Tatum JL, Kelloff GJ, Gillies RJ, Arbeit JM, Brown JM, Chao KS et al (2006) Hypoxia: importance in tumor biology, noninvasive measurement by imaging, and value of its measurement in the management of cancer therapy. Int J Radiat Biol 82(10):699–757
109. Ilangovan G, Li H, Zweier JL, Kuppusamy P (2002) In vivo measurement of tumor redox environment using EPR spectroscopy. Mol Cell Biochem 234–235(1–2):393–398
110. Hou H, Abramovic Z, Lariviere JP, Sentjurc M, Swartz H, Khan N (2010) Effect of a topical vasodilator on tumor hypoxia and tumor oxygen guided radiotherapy using EPR oximetry. Radiat Res 173(5):651–658
111. Hou H, Dong R, Lariviere JP, Mupparaju SP, Swartz HM, Khan N (2011) Synergistic combination of hyperoxygenation and radiotherapy by repeated assessments of tumor pO2 with EPR oximetry. J Radiat Res 52(5):568–574
112. Hou H, Lariviere JP, Demidenko E, Gladstone D, Swartz H, Khan N (2009) Repeated tumor pO(2) measurements by multi-site EPR oximetry as a prognostic marker for enhanced therapeutic efficacy of fractionated radiotherapy. Radiother Oncol 91(1):126–131
113. Goda F, Bacic G, O'Hara JA, Gallez B, Swartz HM, Dunn JF (1996) The relationship between partial pressure of oxygen and perfusion in two murine tumors after X-ray irradiation: a combined gadopentetate dimeglumine dynamic magnetic resonance imaging and in vivo electron paramagnetic resonance oximetry study. Cancer Res 56(14):3344–3349
114. Goda F, O'Hara JA, Rhodes ES, Liu KJ, Dunn JF, Bacic G et al (1995) Changes of oxygen tension in experimental tumors after a single dose of X-ray irradiation. Cancer Res 55(11):2249–2252
115. Ansiaux R, Baudelet C, Jordan BF, Beghein N, Sonveaux P, De Wever J et al (2005) Thalidomide radiosensitizes tumors through early changes in the

tumor microenvironment. Clin Cancer Res 11(2 Pt 1):743–750
116. Jordan BF, Gregoire V, Demeure RJ, Sonveaux P, Feron O, O'Hara J et al (2002) Insulin increases the sensitivity of tumors to irradiation: involvement of an increase in tumor oxygenation mediated by a nitric oxide-dependent decrease of the tumor cells oxygen consumption. Cancer Res 62(12):3555–3561
117. Elas M, Bell R, Hleihel D, Barth ED, McFaul C, Haney CR et al (2008) Electron paramagnetic resonance oxygen image hypoxic fraction plus radiation dose strongly correlates with tumor cure in FSa fibrosarcomas. Int J Radiat Oncol Biol Phys 71(2):542–549
118. Hou H, Mupparaju SP, Lariviere JP, Hodge S, Gui J, Swartz HM et al (2013) Assessment of the changes in 9L and C6 glioma pO2 by EPR oximetry as a prognostic indicator of differential response to radiotherapy. Radiat Res 179(3):343–351
119. Ardenkjaer-Larsen JH, Laursen I, Leunbach I, Ehnholm G, Wistrand LG, Petersson JS et al (1998) EPR and DNP properties of certain novel single electron contrast agents intended for oximetric imaging. J Magn Reson 133(1):1–12
120. Elas M, Ahn KH, Parasca A, Barth ED, Lee D, Haney C et al (2006) Electron paramagnetic resonance oxygen images correlate spatially and quantitatively with Oxylite oxygen measurements. Clin Cancer Res 12(14 Pt 1):4209–4217
121. Matsumoto S, Yasui H, Mitchell JB, Krishna MC (2010) Imaging cycling tumor hypoxia. Cancer Res 70(24):10019–10023
122. Matsumoto S, Hyodo F, Subramanian S, Devasahayam N, Munasinghe J, Hyodo E et al (2008) Low-field paramagnetic resonance imaging of tumor oxygenation and glycolytic activity in mice. J Clin Invest 118(5):1965–1973
123. Matsumoto S, Batra S, Saito K, Yasui H, Choudhuri R, Gadisetti C et al (2011) Antiangiogenic agent sunitinib transiently increases tumor oxygenation and suppresses cycling hypoxia. Cancer Res 71(20):6350–6359
124. Zijlstra WG, Buursma A (1991) Meeuwsen-van der Roest WP. Absorption spectra of human fetal and adult oxyhemoglobin, de-oxyhemoglobin, carboxyhemoglobin, and methemoglobin. Clin Chem 37(9):1633–1638
125. Leow MK (2007) Configuration of the hemoglobin oxygen dissociation curve demystified: a basic mathematical proof for medical and biological sciences undergraduates. Adv Physiol Educ 31(2):198–201
126. Severinghaus JW (1979) Simple, accurate equations for human blood O_2 dissociation computations. J Appl Physiol Respir Environ Exerc Physiol 46(3):599–602
127. van der Sanden BP, Heerschap A, Hoofd L, Simonetti AW, Nicolay K, van der Toorn A et al (1999) Effect of carbogen breathing on the physiological profile of human glioma xenografts. Magn Reson Med 42(3):490–499
128. Baikejiang R, Zhang W, Li C (2017) Diffuse optical tomography for breast cancer imaging guided by computed tomography: a feasibility study. J Xray Sci Technol 25(3):341–355
129. Kim MJ, Su MY, Yu HJ, Chen JH, Kim EK, Moon HJ et al (2016) US-localized diffuse optical tomography in breast cancer: comparison with pharmacokinetic parameters of DCE-MRI and with pathologic biomarkers. BMC Cancer 16:50
130. Gunther JE, Lim EA, Kim HK, Flexman M, Altoe M, Campbell JA et al (2018) Dynamic diffuse optical tomography for monitoring neoadjuvant chemotherapy in patients with breast cancer. Radiology 287(3):778–786
131. Tromberg BJ, Pogue BW, Paulsen KD, Yodh AG, Boas DA, Cerussi AE (2008) Assessing the future of diffuse optical imaging technologies for breast cancer management. Med Phys 35(6):2443–2451
132. Erickson PR, Moor KJ, Werner JJ, Latch DE, Arnold WA, McNeill K (2018) Singlet oxygen phosphorescence as a probe for triplet-state dissolved organic matter reactivity. Environ Sci Technol 52(16):9170–9178
133. DeRosa CA, Samonina-Kosicka J, Fan Z, Hendargo HC, Weitzel DH, Palmer GM et al (2015) Oxygen sensing difluoroboron dinaphthoylmethane polylactide. Macromolecules 48(9):2967–2977
134. Palmer GM, Fontanella AN, Zhang G, Hanna G, Fraser CL, Dewhirst MW (2010) Optical imaging of tumor hypoxia dynamics. J Biomed Opt 15(6):066021
135. Wilson DF, Evans SM, Jenkins WT, Vinogradov SA, Ong E, Dewhirst MW (1998) Oxygen distributions within R3230Ac tumors growing in dorsal flap window chambers in rats. Adv Exp Med Biol 454:603–609
136. Cardenas-Navia LI, Mace D, Richardson RA, Wilson DF, Shan S, Dewhirst MW (2008) The pervasive presence of fluctuating oxygenation in tumors. Cancer Res 68(14):5812–5819
137. Overgaard J, Horsman MR (1996) Modification of hypoxia-induced radioresistance in tumors by the use of oxygen and sensitizers. Semin Radiat Oncol 6(1):10–21
138. Fontanella AN, Schroeder T, Hochman DW, Chen RE, Hanna G, Haglund MM et al (2013) Quantitative mapping of hemodynamics in the lung, brain, and dorsal window chamber-grown tumors using a novel, automated algorithm. Microcirculation 20(8):724–735

Hypoxia-Induced Phenotypes that Mediate Tumor Heterogeneity

3

Jin Qian and Erinn B. Rankin

Abstract

Intratumoral heterogeneity is an important factor contributing to metastasis and therapy resistance. The phenotypic diversity of cancer cells within the tumor microenvironment is strongly influenced by microenvironmental factors such as hypoxia. Clinically, hypoxia and the hypoxia inducible transcription factors HIF-1 and HIF-2 are associated with cancer stem cells, metastasis and drug resistance in multiple tumor types. Experimental models have demonstrated an important functional role for HIF signaling in driving CSC, metastatic and drug resistant phenotypes *in vitro* and *in vivo*. Here we will review recent studies that highlight novel mechanisms by which hypoxia promotes cancer stem cell, metastatic and drug resistant phenotypes.

Keywords

Hypoxia · Metastasis · Stemness · Cancer stem cell · HIF · Therapy resistance · Chemotherapy · Metabolism

3.1 Introduction

It is well established that tumors are composed of a heterogeneous population of tumor cells that are morphologically and phenotypically distinct. The phenotypic diversity of cancer cells within the tumor microenvironment is thought to be a major diver of metastasis and therapy resistance. There are multiple mechanisms driving tumor heterogeneity including genetic, epigenetic and microenvironmental factors [1]. Accumulating evidence suggests that hypoxia is an important microenvironmental factor that influences tumor heterogeneity by promoting cancer stem cell, metastatic and drug resistant phenotypes.

Areas of hypoxia, or low oxygen tensions, are commonly found in solid tumors. Hypoxia develops when there is an imbalance between oxygen delivery and oxygen consumption [2]. Within the tumor microenvironment, oxygen delivery is impaired due to the irregular vasculature characterized by distended capillaries with leaky and sluggish blood flow [2]. In addition, oxygen consumption rates are increased by proliferating tumor cells and infiltrating immune cells. The hypoxia inducible transcription factors HIF-1

J. Qian
Department of Radiation Oncology, Stanford University School of Medicine, Stanford, CA, USA

E. B. Rankin (✉)
Department of Radiation Oncology, Stanford University School of Medicine, Stanford, CA, USA

Department of Obstetrics & Gynecologic Oncology, Stanford University School of Medicine, Stanford, CA, USA
e-mail: erankin@stanford.edu

D. M. Gilkes (ed.), *Hypoxia and Cancer Metastasis*, Advances in Experimental Medicine and Biology 1136, https://doi.org/10.1007/978-3-030-12734-3_3

and HIF-2 coordinate the adaptive cellular response to low oxygen tensions by activating gene expression programs controlling glycolysis, angiogenesis, cell survival, invasion, immune suppression, the cancer stem cell phenotype, metastasis and resistance to conventional therapy [3]. In the presence of oxygen, the α-subunits of HIF-1 and HIF-2 are rapidly degraded through the cooperative actions of prolyl hydroxylase enzymes (PHD1, 2, and 3) and the E3 ubiquitin ligase substrate recognition component VHL [4]. Under hypoxia, HIF-1 and HIF-2 are stabilized and activate the expression of genes containing hypoxia response elements (HRE) [5]. Over 200 genes are activated in response to HIF-1 and HIF-2 that allow cells to survive and adapt to low oxygen tensions.

3.2 Hypoxia and Cancer Stem Cells

Cancer stem cells (CSCs) are a rare subpopulation of tumor cells within the tumor microenvironment that are thought to mediate tumor initiation, metastasis and therapy resistance. Cancer stem cells were first described in AML and have subsequently been identified in many solid tumor types including breast, colorectal, brain, pancreatic and ovarian cancers [6]. Functionally, CSC are defined by their unlimited self-renewal, clonal tumor initiation and clonal long-term repopulation capacity [7]. Recent evidence suggests that CSCs exist in a state of plasticity where they can transition between stem and non-stem cell states. Accumulating data suggests that cellular and molecular factors within the tumor microenvironment play a key role in determining CSC self-renewal, promote their metastatic potential as well as protect them from chemotherapy and radiotherapy [8]. Here we will review recent literature describing how hypoxia and HIF signaling within the tumor microenvironment influences the CSC phenotype.

Clinically, HIF-1 and HIF-2 are highly expressed in CSCs. Immunohistochemical analysis of HIF-1 and HIF-2 with CSC markers such as CD133 or CD44 revealed high levels of HIF-1 and HIF-2 expression within CSC pools in glioma, acute myeloid leukemia (AML), and breast cancer [9–11]. In experimental models, HIF signaling is required to maintain the CSC phenotype *in vitro* and *in vivo*. Samanta et al. observed that paclitaxel or gemcitabine treatment of triple negative breast cancer cell lines induced HIF expression and activity and enriched the breast CSC population [12]. In this model, coadministration of HIF inhibitors reduced the breast CSC resistance to paclitaxel and gemcitabine resulting in increased tumor control [12]. In the polyoma virus middle T antigen (PyMT) transgenic model of breast cancer, genetic inactivation of HIF-1 reduced the number of CD133+ breast CSCs and decreased the CSC pool in limiting dilution studies in vivo [13]. Moreover, Cecil et al. utilized an immunization approach to test the hypothesis that HIF-1 is immunogenic in triple negative breast cancer and active immunization against HIF-1 could impact tumor growth and CSCs in a basal model of breast cancer. Indeed, HIF-1 vaccination reduced the CSC population and growth of basal mammary tumors [14]. Knockdown of HIF-1 or HIF-2 in CSCs is also sufficient to reduce glioma CSC survival and suppress CSC-mediated tumor growth [11]. Similarly, genetic or therapeutic targeting of HIF-1 abrogated the colony-forming unit (cfu) ability of mouse lymphoma and human AML CSCs [10]. In chronic myeloid leukemia stem cells, deletion of HIF-1 inhibits CML progression by impairing cell-cycle progression and inducing apoptosis of LSCs [15]. Collectively, these studies demonstrate that HIF-1 and HIF-2 are highly expressed within human and mouse CSCs where they play an important functional role in mediating CSC growth and survival. These findings suggest that therapeutic strategies targeting HIF-1/HIF-2 may be efficacious in reducing CSCs and promote therapeutic responses to conventional chemo radiation therapy.

3.3 Mechanisms of HIF-Mediated CSC Phenotype

There are multiple mechanisms by which hypoxia and HIF signaling promote CSC growth and survival including upregulation of CSC surface markers, induction of pluripotency factors, stemness pathways, metabolism, epithelial to mesenchymal transition (EMT), and mediating CSC crosstalk with their niche. Here we will review the recent literature identifying mechanisms by which HIF signaling promotes the CSC phenotype.

3.3.1 HIF Signaling and CSC Surface Marker Expression

There has been much interest in the identification of unique markers expressed by CSCs, as these markers have the potential to be exploited as both prognostic factors and therapeutic targets. Currently, there are approximately 30 CSC surface markers that have been identified across multiple tumor types [16]. Most of the CSC cell surface markers are glycoproteins including CD133, CD24 and CD44 [16]. More recently, there has been interest in the identification of metabolic CSC markers. For example, Li et al. identified an increase in lipid unsaturation levels as a marker of ovarian cancer stem cells (CSCs) compared to non-CSCs. Importantly, they found that lipid unsaturation, mediated by the lipid desaturase SCD1, is important to maintain CSC stemness and tumor formation indicating that lipid unsaturation levels may be an functional marker of ovarian CSCs [17]. Future studies are needed to determine if increased unsaturated lipid are found in additional CSC populations.

Given the important role of hypoxia and HIF signaling in maintaining CSCs, multiple CSC markers have been shown to be regulated by hypoxia. CD133 is an cell surface glycoprotein that is associated with CSCs in a variety of cancers including the brain, prostate, liver, pancreas and colon cancer [18]. Multiple studies observed that CD133 expression on CSCs is induced by hypoxia and HIF [19–22]. Molecular analysis of the CD133 (PROM1) promoter revealed that HIF-1 and HIF-2 bind to the PROM1 promoter at an ETS transcription factor binding site. PROM1 expression mediated by HIF likely occurs through an association with ETS-family transcription factors [22]. In support of this notion, immunoprecipitation studies identified a physical interaction between HIF-1 and ETS transcription factor ELK1 [22]. CD24 is another cell surface protein that is associated with a CSC phenotype and plays and particularly important role in the progression of hepatocellular carcinoma and male urothelial cancers [23, 24]. Combined analysis of HIF-1 and CD24 expression in a cohort of 101 human urothelial cancer specimens revealed a statistically significant association with reduced overall survival suggesting an coordinate regulation between HIF-1 and CD24 [25]. Functional studies demonstrated that HIF-1 is both necessary and sufficient for the hypoxic induction of CD24 in multiple cell lines. Moreover, a functional hypoxia-responsive element was identified in the promoter of CD24 demonstrating a direct regulation of CD24 by HIF-1 [25]. Importantly, genetic inactivation of HIF-1 reduced primary and metastatic growth that could be rescued in part by CD24 overexpression [25]. These suggest that the CSC membrane protein CD24 as an important effector of HIF-1 activity. In the glioma stem cell niche, CD44 expression correlates with hypoxia induced gene signatures. CD44 is a glycoprotein transmembrane receptor and CSC markers that binds to osteopontin. Osteopontin binding to CD44 results in cleavage of the CD44 intracellular domain (ICD) that promotes glioma growth and a stem cell-like phenotype in a HIF-2 dependent manner. Mechanistically, the CD44 ICD enhances the transcriptional activity of HIF-2 in a CBP/p300 dependent manner to promote stem cell marker gene expression and facilitate tumor growth [26]. Future studies are needed to elucidate the mechanisms by which the CD44 ICD promotes HIF-2 activity. A recent report suggests that CD44 ICD directly interacts with HIF-2, but not HIF-1, to promote stabilization of the HIF-2 CBP/p300 complex [27]. Collectively, these data identify a functional link between hypoxic signaling and CSC stem cell markers.

3.3.2 HIF Signaling and Pluripotency Transcription Factors

The core pluripotency transcription factors OCT4, SOX2, NANOG and KLF4 are required to maintain embryonic stem cell pluripotency and self-renewal [28, 29]. Similar to embryonic stem cells, CSCs express OCT4, SOX2, NANOG and KLF4 to maintain a stem cell phenotype and tumor growth [30–35].

Hypoxia and HIF signaling play an important role in stem cell proliferation and differentiation during development and cancer. During development, germline inactivation of *Arnt*, the common binding partner for Hif-1α and Hif-2α results in embryonic lethality at E9.5-E10.5 due to defective placental vascularization and trophoblast lineage differentiation [36]. Analysis of trophoblast stem cells revealed that HIF activity is required to induce the expression of Mash2, a bHLH protein that is directs specific trophoblast lineages [36]. Similar results were observed in *Hif-1/Hif-2* double knockout mice demonstrating a critical role for HIF transcriptional activity in trophoblast lineage determination [36]. Covello et al. utilized an elegant genetic knock-in strategy to compare the relative contribution of Hif-1 and Hif-2 on stem cell function during development. Interestingly, *Hif-2* knock-in into the *Hif-1* locus resulted in embryonic lethality at an earlier timepoint (E6.5-E7.5) compared to the *Hif-1/Hif-2* deficient embryos. In this setting, embryonic lethality was associated with developmental patterning defects and increased expression of Oct-4 [37]. This observation lead to the identification of Oct-4 as a novel Hif-2 target. Importantly, knockdown of Oct-4 was sufficient to rescue some of the hematopoietic defects in Hif-2 knock-in cells indicating that Hif-2 modulates stem cell function through Oct-4 during development [37].

In cancer, HIF activity has been shown to regulate the expression of multiple pluripotency factors. In prostate tumors, HIF-1, NANOG, OCT4 and SOX2 staining co-localize indicating a co-regulation of these factors in human tumors [38, 39]. Indeed, HIF signaling can promote the expression of the core pluripotency transcription factors in cancer cells through direct and indirect mechanisms [19, 38]. As described above, a direct role for HIF-2 in the regulation of OCT4 has been demonstrated through the identification of a functional hypoxia response element in the Oct-4 promoter [37]. Hypoxic regulation of Oct4 has been detected in multiple cancer cell lines [38, 40]. Data suggest that HIF-1 and HIF-2 indirectly promote SOX2 expression through the regulation of NOTCH1 signaling [39, 41]. NANOG expression is regulated by HIF in cancer cells through transcriptional and posttranscriptional mechanisms. In breast cancer cells, HIF-1 can promote NANOG expression by upregulating the expression of adenosine receptor 2B (A2BR) resulting in an increase in adenosine signaling and activation of protein kinase C-δ leading to STAT3-mediated activation Nanog transcription [42]. In addition, HIF-mediated increase in the expression of the cystine transporter xCT and the regulatory subunit of glutamate-cysteine ligase (GCLM) results in an increase in intracellular glutathione levels that act to suppress mitogen activated protein kinase kinase (MEK) through copper chelation. MEK-ERK inhibition leads to the nuclear translocation of the FoxO3 transcription factor and activation of its target Nanog [43]. HIFs can also increase NANOG levels in cancer cells by controlling Nanog mRNA stability. HIF-1 transactivation of ALKBH5, an N^6-methyladenosine demethylase, results in HIF dependent demethylation of N^6-methyladenosine of Nanog mRNA [44]. While m^6A modifications of mRNA can control protein binding, mRNA splicing, translation, stability and degradation; m^6A modifications on pluripotency factors such as Nanog results in reduced mRNA stability [45, 46]. Indeed, HIF-mediated ALKBH5 promotes Nanog m^6A demethylation and mRNA stability leading to increased Nanog and enhanced breast cancer stem cell enrichment [44]. Together these data illustrate multiple mechanisms by which HIF activity can promote the CSC phenotype through the activation of pluripotency factor gene expression.

3.3.3 HIF Signaling and CSC Signaling Pathways

Similar to embryonic or tissue stem cells, CSCs activate conserved signaling pathways that promote development and tissue homeostasis including Notch, Hedgehog and Wnt pathways (for a recent review [47]). Hypoxia and the activation of HIF signaling plays an important role in maintaining stem cell signaling in development and in cancer [48]. For example, Notch signaling is an evolutionary conserved pathway controlling cell fate in most metazoan species [49]. Gustafsson et al. discovered that hypoxia promotes an undifferentiated cell state through the activation of Notch signaling. They discovered that the Notch intracellular domain interacts with HIF-1α and HIF-1α is recruited to Notch-responsive promoters under hypoxic conditions to enhance the activation of downstream Notch target genes [50]. In human glioblastoma, hypoxia promotes the tumorigenic capacity of glioma stem cells through the activation of HIF and Notch signaling [11, 51]. HIF-1 may augment Notch signaling under hypoxic conditions by increasing the expression of vasorin, a protein that binds to Notch1 at the cell membrane to prevent its degradation [52]. In AML, HIF promotes CSC self-renewal by inhibiting the Notch negative feedback loop mediated by Hes1 [10]. Thus, there are multiple mechanisms by which HIF signaling controls Notch signaling to maintain a CSC phenotype in tumor cells. In addition to the Notch pathway, hypoxia has also been shown to promote the self-renewal properties of CSCs through activation of the Wnt/β-catenin [53, 54], JAK/STAT3 [55], NF-kB [56], PI3K/AKT and Hippo-YAP pathways [57].

3.3.4 HIF Signaling and CSC Metabolism

Altered metabolism is a hallmark of cancer [58]. There is growing interest in understanding the metabolic heterogeneity of CSCs and how metabolic stressors such as hypoxia affect the CSC phenotype. Samanta et al. observed that metabolic adaptation is required for the hypoxic induction of breast cancer stem cells [59]. Under hypoxic conditions, BCSCs upregulate the expression of multiple genes controlling metabolic adaptation including phosphoglycerate dehydrogenase (PHGDH), an enzyme that shunts glucose metabolites into the serine synthesis pathway. Genetic inactivation of PHGDH resulted in increased oxidant levels, apoptosis, loss of BCSCs and reduced tumorigenic and metastatic potential of BCSCs [59]. Luo et al. recently observed that the transition of breast cancer stem cells (BCSCs) from a quiescent mesenchymal-like (M) state to a proliferative epithelial-like (E) state is regulated by the redox state within the cell. They found that ROS-induced BCSC transition from the M to E state is facilitated by activation of HIF-1 and Notch signaling [60]. This finding is consistent with another report demonstrating that MYC and MCL1 amplification in TNBC promotes CSC enrichment and tumor growth through ROS-mediated accumulation of HIF-1 [61]. These studies identify a molecular link between cellular redox status and CSC maintenance mediated by HIF-1.

3.3.5 HIF Signaling, EMT and CSC

The epithelial to mesenchymal (EMT) program is a key pathway driving the CSC phenotype. In human clinical samples, CSC populations express high levels of the EMT transcription factors (Slug, Snail, Zeb1) or mesenchymal markers (vimentin, N-cadherin and reduced E-cadherin) relative to non-CSC populations. Moreover, experimental manipulation of the EMT program through knockdown of EMT transcription factors such as Snail in CSC populations can significantly reduce the CSC phenotype in vitro and in vivo (for a recent review [1]). Conversely, overexpression of EMT transcription factors such as TWIST1 is sufficient to promote and CSC-like phenotype [62].

Hypoxia is an important microenvironmental factor promoting EMT. Hypoxic signaling promotes EMT in cancer cells through direct and indirect mechanisms. For example, the expression of the EMT transcription factors Twist1, Zeb1/2, and Snail are directly induced by hypoxia in and HIF-dependent manner through the activation of hypoxia response elements (HREs) within regulatory elements of these genes [63–65]. HIF signaling has also been shown to indirectly promote EMT through the activation of Notch, Wnt, TGFββ, integrin-linked kinase (ILK), and the receptor tyrosine kinase AXL (for a recent review [66]). Thus, hypoxia and HIF signaling have the potential to influence CSC populations through the regulation of EMT. In support of this hypothesis, hypoxia induced CSC EMT and tumor formation through HIF-1 mediated activation of Snail in gastric cancer cell lines [67].

3.3.6 HIF Signaling and CSC Crosstalk in their Niche

Similar to normal stem cells, CSCs reside within specialized niches where they rely on stromal derived factors to maintain stemness, protect them from the immune system and facilitate their metastatic potential [8]. For example, cancer associated fibroblasts are an important cellular component of the CSC niche. They secrete factors that promote stem cell signaling pathways such as Wnt, Notch, and Hedgehog [8]. In colorectal cancer, CAFs secrete TGF-B2, which activates the expression of GLI2, a transcription factor involved in the Hedgehog pathway. Interestingly, hypoxia cooperates with CAFs to further induce GLI2 expression in colorectal cancer cells resulting in robust resistance to chemotherapy [68].

Peritoneal metastases are commonly found in ovarian and gastric cancer patients, where CSC rich spheroids present in malignant ascites fluid are abundant sources for peritoneal dissemination, distant metastasis to peritoneal organs and therapy relapse [69]. A recent study suggests that the hypoxic peritoneal milky spots (PMS) at the omentum may create an ideal CSC niche for gastric cancer engraftments [70]. In animal models, gastric CSCs preferentially localized to hypoxic regions of PMS, and these engraftments expressed high levels of stem genes. Moreover, inhibition of HIF-1 in gastric cancer cells significantly reduced peritoneal dissemination and self-renewal [70]. Future studies are needed to understand how hypoxia is induced and maintained within milky spots and whether hypoxia influences other cell compartments of peritoneum surface (macrophages, adipocytes and mesenchymal cells, etc.) to support CSC stemness through paracrine signaling or cell-to-cell contact.

A growing area in CSC research is understanding the immunobiological properties of CSCs and how these properties are modulated by the tumor microenvironment. Cancer stem cells have been shown to inhibit both innate and adaptive immune responses through multiple mechanisms (for a recent review [71]). In addition, the hypoxic microenvironment promotes an immunosuppressive phenotype through inhibition of natural killer cell mediated antitumor responses, promoting resistance to T cell mediated killing, inhibition of macrophage phagocytosis, recruitment of regulatory T cells, and inhibition of T cell effector functions [66]. Recent studies have focused on understanding how hypoxia potentiates CSC immune suppression. In B16-F10 melanoma tumors, hypoxia-induced Nanog promotes a CSC phenotype, but also promotes the infiltration of regulatory T cells through the upregulation of TGF-B1 providing a molecular link between the hypoxic induction of the CSC and immunosuppressive phenotype [72]. In glioblastoma, hypoxia enhanced glioblastoma CSC ability to inhibit T cell proliferation and activation as well as macrophage phagocytosis. The immunosuppressive effects of hypoxia in this model were mediated through the activation of STAT3 and HIF-1 [73]. Another mechanism by which HIF signaling couples immune suppression with a CSC phenotype is through the activation of CD47. CD47 is a cell surface protein that inhibits macrophage mediated phagocytosis by binding to signal regulatory protein SIRPα macrophages [74]. CD47 also regulates the breast cancer CSC phenotype through mechanisms that remain unclear [75]. Zhang and colleagues discovered that HIF-1 directly activates the expression of CD47 to promote a CSC phenotype and inhibit macrophage-mediated phagocytosis [75]. In addition, HIF signaling activates the expression of the immunosuppressive factors CD73 and PDL1 to promote chemotherapy-induced immune evasion in TNBCs [76]. Together, these studies suggest that HIF signaling is an important pathway coupling innate and adaptive immune suppression with a CSC phenotype.

3.4 HIF Signaling, CSC and Metastasis

Metastasis remains the leading cause of cancer related deaths. Metastatic disease is thought to arise from disseminated cancer cells, or cancer stem cells, that retain the ability to reinitiate tumor growth at a distant tissue site [3]. There are multiple proposed models to explain how metastatic stem cells accumulate at distant sites. For example, it is thought that in some cases such as colorectal cancer, the metastatic CSCs originate from the primary tumor where they travel to distant tissue sites and resume their regenerative potential [3]. Another model is metastatic stem cells at distant sites arise through phenotypic plasticity and regain tumor initiating capacity [3]. Regardless of their origin, the tumor microenvironment at both the primary and distant metastatic sites plays an important role in promoting the metastatic phenotype of CSCs [3].

In addition to regulating the CSC phenotype, hypoxia plays an important role in metastasis. Hypoxia and the activation of HIF signaling is associated with metastasis and poor survival in variety of cancer patients [5, 66, 77, 78]. Moreover, preclinical studies have shown that hypoxia and HIF signaling promote multiple steps within the metastatic cascade and have been reviewed extensively elsewhere [66]. Here we will focus our review on recent literature that has identified a link functional link between hypoxic signaling, the CSC metastatic phenotype and metastasis.

As described above, Notch signaling plays an important role in stem cell self-renewal. This pathway is activated by HIF signaling through multiple mechanisms to promote a CSC phenotype. In breast cancer patients, the Notch ligand Jagged 2, correlates with poor overall and metastasis-free survival. Xing and colleagues discovered that breast cancer cells activate Notch signaling under hypoxic conditions to promote metastasis and a CSC phenotype. They found that Jagged 2 expression and activation of Notch are activated within the hypoxic invasive front of breast cancer specimens. Mechanistically, hypoxia increases the expression of Jagged 2 and Notch signaling to promote tumor cell EMT, survival and CSC-like phenotype, indicating that hypoxia may facilitate metastatic and CSC phenotype through the activation of Notch signaling [79]. Another mechanism by which hypoxia may coordinate the metastatic and CSC phenotype is through the activation of CD24 expression. In hepatocellular carcinoma, CD24 is an important functional CSC marker that is required to maintain self-renewal, differentiation and metastasis [24]. Thomas et al. discovered that CD24 is a direct HIF-1 target that promotes HIF-1 driven metastasis in preclinical models of bladder and prostate cancer [25]. Pang et al. recently linked hypoxia with the CSC phenotype and metastasis through the induction of integrin-like kinase (ILK). ILK is a kinase that transmits signals from the B1-integrin and the extracellular matrix to promote tumor cell migration, invasion, proliferation and angiogenesis [80]. Pang et al. found that hypoxia and stiff microenvironments upregulated the expression of B1-integrin and ILK to promote a CSC phenotype in invasive breast cancer cell lines suggesting an link between hypoxia, the ECM and the CSC metastatic phenotype [81]. A recent study utilized a genetic model of pancreatic ductal adenocarcinoma (PDAC) to identify a subpopulation of cancer cells with enhanced metastatic potential. Within this population, they found elevated levels of the transcription factor BLIMP1 and hypoxic gene signatures suggesting that BLIMP1 may be a hypoxia induced gene. Indeed BLIMP1 is a novel HIF-1 target that is required for the metastatic potential of PDAC metastatic stem cells. Interestingly, approximately 35% of hypoxia regulated genes within this metastatic population were regulated by BLIMP1 under hypoxic conditions suggesting that BLIMP1 is an hypoxia regulated factor that potentiates a subset of hypoxia inducible genes within metastatic cell populations. These data suggest that hypoxia promotes the metastatic behavior of PDAC cells through the activation of BLIMP1 [82]. Collectively, these studies identify an important role for hypoxia and HIF signaling in promoting metastatic stem cell phenotypes.

3.5 HIF, CSCs and Chemoresistance

Chemoresistance is a major clinical challenge in the treatment of cancer. Experimental evidence suggests that cancer stem cells are enriched in cancer populations that persist following chemotherapy and facilitate the regrowth of disease in the recurrent setting. There are multiple mechanisms by which CSCs can mediate chemoresistance including upregulation of drug efflux pumps, enhanced DNA-repair capacity, enhanced protection against reactive oxygen species and induction of an quiescent state (for a recent review [7]). Hypoxia is an important microenvironmental factor that facilitates chemoresistance. Clinically, hypoxia and HIF expression is associated with therapeutic resistance or decreased survival following ionizing radiation or chemotherapy [83–85]. There are multiple mechanisms by which hypoxia and HIF signaling promote chemoresistance including upregulation of drug efflux, induction of autophagy, hypoxia-driven selection of tumor cells with reduced apoptotic capacity and inhibition of DNA damage, metabolic reprogramming EMT and the CSC phenotype (for a recent review [86]). Here we will focus our discussion on recent studies that have identified a role of HIF in promoting chemoresistance associated with a CSC phenotype.

Advanced stage colorectal cancers are commonly resistant to chemotherapy. Cancer stem cells have been implicated in colorectal cancer relapse. Hypoxia is also known to promote chemoresistance in colorectal cancer cells although the mechanisms by which this occurs remain unclear. Tang et al. propose that one mechanism by which hypoxia facilitates chemoresistance in colorectal cancer occurs through an interaction with the CSC niche. Cancer associated fibroblasts are a key cellular component of the colorectal CSC niche and produce multiple factors including Wnts, HGF, and IL-11 to promote colorectal CSCs and metastasis. Under hypoxic conditions, CAFs secrete the growth factor TGF-B2 leading to the induction of the hedgehog transcription factor GLI2 in colorectal cancer cells. Notably, GLI2 expression in colorectal cancer cells is required for chemoresistance induced by CAFs and hypoxia. Clinically, HIF-1/TGF-B2/GLI2 expression are associated with relapse and poor patient survival suggesting that targeting these pathways in combination may be efficacious to sensitize resistant colorectal tumors to chemotherapy [68]. Future studies are needed to determine whether hypoxic signaling in additional stromal cell populations may enhance CSC and chemoresistance phenotypes.

Triple negative breast cancer (TNBC) is commonly treated with cytotoxic chemotherapeutic agents given the lack of known molecular targets. The development of recurrent chemoresistant disease is a leading cause of mortality in women diagnosed with TNBC. Cancer stem cells are thought to be a rare population of cells in TNBC patients that are resistant to chemotherapy and drive relapse following adjuvant chemotherapy [87]. Current research efforts are focused on elucidating the mechanisms by which breast cancer stem cells facilitate recurrence and metastasis. Multiple groups have observed that chemotherapy induces the normoxic expression of HIF-1 and/or HIF-2 to drive breast cancer stem cell enrichment and drug resistance [12, 43, 88, 89]. There are likely multiple mechanisms by which chemotherapy induces the normoxic accumulation of HIF. Cao et al. found that doxorubicin induced the expression of HIF in a STAT1-dependent manner. STAT1 activation by chemotherapy resulted in the activation of iNOS and the accumulation of intracellular nitric oxide, which can impair the normoxic degradation of HIF-1 [88]. Another mechanism by which HIF-signaling is activated in TNBC following chemotherapy is mediated through the induction of MYC and

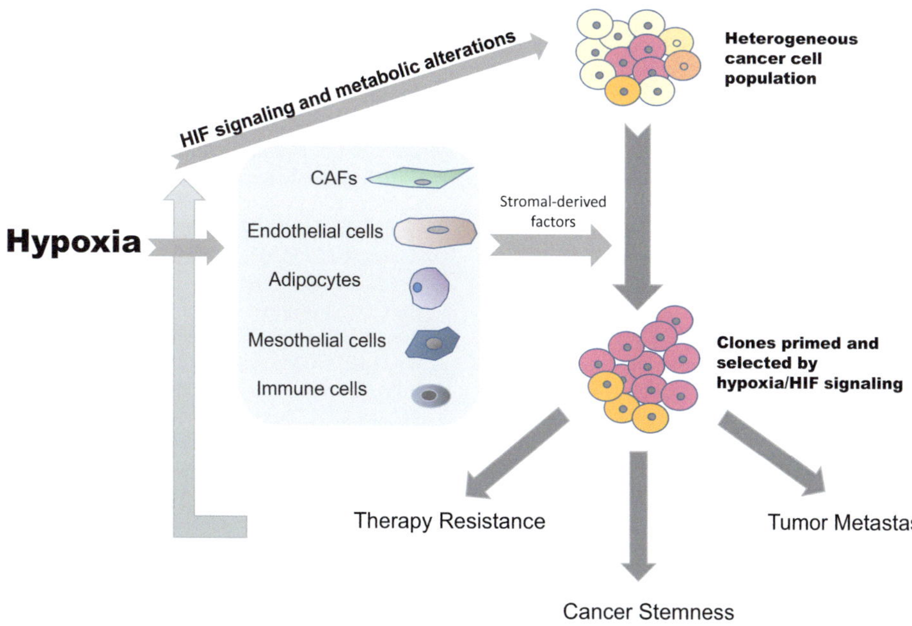

Fig. 3.1 Mechanisms of hypoxia induced tumor heterogeneity. The hypoxic tumor microenvironment facilitates cancer stem cell, metastatic and therapy resistant phenotypes through multiple mechanisms. Both HIF-dependent and independent mechanisms drive the selection of cancer stem cells within the hypoxic tumor microenvironment that contribute to metastatic disease and therapy relapse. Additionally, the selection of chemoresistant clones can stabilize HIF under normoxic conditions to further promote stemness and metastatic phenotypes

MCL1. MYC and MCL1 are commonly coamplified in drug resistant TNBC following neoadjuvant chemotherapy. MYC and MCL1 cooperatively promote mitochondrial oxidative phosphorylation and the generation of ROS. In this setting, increased ROS enhances the normoxic stabilization of HIF-1 to promote CSC enrichment in chemotherapy resistant TNBC [61]. Once stabilized, HIF promotes the BCSC phenotype and chemoresistance through multiple mechanisms including increased IL-6 and IL-8 signaling, increased expression of multidrug resistance 1 gene expression and the activation of pluripotency factors [12, 43, 89].

3.6 Conclusion

Together the findings presented in this review highlight the central role for hypoxia and HIF signaling in coordinating CSC, chemoresistant and metastatic phenotypes in cancer (Fig. 3.1). With the clinical development of HIF inhibitors, it will be important to investigate the combination of these agents with chemotherapy and/or targeted therapies aimed at the inhibition of CSC signaling pathways.

Acknowledgements This work was supported by the Office of the Assistant Secretary of Defense for Health Affairs through the Department of Defense Ovarian Cancer Research Program under Award No. W81XWH-15-1-0097 (EBR). We apologize to those colleagues whose work we could not cite due to space constraints.

References

1. Shibue T, Weinberg RA (2017) EMT, CSCs, and drug resistance: the mechanistic link and clinical implications. Nat Rev Clin Oncol 14:611–629. https://doi.org/10.1038/nrclinonc.2017.44
2. Brown JM, Giaccia AJ (1998) The unique physiology of solid tumors: opportunities (and problems) for cancer therapy. Cancer Res 58:1408–1416

3. Oskarsson T, Batlle E, Massague J (2014) Metastatic stem cells: sources, niches, and vital pathways. Cell Stem Cell 14:306–321. https://doi.org/10.1016/j.stem.2014.02.002
4. Jaakkola P et al (2001) Targeting of HIF-alpha to the von Hippel-Lindau ubiquitylation complex by O2-regulated prolyl hydroxylation. Science 292: 468–472. doi:https://doi.org/10.1126/science.1059796 1059796 [pii]
5. Rankin EB, Giaccia AJ (2008) The role of hypoxia-inducible factors in tumorigenesis. Cell Death Differ 15: 678–685. doi: cdd200821 [pii] 10.1038/cdd.2008.21
6. Bonnet D, Dick JE (1997) Human acute myeloid leukemia is organized as a hierarchy that originates from a primitive hematopoietic cell. Nat Med 3:730–737
7. Batlle E, Clevers H (2017) Cancer stem cells revisited. Nat Med 23:1124–1134. https://doi.org/10.1038/nm.4409
8. Plaks V, Kong N, Werb Z (2015) The cancer stem cell niche: how essential is the niche in regulating stemness of tumor cells? Cell Stem Cell 16:225–238. https://doi.org/10.1016/j.stem.2015.02.015
9. Oliveira-Costa JP et al (2011) Differential expression of HIF-1alpha in CD44+CD24−/low breast ductal carcinomas. Diagn Pathol 6:73. https://doi.org/10.1186/1746-1596-6-73
10. Wang Y, Liu Y, Malek SN, Zheng P, Liu Y (2011) Targeting HIF1alpha eliminates cancer stem cells in hematological malignancies. Cell Stem Cell 8:399–411. https://doi.org/10.1016/j.stem.2011.02.006
11. Li Z et al (2009) Hypoxia-inducible factors regulate tumorigenic capacity of glioma stem cells. Cancer Cell 15:501–513. https://doi.org/10.1016/j.ccr.2009.03.018
12. Samanta D, Gilkes DM, Chaturvedi P, Xiang L, Semenza GL (2014) Hypoxia-inducible factors are required for chemotherapy resistance of breast cancer stem cells. Proc Natl Acad Sci USA 111:E5429–E5438. https://doi.org/10.1073/pnas.1421438111
13. Schwab LP et al (2012) Hypoxia-inducible factor 1alpha promotes primary tumor growth and tumor-initiating cell activity in breast cancer. Breast Cancer Res 14:R6. https://doi.org/10.1186/bcr3087
14. Cecil DL et al (2017) Immunization against HIF-1alpha inhibits the growth of basal mammary tumors and targets mammary stem cells in vivo. Clin Cancer Res 23:3396–3404. https://doi.org/10.1158/1078-0432.CCR-16-1678
15. Zhang H, Li H, Xi HS, Li S (2012) HIF1alpha is required for survival maintenance of chronic myeloid leukemia stem cells. Blood 119:2595–2607. https://doi.org/10.1182/blood-2011-10-387381
16. Mallard BW, Tiralongo J (2017) Cancer stem cell marker glycosylation: nature, function and significance. Glycoconj J 34:441–452. https://doi.org/10.1007/s10719-017-9780-9
17. Li J et al (2017) Lipid desaturation is a metabolic marker and therapeutic target of ovarian cancer stem cells. Cell Stem Cell 20:303–314 e305. https://doi.org/10.1016/j.stem.2016.11.004
18. Glumac PM, LeBeau AM (2018) The role of CD133 in cancer: a concise review. Clin Transl Med 7:18. https://doi.org/10.1186/s40169-018-0198-1
19. Bar EE, Lin A, Mahairaki V, Matsui W, Eberhart CG (2010) Hypoxia increases the expression of stem-cell markers and promotes clonogenicity in glioblastoma neurospheres. Am J Pathol 177:1491–1502. https://doi.org/10.2353/ajpath.2010.091021
20. Soeda A et al (2009) Hypoxia promotes expansion of the CD133-positive glioma stem cells through activation of HIF-1alpha. Oncogene 28:3949–3959. https://doi.org/10.1038/onc.2009.252
21. Iida H, Suzuki M, Goitsuka R, Ueno H (2012) Hypoxia induces CD133 expression in human lung cancer cells by up-regulation of OCT3/4 and SOX2. Int J Oncol 40:71–79. https://doi.org/10.3892/ijo.2011.1207
22. Ohnishi S et al (2013) Hypoxia-inducible factors activate CD133 promoter through ETS family transcription factors. PLoS One 8:e66255. https://doi.org/10.1371/journal.pone.0066255
23. Overdevest JB et al (2012) CD24 expression is important in male urothelial tumorigenesis and metastasis in mice and is androgen regulated. Proc Natl Acad Sci USA 109:E3588–E3596. https://doi.org/10.1073/pnas.1113960109
24. Lee TK et al (2011) CD24(+) liver tumor-initiating cells drive self-renewal and tumor initiation through STAT3-mediated NANOG regulation. Cell Stem Cell 9:50–63. https://doi.org/10.1016/j.stem.2011.06.005
25. Thomas S et al (2012) CD24 is an effector of HIF-1-driven primary tumor growth and metastasis. Cancer Res 72:5600–5612. https://doi.org/10.1158/0008-5472.CAN-11-3666
26. Pietras A et al (2014) Osteopontin-CD44 signaling in the glioma perivascular niche enhances cancer stem cell phenotypes and promotes aggressive tumor growth. Cell Stem Cell 14:357–369. https://doi.org/10.1016/j.stem.2014.01.005
27. Johansson E et al (2017) CD44 interacts with HIF-2alpha to modulate the hypoxic phenotype of perinecrotic and perivascular glioma cells. Cell Rep 20:1641–1653. https://doi.org/10.1016/j.celrep.2017.07.049
28. Boiani M, Scholer HR (2005) Regulatory networks in embryo-derived pluripotent stem cells. Nat Rev Mol Cell Biol 6:872–884. https://doi.org/10.1038/nrm1744
29. Boyer LA et al (2005) Core transcriptional regulatory circuitry in human embryonic stem cells. Cell 122:947–956. https://doi.org/10.1016/j.cell.2005.08.020
30. Ben-Porath I et al (2008) An embryonic stem cell-like gene expression signature in poorly differentiated aggressive human tumors. Nat Genet 40:499–507. https://doi.org/10.1038/ng.127
31. Hu T et al (2008) Octamer 4 small interfering RNA results in cancer stem cell-like cell apoptosis. Cancer

Res 68:6533–6540. https://doi.org/10.1158/0008-5472.CAN-07-6642
32. Sarig R et al (2010) Mutant p53 facilitates somatic cell reprogramming and augments the malignant potential of reprogrammed cells. J Exp Med 207:2127–2140. https://doi.org/10.1084/jem.20100797
33. Hu G et al (2009) A genome-wide RNAi screen identifies a new transcriptional module required for self-renewal. Genes Dev 23:837–848. https://doi.org/10.1101/gad.1769609
34. Ikushima H et al (2009) Autocrine TGF-beta signaling maintains tumorigenicity of glioma-initiating cells through Sry-related HMG-box factors. Cell Stem Cell 5:504–514. https://doi.org/10.1016/j.stem.2009.08.018
35. Jeter CR et al (2009) Functional evidence that the self-renewal gene NANOG regulates human tumor development. Stem Cells 27:993–1005. https://doi.org/10.1002/stem.29
36. Cowden Dahl KD et al (2005) Hypoxia-inducible factors 1alpha and 2alpha regulate trophoblast differentiation. Mol Cell Biol 25:10479–10491. https://doi.org/10.1128/MCB.25.23.10479-10491.2005
37. Covello KL et al (2006) HIF-2alpha regulates Oct-4: effects of hypoxia on stem cell function, embryonic development, and tumor growth. Genes Dev 20:557–570. https://doi.org/10.1101/gad.1399906
38. Mathieu J et al (2011) HIF induces human embryonic stem cell markers in cancer cells. Cancer Res 71:4640–4652. https://doi.org/10.1158/0008-5472.CAN-10-3320
39. Bae KM, Dai Y, Vieweg J, Siemann DW (2016) Hypoxia regulates SOX2 expression to promote prostate cancer cell invasion and sphere formation. Am J Cancer Res 6:1078–1088
40. Kumar SM et al (2012) Acquired cancer stem cell phenotypes through Oct4-mediated dedifferentiation. Oncogene 31:4898–4911. https://doi.org/10.1038/onc.2011.656
41. Seo EJ et al (2016) Hypoxia-NOTCH1-SOX2 signaling is important for maintaining cancer stem cells in ovarian cancer. Oncotarget 7:55624–55638. https://doi.org/10.18632/oncotarget.10954
42. Lan J et al (2018) Hypoxia-inducible factor 1-dependent expression of adenosine receptor 2B promotes breast cancer stem cell enrichment. Proc Natl Acad Sci USA. https://doi.org/10.1073/pnas.1809695115
43. Lu H et al (2015) Chemotherapy triggers HIF-1-dependent glutathione synthesis and copper chelation that induces the breast cancer stem cell phenotype. Proc Natl Acad Sci USA 112:E4600–E4609. https://doi.org/10.1073/pnas.1513433112
44. Zhang C et al (2016) Hypoxia induces the breast cancer stem cell phenotype by HIF-dependent and ALKBH5-mediated m(6)A-demethylation of NANOG mRNA. Proc Natl Acad Sci USA 113:E2047–E2056. https://doi.org/10.1073/pnas.1602883113
45. Meyer KD, Jaffrey SR (2014) The dynamic epitranscriptome: N6-methyladenosine and gene expression control. Nat Rev Mol Cell Biol 15:313–326. https://doi.org/10.1038/nrm3785
46. Geula S et al (2015) Stem cells. m6A mRNA methylation facilitates resolution of naive pluripotency toward differentiation. Science 347:1002–1006. https://doi.org/10.1126/science.1261417
47. Takebe N et al (2015) Targeting Notch, hedgehog, and Wnt pathways in cancer stem cells: clinical update. Nat Rev Clin Oncol 12:445–464. https://doi.org/10.1038/nrclinonc.2015.61
48. Keith B, Simon MC (2007) Hypoxia-inducible factors, stem cells, and cancer. Cell 129:465–472. https://doi.org/10.1016/j.cell.2007.04.019
49. Artavanis-Tsakonas S, Rand MD, Lake RJ (1999) Notch signaling: cell fate control and signal integration in development. Science 284:770–776
50. Gustafsson MV et al (2005) Hypoxia requires notch signaling to maintain the undifferentiated cell state. Dev Cell 9:617–628. https://doi.org/10.1016/j.devcel.2005.09.010
51. Qiang L et al (2012) HIF-1alpha is critical for hypoxia-mediated maintenance of glioblastoma stem cells by activating Notch signaling pathway. Cell Death Differ 19:284–294. https://doi.org/10.1038/cdd.2011.95
52. Man J et al (2018) Hypoxic induction of vasorin regulates Notch1 turnover to maintain glioma stem-like cells. Cell Stem Cell 22:104–118 e106. https://doi.org/10.1016/j.stem.2017.10.005
53. Dong HJ et al (2016) The wnt/beta-catenin signaling/Id2 cascade mediates the effects of hypoxia on the hierarchy of colorectal-cancer stem cells. Sci Rep 6:22966. https://doi.org/10.1038/srep22966
54. Giambra V et al (2015) Leukemia stem cells in T-ALL require active Hif1alpha and wnt signaling. Blood 125:3917–3927. https://doi.org/10.1182/blood-2014-10-609370
55. Almiron Bonnin DA et al (2018) Secretion-mediated STAT3 activation promotes self-renewal of glioma stem-like cells during hypoxia. Oncogene 37:1107–1118. https://doi.org/10.1038/onc.2017.404
56. Qin J et al (2017) Hypoxia-inducible factor 1 alpha promotes cancer stem cells-like properties in human ovarian cancer cells by upregulating SIRT1 expression. Sci Rep 7:10592. https://doi.org/10.1038/s41598-017-09244-8
57. Kim H, Lin Q, Glazer PM, Yun Z (2018) The hypoxic tumor microenvironment in vivo selects the cancer stem cell fate of breast cancer cells. Breast Cancer Res 20:16. https://doi.org/10.1186/s13058-018-0944-8
58. Hanahan D, Weinberg RA (2011) Hallmarks of cancer: the next generation. Cell 144:646–674. doi:S0092-8674(11)00127-9 [pii] 10.1016/j.cell.2011.02.013
59. Samanta D et al (2016) PHGDH expression is required for mitochondrial redox homeostasis, breast Cancer stem cell maintenance, and lung metastasis. Cancer Res 76:4430–4442. https://doi.org/10.1158/0008-5472.CAN-16-0530

60. Luo M et al (2018) Targeting breast Cancer stem cell state equilibrium through modulation of redox signaling. Cell Metab 28:69–86 e66. https://doi.org/10.1016/j.cmet.2018.06.006
61. Lee KM et al (2017) MYC and MCL1 cooperatively promote chemotherapy-resistant breast Cancer stem cells via regulation of mitochondrial oxidative phosphorylation. Cell Metab 26:633–647 e637. https://doi.org/10.1016/j.cmet.2017.09.009
62. Schmidt JM et al (2015) Stem-cell-like properties and epithelial plasticity arise as stable traits after transient Twist1 activation. Cell Rep 10:131–139. https://doi.org/10.1016/j.celrep.2014.12.032
63. Krishnamachary B et al (2006) Hypoxia-inducible factor-1-dependent repression of E-cadherin in von hippel-Lindau tumor suppressor-null renal cell carcinoma mediated by TCF3, ZFHX1A, and ZFHX1B. Cancer Res 66:2725–2731. doi:66/5/2725 [pii] 10.1158/0008-5472.CAN-05-3719
64. Yang MH et al (2008) Direct regulation of TWIST by HIF-1alpha promotes metastasis. Nat Cell Biol 10: 295–305. doi:ncb1691 [pii] 10.1038/ncb1691
65. Imai T et al (2003) Hypoxia attenuates the expression of E-cadherin via up-regulation of SNAIL in ovarian carcinoma cells. Am J Pathol 163: 1437–1447. doi:S0002-9440(10)63501-8 [pii] 10.1016/S0002-9440(10)63501-8
66. Rankin EB, Giaccia AJ (2016) Hypoxic control of metastasis. Science 352:175–180. https://doi.org/10.1126/science.aaf4405
67. Yang SW et al (2017) HIF-1alpha induces the epithelial-mesenchymal transition in gastric cancer stem cells through the Snail pathway. Oncotarget 8:9535–9545. https://doi.org/10.18632/oncotarget.14484
68. Tang YA et al (2018) Hypoxic tumor microenvironment activates GLI2 via HIF-1alpha and TGF-beta2 to promote chemoresistance in colorectal cancer. Proc Natl Acad Sci USA 115:E5990–E5999. https://doi.org/10.1073/pnas.1801348115
69. Lupia M, Cavallaro U (2017) Ovarian cancer stem cells: still an elusive entity? Mol Cancer 16:64. https://doi.org/10.1186/s12943-017-0638-3
70. Miao ZF et al (2014) Peritoneal milky spots serve as a hypoxic niche and favor gastric cancer stem/progenitor cell peritoneal dissemination through hypoxia-inducible factor 1alpha. Stem Cells 32:3062–3074. https://doi.org/10.1002/stem.1816
71. Maccalli C, Rasul KI, Elawad M, Ferrone S (2018) The role of cancer stem cells in the modulation of anti-tumor immune responses. Semin Cancer Biol. https://doi.org/10.1016/j.semcancer.2018.09.006
72. Hasmim M et al (2013) Cutting edge: hypoxia-induced Nanog favors the intratumoral infiltration of regulatory T cells and macrophages via direct regulation of TGF-beta1. J Immunol 191:5802–5806. https://doi.org/10.4049/jimmunol.1302140
73. Wei J et al (2011) Hypoxia potentiates glioma-mediated immunosuppression. PLoS One 6:e16195. https://doi.org/10.1371/journal.pone.0016195
74. Chao MP, Weissman IL, Majeti R (2012) The CD47-SIRPalpha pathway in cancer immune evasion and potential therapeutic implications. Curr Opin Immunol 24:225–232. https://doi.org/10.1016/j.coi.2012.01.010
75. Zhang H et al (2015) HIF-1 regulates CD47 expression in breast cancer cells to promote evasion of phagocytosis and maintenance of cancer stem cells. Proc Natl Acad Sci USA 112:E6215–E6223. https://doi.org/10.1073/pnas.1520032112
76. Samanta D et al (2018) Chemotherapy induces enrichment of CD47(+)/CD73(+)/PDL1(+) immune evasive triple-negative breast cancer cells. Proc Natl Acad Sci USA 115:E1239–E1248. https://doi.org/10.1073/pnas.1718197115
77. Schindl M et al (2002) Overexpression of hypoxia-inducible factor 1alpha is associated with an unfavorable prognosis in lymph node-positive breast cancer. Clin Cancer Res 8:1831–1837
78. Yamamoto Y et al (2008) Hypoxia-inducible factor 1alpha is closely linked to an aggressive phenotype in breast cancer. Breast Cancer Res Treat 110:465–475. https://doi.org/10.1007/s10549-007-9742-1
79. Xing F et al (2011) Hypoxia-induced Jagged2 promotes breast cancer metastasis and self-renewal of cancer stem-like cells. Oncogene 30:4075–4086. https://doi.org/10.1038/onc.2011.122
80. Hannigan G, Troussard AA, Dedhar S (2005) Integrin-linked kinase: a cancer therapeutic target unique among its ILK. Nat Rev Cancer 5:51–63. https://doi.org/10.1038/nrc1524
81. Pang MF et al (2016) Tissue stiffness and hypoxia modulate the integrin-linked kinase ILK to control breast Cancer stem-like cells. Cancer Res 76:5277–5287. https://doi.org/10.1158/0008-5472.CAN-16-0579
82. Chiou SH et al (2017) BLIMP1 induces transient metastatic heterogeneity in pancreatic Cancer. Cancer Discov 7:1184–1199. https://doi.org/10.1158/2159-8290.CD-17-0250
83. Vergis R et al (2008) Intrinsic markers of tumour hypoxia and angiogenesis in localised prostate cancer and outcome of radical treatment: a retrospective analysis of two randomised radiotherapy trials and one surgical cohort study. Lancet Oncol 9:342–351. https://doi.org/10.1016/S1470-2045(08)70076-7
84. Generali D et al (2006) Hypoxia-inducible factor-1alpha expression predicts a poor response to primary chemoendocrine therapy and disease-free survival in primary human breast cancer. Clin Cancer Res 12:4562–4568. https://doi.org/10.1158/1078-0432.CCR-05-2690
85. Koukourakis MI et al (2002) Hypoxia-inducible factor (HIF1A and HIF2A), angiogenesis, and chemoradiotherapy outcome of squamous cell head-and-neck cancer. Int J Radiat Oncol Biol Phys 53:1192–1202
86. Wilson WR, Hay MP (2011) Targeting hypoxia in cancer therapy. Nat Rev Cancer 11:393–410. https://doi.org/10.1038/nrc3064

87. Yu KD et al (2013) Identification of prognosis-relevant subgroups in patients with chemoresistant triple-negative breast cancer. Clin Cancer Res 19:2723–2733. https://doi.org/10.1158/1078-0432.CCR-12-2986
88. Cao Y et al (2013) Tumor cells upregulate normoxic HIF-1alpha in response to doxorubicin. Cancer Res 73:6230–6242. https://doi.org/10.1158/0008-5472.CAN-12-1345
89. Lu H et al (2017) Chemotherapy-induced Ca(2+) release stimulates breast Cancer stem cell enrichment. Cell Rep 18:1946–1957. https://doi.org/10.1016/j.celrep.2017.02.001

Hypoxia and the Tumor Secretome

4

Ying Liu, Gabrielle E. Ciotti, and T. S. Karin Eisinger-Mathason

Abstract

Metastasis remains the leading cause of cancer-related deaths. To date, there are no specific treatments targeting disseminated disease. New therapeutic options will become available only if we enhance our understanding of mechanisms underlying metastatic spread. A large body of literature shows that the metastatic potential of tumor cells is strongly influenced by microenvironmental cues such as low oxygen (hypoxia). Clinically, hypoxia is a hallmark of most solid tumors and is associated with increased metastasis and poor survival in a variety of cancer types. Mechanistically, hypoxia influences multiple steps within the metastatic cascade and particularly impacts the interactions between tumor cells and host stroma at both primary and secondary sites. Here we review current evidence for a hypoxia-induced tumor secretome and its impact on metastatic progression. These studies have identified potential biomarkers and therapeutic targets that could be integrated into strategies for preventing and treating metastatic disease.

Keywords

Hypoxia · Tumor microenvironment · Metastasis · Secretome · Pro-survival · Blood vessel formation · Immune evasion · Motility · ECM remodeling · Vascular permeability · Premetastatic niche

4.1 Introduction

Metastasis is a complex process that selects for highly aggressive tumor cells able to disseminate from their primary tumor, survive in the circulation, and grow in foreign environments. This process involves a reciprocal interplay between tumor cells and host stroma at both primary and secondary sites, and the metastatic potential is profoundly influenced by microenvironmental factors such as hypoxia. Among the diverse ways by which hypoxia promotes metastasis, the regulation of tumor cells' secretome is emerging as a key mechanism. Hypoxia-induced tumor secretome impacts almost all steps in the metastatic cascade and presents significant opportunities for therapeutic intervention (Fig. 4.1).

The term 'secretome' was first coined in 2000 in a study on bacterial secretory pro-

Y. Liu · G. E. Ciotti · T. S. K. Eisinger-Mathason (✉)
The Abramson Family Cancer Research Institute, Department of Pathology and Laboratory Medicine, University of Pennsylvania Perelman School of Medicine, Philadelphia, PA, USA
e-mail: karineis@pennmedicine.upenn.edu

D. M. Gilkes (ed.), *Hypoxia and Cancer Metastasis*, Advances in Experimental Medicine and Biology 1136,
https://doi.org/10.1007/978-3-030-12734-3_4

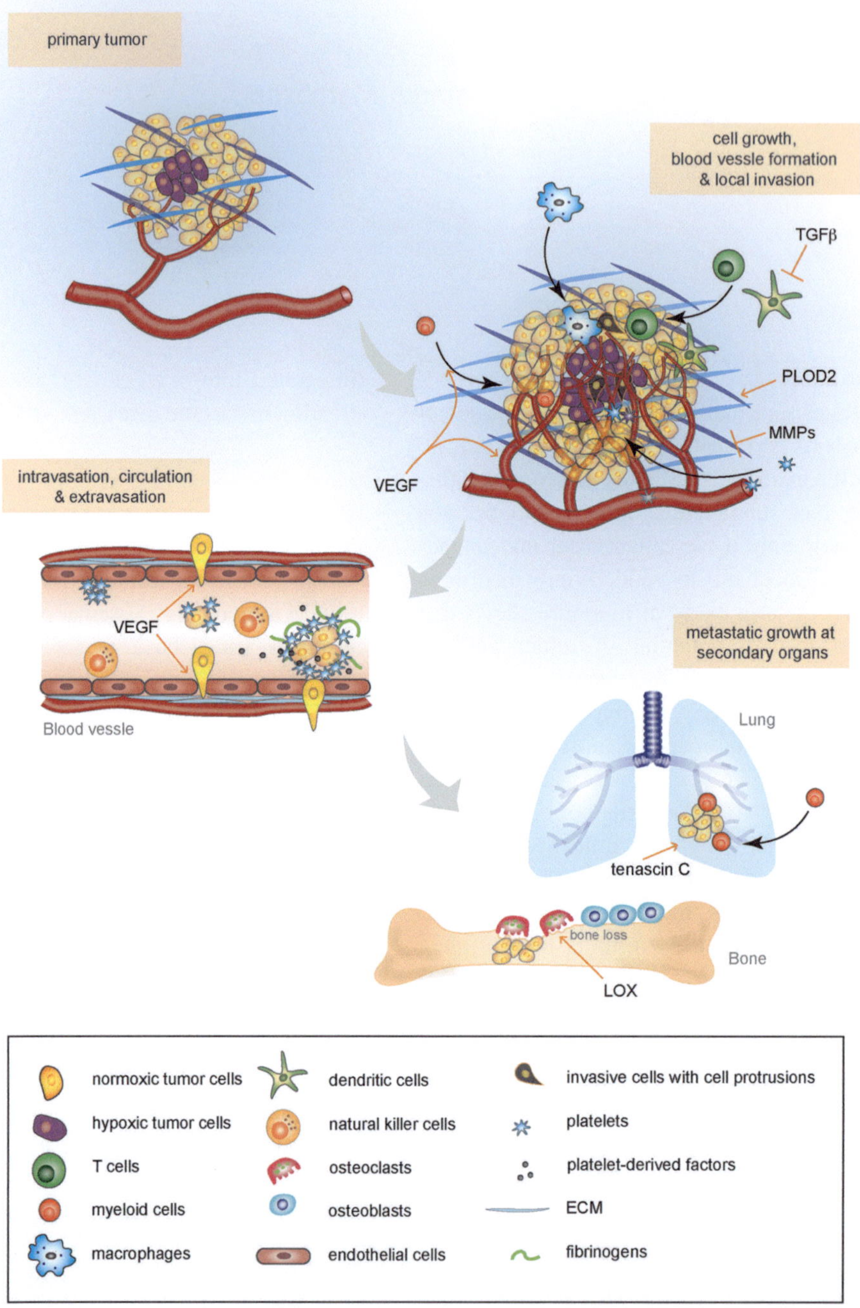

Fig. 4.1 Hypoxia Induced Tumor Secretome in the Metastatic Cascade Hypoxia promotes primary tumor cell growth, blood vessel formation, immune escape, and local invasion by up-regulating the expression and secretion of factors such as VEGF, TGFβ, MMPs, and PLOD2. As tumor cells intravasate into and extravasate out of the blood vessels, hypoxia increases endothelial permeability and facilitates transendothelial migration. Hypoxia also protects disseminating tumor cells against immunolysis through factors such as ADP that can activate and recruit platelets to shield malignant cells. In distant target organs, hypoxia fosters the formation of a premetastatic niche and the metastatic outgrowth through the secretion of factors such as LOX and tenascin C. Abbreviation: *VEGF* vascular endothelial growth factor, *TGF*β transforming growth factor β, *MMPs* matrix metalloproteinases, *PLOD2* Procollagen-lysine,2-oxoglutarate 5-dioxygenase 2, *LOX* lysyl oxidase

cesses, and was later re-defined as 'the global group of proteins secreted into the extracellular space by a cell, tissue, organ, or organism through known and unknown mechanisms involving constitutive and regulated secretory organelles [1, 2]. Secreted proteins can be released into the extracellular space through at least two mechanisms: the classical and non-classical secretory pathways. In the former pathway, proteins targeted for extracellular release are synthesized as protein precursors that usually contain N-terminal signal sequences; these signal sequences allow trafficking to the rough endoplasmic reticulum (ER) and subsequently to the Golgi apparatus, following which the proteins are released into the extracellular space in Golgi-derived secretory vesicles [3, 4]. In the non-classical secretory pathways, proteins can be exported through ER/Golgi-independent mechanisms such as exosomes and ectosomes (a.k.a. shedding vesicles, microvesicles, exosome-like vesicles, nanoparticles, microparticles, and oncosomes) [5]. In the exosomal pathway, cargos are carried in exosomes (30–120 nm diameter) derived from the exocytosis of multivesicular bodies (MVB); whereas in ectosome shedding, cargos are released in ectosomes (100–1000 nm diameter) generated by plasma membrane budding and pinching off [5–7]. Importantly, it is now evident that contents released by these alternative mechanisms include not only proteins but also non-protein cargos such as lipids, micro-RNAs (miRNA), and messenger RNAs (mRNA). Here we use the term 'secretome' to describe all kinds of molecules released by tumor cells regardless of secretory pathways. We will provide an overview of the hypoxia-induced tumor secretome and discuss the mechanisms by which the secreted factors 1) allow tumor cells to survive and adapt to low oxygen tensions, and 2) confer changes in the microenvironment that dramatically affect immune surveillance, invasion/migration, and metastatic colonization. Finally, we will discuss how the hypoxia-induced tumor secretome may guide new areas of research into therapeutic strategies designed to prevent and treat metastasis.

4.2 Acquisition of Pro-metastatic Traits at the Primary Site

4.2.1 Promotion of Cell Survival

Hypoxia is a hallmark of the microenvironment of most solid tumors. In the harsh hypoxic milieu unfavorable for rapid tumor expansion, tumor cells rewire their signaling programs to permit cell survival and raise metastatic propensity. One way to achieve this is to secrete proteins to the tumor microenvironment to activate pro-proliferative and/or anti-apoptotic pathways *via* autocrine mechanisms. For instance, in breast cancer cells, hypoxia induces the secretion of heat shock protein-90α (HSP90α) that helps tumor cells survive the hostile environment through the LDL receptor-related protein-1 (LRP1) receptor [8]. As the HSP90 family proteins are upregulated in a variety of tumor types [9–11], the HSP90-LRP1 pathway and its pro-survival role are likely not only limited to breast cancer but may also be implicated in other cancer types. In colon cancer, hypoxia induces the secretion of periostin. Periostin is overexpressed in more than 80% of human colon cancers examined, with the highest expression in metastatic tumors. Molecularly, secreted periostin enhances colon cancer growth by activating the protein kinase B (PKB/AKT) signaling pathway through integrin αvβ3 [12]. Another integrin-interacting protein secreted by tumors in response to hypoxia is the matrix glycoprotein osteopontin (OPN). OPN has been shown to facilitate metastasis and has prognostic value in several types of malignancies, including gastric and ovarian cancers. Mechanistically, similar to periostin, OPN activates AKT to enhance cell survival, and it also promotes angiogenesis [13, 14].

In addition to proteins, the secretion of exosomes represents an important and effective mechanism for communication between tumor cells and their microenvironment to enhance survival. In bladder cancer, hypoxia induces the tumor cells to secrete oncogenic long non-coding RNA-urothelial cancer-associated 1 (UCA1) through exosomes. Highly expressed in bladder cancer tissues, lnRNA-UCA1 promotes cell proliferation by regulating several downstream targets, including cAMP response element-binding protein (CREB), chromatin remodeling factor BRG1, phosphoinositide 3-kinase (PI3K), AKT and WNT pathways [15]. Similarly, in prostate cancer, hypoxia upregulates exosomes rich in lactate; inhibition of the exosome release impaired the survival of the prostate tumor cells, although the underlying mechanism is unclear [16].

4.2.2 Induction of Blood Vessel Formation

Blood vessel formation involves multiple steps. During embryogenesis, blood vessel formation is initiated *de novo* by vasculogenesis where bone marrow-derived endothelial progenitor cells (EPC) form an immature vascular plexus [17]. EPCs differentiate into endothelial cells (ECs), which interact with the extracellular matrix (ECM) creating a tight barrier between the blood and tissue. Following vasculogenesis, angiogenesis occurs, where new blood vessels are generated from pre-existing vasculature. There are two types of angiogenesis: sprouting and intussuscepted angiogenesis. Both types occur in utero and in adults and are thought to occur in essentially all tissues and organs [17]. Sprouting angiogenesis is initiated when oxygen sensing machineries detect a state of hypoxia and thus demand new blood vessel formation to meet the metabolic requirements of parenchymal cells [18]. As implied by its name, sprouting angiogenesis is characterized by sprouts composed of endothelial cells that migrate across ECM in response to pro-angiogenic signals, allowing for vessel sprouting to occur [17]. Sprouting angiogenesis can, therefore, add blood vessels to parts of tissues previously containing no blood vessels. In contrast, in intussuscepted angiogenesis, the capillary wall extends into the lumen of an existing vessel, forms a pillar that expands and splits a single vessel into two. Post angiogenesis, newly formed vessels then undergo maturation by interacting with smooth muscle cells and pericytes [17].

As a tumor rapidly proliferates, it outgrows the blood supply leaving regions within the tumor low in nutrients and oxygen. Tumor cells, in response to hypoxia and hypoxia-inducible factor (HIF) signaling, produce and release pro-angiogenic growth factors and enzymes that affect all steps of blood vessel formation, rapidly generating abundant but disordered blood vessels. Hypoxic tumor cells secrete vascular endothelial growth factor (VEGF), which is a highly specific and potent mitogen for endothelial cells; VEGF contributes to the EPCs' recruitment from the bone marrow and the induction of their differentiation into ECs [19–23]. Several other HIF targets, including the stromal-derived factor-1 (SDF-1), fibroblast growth factors (FGFs), and monocyte chemoattractant protein-1 (MCP-1), also act as chemoattractive signals and recruit bone marrow-derived cells to sites of vasculogenesis [24–26]. Overexpressed in breast cancer and glioma, stem cell factor (SCF) is another HIF-1α transcriptional target that mediates neovascularization by enhancing EC survival, migration, and EPC mobilization [27–30]. Hypoxic tumors also secrete enzymes such as matrix metalloproteinases (MMPs) to split the pre-existing vessels to allow new vessels to form. Finally, hypoxia supports vessel maturation *via* induction and secretion of angiopoietin 1 (ANG-1), platelet-derived growth factor (PDGF), and transforming growth factor β1 (TGF-β1) by recruiting supporting cells such as smooth muscle cells and pericytes and thereby creating mature and stable blood vessels [31–33].

In addition to increasing pro-angiogenic factors, hypoxic tumors also reduce the production of angiogenesis inhibitors. For instance, hypoxic tumor cells decrease the levels of thrombospondin, a secreted adhesive glycoprotein that is anti-angiogenic; attenuation of thrombospondin expression thus removes an important brake on angiogenesis [34].

Recent studies have highlighted the role of hypoxia-induced exosomes in angiogenesis. In glioblastoma multiforme (GBM), exosomes mediate hypoxia-dependent stimulation of angiogenesis through the activation of paracrine signaling in endothelial cells and pericytes; hypoxia-resistant multiple myeloma (HR-MM) cells release and transfer exosomal miR-135b to endothelial cells to enhance angiogenesis [35]. Hypoxic lung cancer cells produce exosomes containing miR-23a that can increase angiogenesis by targeting prolyl hydroxylase and tight junction protein zonula occludens 1 (ZO-1) [36]. Also, exosomes derived from hypoxic leukemia cells enhances the tube formation in human umbilical vein endothelial cells (HUVECs) *via* miR-210 [37]. Moreover, hypoxic non-small cell lung cancers (NSCLCs) produces exosomal miR-494 that promotes angiogenesis by downregulating phosphatase and tensin homolog (PTEN) and activating the AKT/eNOS pathway in ECs [38].

Notably, the transdifferentiation of tumor cells into endothelial cells has emerged as another mechanism of tumor-induced angiogenesis. Several studies demonstrate that stem cell-like cells in glioblastoma, which frequently contains hypoxic regions, can transdifferentiate into vascular endothelial cells [39–41]. Interestingly, hypoxia appears to be an important factor in this transdifferentiation process that is independent of VEGF [41]. Later studies revealed that the endothelial transdifferentiation of tumor cells is mediated through the Twist1-Jagged1/KLF4 axis [42]. Inhibition of the γ-secretase, which is responsible for the cleaving and release of membrane-bound Jagged1 to extracellular space as a free ligand, blocks the transdifferentiation [42]; therefore, it is an interesting speculation that hypoxia may promote this process by producing soluble Jagged1.

4.2.3 Immune Escape During Invasion

During dissemination, tumor cells commonly escape the immune surveillance using a complex set of mechanisms that prevent the immune system from mounting an effective response. Hypoxic microenvironment induces tumor cells to produce chemoattractants and other soluble factors that recruit myeloid-derived suppressor cells (MDSC), T cells, and macrophages to tumor sites.

Through the secretion of chemokines and cytokines, including CC-chemokine ligand 5 (CCL5), C-X-C motif chemokine ligand 12 (CXCL12 or SDF-1), VEGF, and endothelins, hypoxic tumor cells recruit MDSCs that are potent suppressors of CD8+ T cells, natural killer (NK) cells, dendritic cells, and macrophages. [43–50] Similarly, the secreted VEGF, CCL22, CCL28, and TGF-β, inhibit effector T-cell response and promote the generation and recruitment of immunosuppressive Tregs [51–53]. Once in the tumor microenvironment, Tregs promote immune tolerance and angiogenesis and thus facilitate metastasis. Hypoxic tumor-secreted interleukin 10 (IL10) and TGF-β, induce the differentiation of tumor-associated macrophages (TAM) into immunosuppressive M2 macrophages [54]. Moreover, the production of VEGF and TGF-β the maturation and function of dendritic cells (DCs) [53, 55]. VEGF can also increase the expression of the programmed death ligand 1 (PD-L1) in DC, which is inhibitory for T-cell function [56].

Hypoxia was also shown to induce immunosuppression by upregulating COX-2 expression and subsequently prostaglandin E2 (PGE2) production in tumor cells. COX-2 is a proinflammatory enzyme that converts arachidonic acid into PGE2. The latter causes immunosuppression by promoting adenosine-cAMP signaling in activated T cells and thereby restrains their effector function [57]. PGE2 secreted by tumor cells can also inhibit the maturation of DCs [58]. Also, PGE2 supports the differentiation of Tregs and enhances their suppressive activity [59]. Finally, PGE2 can stimulate the immunosuppressive functions of MDSCs by binding to EP-4 receptors on these cells [60].

In addition, HIF-1α activation is associated with an increase in the expression of metalloproteinase ADAM10; this enzyme is required for the hypoxia-induced shedding of MHC class I chain-related molecule A (MICA), a ligand that triggers the cytolytic action of immune effectors [61]. Removal of MICA from the tumor cell surface thus leads to tumor cell resistance to lysis mediated by innate immune effectors. Exosomal microRNAs have emerged as a new way that cancer cells use to evade the immune surveillance. In epithelial ovarian cancer, hypoxia induces increased expression of exosomes rich in miR-940, and thereby promotes macrophage polarization towards an immunosuppressive M2 phenotype [62]. Similarly, in pancreatic cancer, hypoxia induces exosomal miR-301a, and thus mediates M2 macrophage polarization *via* PTEN/PI3Kγ [63]. Also, hypoxic tumor-derived microvesicles can transfer TGF-β1 and miR-23a to NK cells, which causes the latter to decrease the expression of the activating receptor NKG2D and CD107a respectively, thereby inhibiting the NK-cell function [64]. Moreover, in nasopharyngeal carcinoma (NPC) cells, hypoxia induces exosomal miR-24-3p, which inhibits T cell proliferation and Th1 and Th17 differentiation, and promotes the differentiation of Tregs [65].

4.2.4 Local Invasion Through ECM Remodeling and Cell Motility Enhancement

One of the initial phases in the metastatic cascade is the local invasion of tumor cells from the primary lesion site into the adjacent tissue parenchyma. Invasion is a multi-step process: first, the migrating cell polarizes and elongates; a cell protrusion (lamellipodia, filopodia, pseudopod or invadopodia) then forms and attaches to the ECM substrate; next, the leading edge of the cell or the entire cell body contracts, thereby generating a traction force that pulls the cell body and the tailing edge to gradually glide forward [66].

Tumor cells utilize a variety of mechanisms to promote local invasion and most of these mechanisms are regulated by hypoxia and hypoxic tumor-derived proteins or factors. For instance, in response to hypoxia, tumor cells increase the expression and secretion of extracellular proteases such as MMPs and cathepsins that break down the ECM to promote tumor invasion [67, 68]. In multiple tumor models, hypoxia upregulates the expression and secretion of the matrix-modifying enzyme lysyl oxidases (LOX). Secreted LOX is required for the focal adhesion between cell and matrix at the leading edge of invasive pseudopod protrusions, and this adhesion is critical for the acquisition of cell motility [69]. Another collagen modification enzyme and HIF1α target, procollagen-lysine, 2-oxoglutarate 5-dioxygenase (PLOD2), originally thought to reside exclusively within the cell, can also be secreted [70]. Earlier works including ours demonstrate that PLOD2 hydroxylates telopeptide lysine residues on collagen to promote the formation of collagen crosslinks that, upon release to extracellular space, facilitate metastasis in multiple tumor models [71, 72]. Recently, using cells lines and human lung cancer samples, it has been shown that PLOD2 can also be secreted by tumor cells; secreted PLOD2 closely associates with type I collagen fibrils, indicating that this enzyme is capable of modifying collagen in the extracellular space [70]. It is intriguing that PLOD2 performs the same enzymatic function both intracellularly and extracellularly. It remains to be studied whether secreted PLOD2 plays other moonlighting roles in addition to its lysyl hydroxylase activity to promote invasion and later steps in metastasis.

In addition to the direct effects on ECM, HIF signaling in tumor cells also induces the expression of chemokines and cytokines. These chemical signals then recruit macrophages and mesenchymal stem cells (MSCs) into the tumor microenvironment to support tumor cell invasion and migration [73, 74]. Moreover, hypoxic tumors can recruit platelets through secreted adenosine diphosphate [75]. The communication between platelets and tumor cells may allow the loading of platelets with tumor-derived factors, which could further promote tumor cell migration and metastasis.

Moreover, hypoxia can also induce tumor cells to secrete factors promoting the formation of invasive cell protrusions. It has been demonstrated that in multiple cancer cell lines, hypoxia increases protease-dependent shedding of the membrane-anchored ligand heparin-binding EGF-like growth factor (HB-EGF) through activation of Notch signaling. Soluble HB-EGF, in turn, can induce the formation of invadopodia in an autonomous manner by activating the epidermal growth factor (EGF) receptor [76].

Finally, an increasing number of studies provide novel evidence that under hypoxic conditions, the migration and invasion ability of cancer cells can be enhanced by hypoxia-induced exosomes. In prostate cancer, hypoxia-induced exosomes promote the invasiveness of prostate cancer cells by targeting the adherence junction; these exosomes cause the loss of E-cadherin in PC3 cells along with an increase in cytoplasmic and nuclear β-catenin expression [77]. Further, hypoxia upregulates exosomes rich in miR-21 in oral squamous cell carcinoma (OSCC) cells; the exosomes derived from hypoxic OSCC cells, in turn, downregulate a pool of genes in normoxic cells and induce their epithelial-mesenchymal-transition (EMT) [78].

4.3 Distant Metastasis

4.3.1 Immune Evasion in the Circulation

Once in the circulation, tumor cells face many survival challenges, including immunological attack, shear forces, and apoptosis. Less than 0.1% of circulating tumor cells (CTCs) survive while the majority of them die [79]. One way to evade the immune system is to interact with platelets. Tumor cells cause tumor cell-induced platelet aggregation (TCIPA) by direct contact or through the release of agonistic mediators, such as ADP, thrombin, thromboxane A2 (TXA2) and tumor-associated proteinases [80–82]. As hypoxia leads to increased ADP and tissue factor-mediated thrombin production by tumor cells [75, 83], intratumoral hypoxia likely plays an important role in TCIPA formation. Platelets are activated in TCIPA and attach to the surface of CTCs through p-selectin and a GPIIb-IIIa-fibrinogen bridge [84].

Work in recent years has revealed that the tumor-platelet interaction protects the tumor cells in a number of ways. Adhered platelets can prevent tumor cells from being recognized and lysed by NK cells [85, 86]. This protective effect can be mediated by soluble factors derived from platelets, including TGF-β and PDGF that inhibit NK cell activity, and by the physical shield composed of a platelet cloak and deposited fibrinogen around cancer cells [87–89]. In addition to NK cells, platelet-derived VEGF may also inhibit the maturation of dendritic cells, the major antigen-presenting cells in the immune system [90]. Furthermore, activated platelets can transfer the major histocompatibility complex (MHC) to CTCs, which in turn can mimic host cells and escape immune surveillance [91].

Hypoxic signaling in primary tumors can also educate neutrophils via CCL2 secretion, giving rise to tumor-educated neutrophils [92]. In normal physiology, neutrophils are the first line of defense against infections. However, in metastasis, neutrophils play multifactorial roles, and whether they promote or curb disease progression is context-dependent. Although neutrophils infiltrated to primary tumors exert immunosuppressive functions, tumor-educated neutrophils appear to accumulate in the circulation and in the lungs of tumor-bearing mice prior to metastatic progression and have been found to prevent cancer cells from seeding in the lungs [92]. Therefore, engineering neutrophils by entraining them with CCL2 may be exploited therapeutically for the prevention of metastatic disease.

4.3.2 Endothelium Permeability

Intravasation and extravasation during metastasis require tumor cells to pass through the endothelial wall in a process termed transendothelial migration (TEM) [93]. HIF activity in tumor cells results in the release of factors that modulate endothelial cell-endothelial cell and endothelial cell-tumor cell interactions to permit TEM. HIF strengthens tumor cell-endothelial cell interactions through the activation of L1 cell adhesion molecule (L1CAM) [94]. Hypoxia-induced secretion of VEGF, angiopoietin 2 (ANG2), angiopoietin-like 4 (ANGPTL4), MMPs, ADAM12, placental growth factor (PGF), and CCL2 in multiple tumor models are capable of destabilizing the vascular wall prior to the arrival of tumor cells, thereby facilitating the subsequent intravasation and extravasation [93, 95–98]. This priming effect by the same molecules that also promote angiogenesis and invasion demonstrates that traits advantageous in the early phase of primary tumor invasion may also be useful at later steps in the metastasis cascade.

As noted before, platelets can adhere to tumor cells and exert protective effects. Once activated by cancer cells, platelets can also signal to nearby endothelial cells. Tumor cells elicit ATP secretion from activated platelets, which can increase the vasculature permeability by targeting P2Y2 receptors on endothelial cells [99].

Chemokines regulated by HIF signaling can also weaken the barrier integrity of lymphatic walls. For instance, in breast cancer, IL-6 secretion by hypoxic tumors activates signal transducer and activator of transcription 3 (STAT3) signaling in lymphatic endothelial cells localized within the lung and lymph node; the STAT3 signaling, in turn, promotes lung vascular permeability, lymph node angiogenesis and recruitment of tumor cells into the lymphatic system [100].

4.3.3 Metastatic Colonization Through Microenvironment Remodeling

The vast majority of cancer cells that extravasate into the distant "foreign" parenchyma will die, but a minority of these cells may enter dormancy and survive for months to decades. From this dormant population, a small number of cancer cells may revive and establish a full-fledged tumor at the distant site in a process termed the "metastatic colonization". Although hypoxic tumor cells will have been exposed to the physiologically normal amount of oxygen by the time they leave the primary site, factors they secrete prior to dissemination can establish a premetastatic niche at the distant locations prior to tumor arrival. Therefore, the early molecular changes at the premetastatic niche due to the hypoxic primary tumor are one of the rate-limiting steps of metastasis. Understanding the mechanism of hypoxia-induced tumor secretome in the regulation of the premetastatic niche is likewise an area of intense investigation.

Hypoxic tumors secret multiple types of ECM remodeling enzymes; these enzymes are critical for establishing a second home favorable for tumor growth. In breast cancer, the hypoxia-induced secretion of LOX is significantly associated with bone-tropism and disease relapse. Mechanistically, LOX is secreted by hypoxic ER^-breast cancer cells and it disrupts the balance between bone formation and destruction such that there is a greater overall bone loss. The site(s) of damaged bone presents a pre-metastatic niche favorable for disseminated breast cancer cells, thereby facilitating the formation of bone metastasis [101]. Similarly, tumor secretion of LOX-like proteins, downstream of hypoxia and HIF signaling, also supports metastatic colonization. For example, LOXL2 has been shown to play a critical role for niche formation in hepatocellular carcinoma [102].

In addition to the matrix remodeling enzymes, HIF activation also promotes the secretion of other proteins to facilitate colonization. In breast cancer cells, hypoxia induces the secretion of parathyroid hormone-related protein (PTHrP) that preconditions the bone marrow microenvironment for colonization [103, 104]. In breast and pancreatic cancers, tumor-derived Angpt2 appears to foster the metastatic colonization by enhancing the ability of infiltrating myeloid cells to promote the vascularization of metastatic nodules [105]. The ECM protein tenascin C, secreted by hypoxic breast cancer cells, promotes metastatic colonization in the lung *via* perturbation of the Notch and Wnt signaling [106]. Finally, in colon cancer, metastatic tumor cells in liver adapt to the metabolic stress within the hypoxic milieu by releasing creatine kinase brain-type (CKB) into the extracellular space; CKB generates phosphocreatine in the microenvironment, which can be imported back to the tumor cells and used as a source of ATP production [107, 108].

4.4 Concluding Remarks

The studies described above demonstrate that systemic cellular and molecular signals through hypoxia-regulated tumor secretome profoundly

Table 4.1 Hypoxia-induced secreted factors with effects at the primary site

Functions	Secreted molecules
Growth/survival	HSP90α lactate, lnRNA-UCA1, OPN, periostin
Blood vessel formation	ANG-1, FGFs, miR-23a, miR-135b, miR-210, miR-494, MMPs, OPN, PDGF, SCF, SDF-1, MCP-1, TGF-β1, VEGF
Transdifferentiation	Jagged1, miR-21
Immune evasion	CCL5, CCL22, CCL28, CXCL12, endothelins, IL10, MICA, miR-23a, miR-24-3p, miR-301a, miR-940, PGE2, SDF-1, TGF-β1, VEGF
Invasion and migration	ADP, β-catenin, cathepsins, HB-EGF, LOX, MMPs, PLOD2

ADP adenosine diphosphate, *ANG-1* angiopoietin 1, *CCL5* CC-chemokine ligand 5, *CCL22* CC-chemokine ligand 22, *CCL28* CC-chemokine ligand 28, *CXCL12 or SDF-1* C-X-C motif chemokine ligand 12, *FGFs* fibroblast growth factors, *HB-EGF* heparin-binding EGF-like growth factor, *HSP90α* heat shock protein-90α, *IL10* interleukin 10, *lnRNA-UCA1* long non-coding RNA-urothelial cancer-associated 1, *LOX* lysyl oxidases, *MCP-1* monocyte chemoattractant protein-1, *MICA* MHC class I chain-related molecule A, *MMPs* matrix metalloproteinases, *OPN* osteopontin *PDGF* platelet-derived growth factor, *PGE2* prostaglandin E2, *PLOD2* procollagen-lysine, 2-oxoglutarate 5-dioxygenase, *SCF* stem cell factor, *SDF-1* stromal-derived factor-1, *TGF-β* transforming growth factor β1, *VEGF* vascular endothelial growth factor

Table 4.2 Hypoxia-induced secreted factors with effects at the distant site

Functions	Secreted molecules
Immune evasion	ADP, CCL2, proteinases, thrombin, TXA2
Intravasation and extravasation	ADAM12, ANG2, ANGPTL4, ATP, CCL2, IL-6; MMPs, PGF, VEGF
Metastatic colonization	ANG2, CKB, LOX, tenascin C, PTHrP

ADAM12 disintegrin snd metalloproteinase domain-containing protein 12, *ANG2* angiopoietin 2, *ANGPTL4* angiopoietin-like 4, *ATP* adenosine triphosphate, *CCL2* CC-chemokine ligand 2, *CKB* creatine kinase brain-type, *IL-6* interleukin 6, *MMPs* matrix metalloproteinases, *LOX* lysyl oxidases, *PGF* placental growth factor, *PTHrP* parathyroid hormone-related protein, *TXA2* thromboxane A2

influence both the early and late stages of metastasis. Within the primary tumor, factors secreted by hypoxic tumor cells promote survival and blood vessel formation, increasing the propensity of tumor cells to metastasize (Table 4.1). During local invasion and transendothelial migration, hypoxic tumor-secreted factors enhance tumor cell motility by ECM remodeling and facilitate the entry into and exit from the vasculature by weakening barrier integrity. During late phases of metastasis, factors derived from hypoxic primary tumor promote the establishment of a pro-metastatic environment within the secondary organs (Table 4.2). At multiple steps in the metastatic process, hypoxia-regulated tumor secretome assists tumor cells in immune escape by promoting immune suppression and tumor resistance. In addition, enzymes secreted by secondary tumors in response to hypoxia play an important role in selecting for metastatic cells that can survive the metabolic stress. However, the secretome of hypoxic tumors and how it impacts metastasis remains understudied and warrants continued investigation. As the list of tumor-derived cytokines, proteins, extracellular particles expands, so will the opportunities to discover novel biomarkers and therapeutic targets.

The outstanding questions are: Are the types of tumor secretome associated with specific locations of distant metastasis? Given the direct and indirect tumor-platelet interactions associated with hypoxia, do platelets act as the message carriers for tumor-derived factors, such that the primary tumor can efficiently relay information to the secondary organs? What are the molecular mechanisms within the hypoxic tumor that regulate exosome content and release? Similarly, what are the mechanisms of target cell selection by hypoxic tumor-derived exosomes? Elucidating these questions is an essential prerequisite for the development of treatment strategies.

Acknowledgments This work was supported by NIH Grant 1U54CA210173-01. We apologize to other researchers whose work we could not cite owing to space constraints.

References

1. Tjalsma H, Bolhuis A, Jongbloed JD, Bron S, van Dijl JM (2000) Signal peptide-dependent protein transport in Bacillus subtilis: a genome-based survey of the secretome. Microbiol Mol Biol Rev 64:515–547
2. Chenau J, Michelland S, Seve M (2008) Le sécrétome: définitions et intérêt biomédical. La Revue de Médecine Interne 29:606–608
3. Palade G (1975) Intracellular aspects of the process of protein synthesis. Science 189:867–867
4. Ferro-Novick S, Brose N (2013) Traffic control system within cells. Nature 504:98–98
5. Cocucci E, Meldolesi J (2015) Ectosomes and exosomes: shedding the confusion between extracellular vesicles. Trends Cell Biol 25:364–372
6. Vlassov AV, Magdaleno S, Setterquist R, Conrad R (2012) Exosomes: current knowledge of their composition, biological functions, and diagnostic and therapeutic potentials. Biochim Biophys Acta **1820**:940–948
7. Cocucci E, Racchetti G, Meldolesi J (2009) Shedding microvesicles: artefacts no more. Trends Cell Biol 19:43–51
8. Dong H et al (2016) Breast Cancer MDA-MB-231 cells use secreted heat shock protein-90alpha (Hsp90α) to survive a hostile hypoxic environment. Sci Rep 6:20605
9. Ehrenfried JA, Herron BE, Townsend CM, Evers BM (1995) Heat shock proteins are differentially expressed in human gastrointestinal cancers. Surg Oncol 4:197–203
10. Li CF et al (2008) Heat shock protein 90 overexpression independently predicts inferior disease-free survival with differential expression of the and isoforms in gastrointestinal stromal tumors. Clin Cancer Res 14:7822–7831
11. Gress TM et al (1994) Differential expression of heat shock proteins in pancreatic carcinoma. Cancer Res 54:547–551
12. Bao S et al (2004) Periostin potently promotes metastatic growth of colon cancer by augmenting cell survival via the Akt/PKB pathway. Cancer Cell 5:329–339
13. Song G et al (2009) Osteopontin promotes gastric cancer metastasis by augmenting cell survival and invasion through Akt-mediated HIF-1alpha up-regulation and MMP9 activation. J Cell Mol Med 13:1706–1718
14. Song G et al (2008) Osteopontin promotes ovarian cancer progression and cell survival and increases HIF-1alpha expression through the PI3-K/Akt pathway. Cancer Sci 99:1901–1907
15. Xue M et al (2017) Hypoxic exosomes facilitate bladder tumor growth and development through transferring long non-coding RNA-UCA1. Mol Cancer 16:143
16. Panigrahi GK et al (2018) Hypoxia-induced exosome secretion promotes survival of African-American and Caucasian prostate cancer cells. Sci Rep 8:290
17. Adair TH, Montani J-P (2010) Angiogenesis. In: Colloquium series on integrated systems physiology: from molecule to function, vol. 2, pp 1–84
18. Krock BL, Skuli N, Simon MC (2012) Hypoxia-induced angiogenesis: good and evil. Genes Cancer 2:1117–1133
19. Li B et al (2006) VEGF and PlGF promote adult vasculogenesis by enhancing EPC recruitment and vessel formation at the site of tumor neovascularization. FASEB J 20:1495–1497
20. Compernolle V et al (2002) Loss of HIF-2alpha and inhibition of VEGF impair fetal lung maturation, whereas treatment with VEGF prevents fatal respiratory distress in premature mice. Nat Med 8:702–710
21. Rankin EB et al (2008) Hypoxia-inducible factor-2 regulates vascular tumorigenesis in mice. Oncogene 27:5354–5358
22. Liu Y, Cox SR, Morita T, Kourembanas S (1995) Hypoxia regulates vascular endothelial growth factor gene expression in endothelial cells. Identification of a 5′ enhancer. *Circ Res* 77:638–643
23. Levy AP, Levy NS, Wegner S, Goldberg MA (1995) Transcriptional regulation of the rat vascular endothelial growth factor gene by hypoxia. J Biol Chem 270:13333–13340
24. Ceradini DJ et al (2004) Progenitor cell trafficking is regulated by hypoxic gradients through HIF-1 induction of SDF-1. Nat Med 10:858–864
25. Reynolds LP, Redmer DA (1998) Expression of the angiogenic factors, basic fibroblast growth factor and vascular endothelial growth factor, in the ovary. J Anim Sci 76:1671
26. Hong KH (2005) Monocyte chemoattractant protein-1-induced angiogenesis is mediated by vascular endothelial growth factor-A. Blood 105:1405–1407
27. Matsui J, Wakabayashi T, Asada M, Yoshimatsu K, Okada M (2004) Stem cell factor/c-kit signaling promotes the survival, migration, and capillary tube formation of human umbilical vein endothelial cells. J Biol Chem 279:18600–18607
28. Han ZB et al (2008) Hypoxia-inducible factor (HIF)-1 directly enhances the transcriptional activity of stem cell factor (SCF) in response to hypoxia and epidermal growth factor (EGF). Carcinogenesis 29:1853–1861
29. Sun L et al (2006) Neuronal and glioma-derived stem cell factor induces angiogenesis within the brain. Cancer Cell 9:287–300
30. Litz J (2006) Imatinib inhibits c-Kit-induced hypoxia-inducible factor-1 activity and vascular endothelial growth factor expression in small cell lung cancer cells. Mol Cancer Ther 5:1415–1422
31. Hellstrom M, Kalen M, Lindahl P et al (1999) Role of PDGF-B and PDGFR-beta in recruitment of vascular smooth muscle cells and pericytes during

embryonic blood vessel formation in the mouse. Development 126:3047–3055
32. Iivanainen E, Nelimarkka L, Elenius V et al (2003) Angiopoietin-regulated recruitment of vascular smooth muscle cells by endothelial-derived heparin binding EGF-like growth factor. FASEB J 17:1609–1621
33. Murakami M (2012) Signaling required for blood vessel maintenance: molecular basis and pathological manifestations. Int J Vasc Med 2012:293641
34. Laderoute KR et al (2000) Opposing effects of hypoxia on expression of the angiogenic inhibitor thrombospondin 1 and the angiogenic inducer vascular endothelial growth factor. Clin Cancer Res 6:2941–2950
35. Umezu T et al (2014) Exosomal miR-135b shed from hypoxic multiple myeloma cells enhances angiogenesis by targeting factor-inhibiting HIF-1. Blood 124:3748–3757
36. Hsu Y-L et al (2017) Hypoxic lung cancer-secreted exosomal miR-23a increased angiogenesis and vascular permeability by targeting prolyl hydroxylase and tight junction protein ZO-1. Oncogene 36:4929–4942
37. Tadokoro H, Umezu T, Ohyashiki K, Hirano T, Ohyashiki JH (2013) Exosomes derived from hypoxic leukemia cells enhance tube formation in endothelial cells. J Biol Chem 288:34343–34351
38. Mao G et al (2015) Tumor-derived microRNA-494 promotes angiogenesis in non-small cell lung cancer. Angiogenesis 18:373–382
39. Wang R et al (2010) Glioblastoma stem-like cells give rise to tumour endothelium. Nature 468:829–833
40. Ricci-Vitiani L et al (2010) Tumour vascularization via endothelial differentiation of glioblastoma stem-like cells. Nature 468:824–828
41. Soda Y et al (2011) Transdifferentiation of glioblastoma cells into vascular endothelial cells. Proc Natl Acad Sci USA 108:4274–4280
42. Chen H-F et al (2014) Twist1 induces endothelial differentiation of tumour cells through the Jagged1-KLF4 axis. Nat Commun 5:4697
43. Du R et al (2008) HIF1α induces the recruitment of bone marrow-derived vascular modulatory cells to regulate tumor angiogenesis and invasion. Cancer Cell 13:206–220
44. Lin S et al (2012) Chemokine C-C motif receptor 5 and C-C motif ligand 5 promote cancer cell migration under hypoxia. Cancer Sci 103:904–912
45. Leek RD et al (2000) Macrophage infiltration is associated with VEGF and EGFR expression in breast cancer. J Pathol 190:430–436
46. Grimshaw MJ, Wilson JL, Balkwill FR (2002) Endothelin-2 is a macrophage chemoattractant: implications for macrophage distribution in tumors. Eur J Immunol 32:2393–2400
47. Grimshaw MJ (2007) Endothelins and hypoxia-inducible factor in cancer. Endocr Relat Cancer 14:233–244
48. Gabrilovich D (2004) Mechanisms and functional significance of tumour-induced dendritic-cell defects. Nat Rev Immunol 4:941–952
49. Kusmartsev S, Nefedova Y, Yoder D, Gabrilovich DI (2004) Antigen-specific inhibition of CD8+ T cell response by immature myeloid cells in cancer is mediated by reactive oxygen species. J Immunol 172:989–999
50. Mantovani A (2010) The growing diversity and spectrum of action of myeloid-derived suppressor cells. Eur J Immunol 40:3317–3320
51. Fridman WH et al (2011) Prognostic and predictive impact of intra- and peritumoral immune infiltrates. Cancer Res 71:5601–5605
52. Facciabene A et al (2011) Tumour hypoxia promotes tolerance and angiogenesis via CCL28 and Treg cells. Nature 475:226–230
53. Wrzesinski SH, Wan YY, Flavell RA (2007) Transforming growth factor-beta and the immune response: implications for anticancer therapy. Clin Cancer Res 13:5262–5270
54. Hao N-B et al (2012) Macrophages in tumor microenvironments and the progression of tumors. Clin Dev Immunol 2012:948098–948011
55. Gabrilovich DI et al (1996) Production of vascular endothelial growth factor by human tumors inhibits the functional maturation of dendritic cells. Nat Med 2:1096–1103
56. Curiel TJ et al (2003) Blockade of B7-H1 improves myeloid dendritic cell–mediated antitumor immunity. Nat Med 9:562–567
57. Whiteside TL, Mandapathil M, Schuler P (2011) The role of the adenosinergic pathway in immunosuppression mediated by human regulatory T cells (Treg). *Curr Med Chem* 18:5217–5223
58. Yang L et al (2003) Cancer-associated immunodeficiency and dendritic cell abnormalities mediated by the prostaglandin EP2 receptor. J Clin Investig 111:727–735
59. Serafini P, Mgebroff S, Noonan K, Borrello I (2008) Myeloid-derived suppressor cells promote cross-tolerance in B-cell lymphoma by expanding regulatory T cells. Cancer Res 68:5439–5449
60. Sinha P, Clements VK, Fulton AM, Ostrand-Rosenberg S (2007) Prostaglandin E2 promotes tumor progression by inducing myeloid-derived suppressor cells. Cancer Res 67:4507–4513
61. Barsoum IB et al (2011) Hypoxia induces escape from innate immunity in Cancer cells via increased expression of ADAM10: role of nitric oxide. Cancer Res 71:7433–7441
62. Chen X et al (2017) Exosomes derived from hypoxic epithelial ovarian cancer deliver microRNA-940 to induce macrophage M2 polarization. Oncol Rep 38:522–528
63. Wang X et al (2018) Hypoxic tumor-derived exosomal miR-301a mediates M2 macrophage polarization via PTEN/PI3Kγ to promote pancreatic cancer metastasis. *Cancer Res* 78(16):4586–4598. https://doi.org/10.1158/0008-5472.CAN-17-3841

64. Berchem G et al (2016) Hypoxic tumor-derived microvesicles negatively regulate NK cell function by a mechanism involving TGF-β and miR23a transfer. Oncoimmunology 5:e1062968
65. Ye S-B et al (2016) Exosomal miR-24-3p impedes T-cell function by targeting FGF11and serves as a potential prognostic biomarker for nasopharyngeal carcinoma. J Pathol 240:329–340
66. Friedl P, Wolf K (2003) Tumour-cell invasion and migration: diversity and escape mechanisms. Nat Rev Cancer 3:362–374
67. Sudhan DR, Siemann DW (2013) Cathepsin L inhibition by the small molecule KGP94 suppresses tumor microenvironment enhanced metastasis associated cell functions of prostate and breast cancer cells. Clin Exp Metastasis 30:891–902
68. Krishnamachary B et al (2003) Regulation of colon carcinoma cell invasion by hypoxia-inducible factor 1. Cancer Res 63:1138–1143
69. Erler JT et al (2009) Hypoxia-induced lysyl oxidase is a critical mediator of bone marrow cell recruitment to form the premetastatic niche. Cancer Cell 15:35–44
70. Chen Y et al (2016) Lysyl hydroxylase 2 is secreted by tumor cells and can modify collagen in the extracellular space. J Biol Chem 291:25799–25808
71. Eisinger-Mathason TSK et al (2013) Hypoxia-dependent modification of collagen networks promotes sarcoma metastasis. Cancer Discov 3:1190–1205
72. Gilkes DM et al (2013) Procollagen lysyl hydroxylase 2 is essential for hypoxia-induced breast cancer metastasis. Mol Cancer Res 11:456–466
73. Chaturvedi P, Gilkes DM, Takano N, Semenza GL (2014) Hypoxia-inducible factor-dependent signaling between triple-negative breast cancer cells and mesenchymal stem cells promotes macrophage recruitment. Proc Natl Acad Sci 111:E2120–E2129
74. Chaturvedi P et al (2013) Hypoxia-inducible factor-dependent breast cancer-mesenchymal stem cell bidirectional signaling promotes metastasis. J Clin Invest 123:189–205
75. Haemmerle M et al (2016) FAK regulates platelet extravasation and tumor growth after antiangiogenic therapy withdrawal. J Clin Invest 126:1885–1896
76. Díaz B, Yuen A, Iizuka S, Higashiyama S, Courtneidge SA (2013) Notch increases the shedding of HB-EGF by ADAM12 to potentiate invadopodia formation in hypoxia. J Cell Biol 201:279–292
77. Ramteke A et al (2015) Exosomes secreted under hypoxia enhance invasiveness and stemness of prostate cancer cells by targeting adherens junction molecules. Mol Carcinog 54:554–565
78. Li L et al (2016) Exosomes derived from hypoxic Oral squamous cell carcinoma cells deliver miR-21 to normoxic cells to elicit a Prometastatic phenotype. Cancer Res 76:1770–1780
79. Fidler IJ (1970) Metastasis: quantitative analysis of distribution and fate of tumor emboli labeled with 125 I-5-iodo-2′-deoxyuridine. J Natl Cancer Inst 45:773–782
80. Zucchella M et al (1989) Human tumor cells cultured 'in vitro' activate platelet function by producing ADP or thrombin. Haematologica 74:541–545
81. Bastida E, Ordinas A, Giardina SL, Jamieson GA (1982) Differentiation of platelet-aggregating effects of human tumor cell lines based on inhibition studies with apyrase, hirudin, and phospholipase. Cancer Res 42:4348–4352
82. Pinto S et al (1993) Increased thromboxane A2 production at primary tumor site in metastasizing squamous cell carcinoma of the larynx. Prostaglandins Leukot Essent Fatty Acids 49:527–530
83. Monteiro RQ et al (2016) Hypoxia regulates the expression of tissue factor pathway signaling elements in a rat glioma model. Oncol Lett 12:315–322
84. Gong L, Cai Y, Zhou X, Yang H (2012) Activated platelets interact with lung cancer cells through P-selectin glycoprotein ligand-1. Pathol Oncol Res 18:989–996
85. Nieswandt B, Hafner M, Echtenacher B, Männel DN (1999) Lysis of tumor cells by natural killer cells in mice is impeded by platelets. Cancer Res 59:1295–1300
86. Gay LJ, Felding-Habermann B (2011) Contribution of platelets to tumour metastasis. Nat Rev Cancer 11:123–134
87. Palumbo JS, Degen JL (2007) Mechanisms linking tumor cell-associated procoagulant function to tumor metastasis. Thromb Res 120(Suppl 2):S22–S28
88. Kopp H-G, Placke T, Salih HR (2009) Platelet-derived transforming growth factor-beta down-regulates NKG2D thereby inhibiting natural killer cell antitumor reactivity. Cancer Res 69:7775–7783
89. Labelle M, Hynes RO (2012) The initial hours of metastasis: the importance of cooperative host-tumor cell interactions during Hematogenous dissemination. Cancer Discov 2:1091–1099
90. Palucka K, Banchereau J (2012) Cancer immunotherapy via dendritic cells. Nat Rev Cancer 12:265–277
91. Placke T et al (2012) Platelet-derived MHC class I confers a Pseudonormal phenotype to Cancer cells that subverts the antitumor reactivity of natural killer immune cells. Cancer Res 72:440–448
92. Granot Z et al (2011) Tumor entrained neutrophils inhibit seeding in the premetastatic lung. Cancer Cell 20:300–314
93. Reymond N, d'Água BB, Ridley AJ (2013) Crossing the endothelial barrier during metastasis. Nat Rev Cancer 13:858–870
94. Zhang H et al (2011) HIF-1-dependent expression of angiopoietin-like 4 and L1CAM mediates vascular metastasis of hypoxic breast cancer cells to the lungs. Oncogene 31:1757–1770
95. Gupta GP et al (2007) Mediators of vascular remodelling co-opted for sequential steps in lung metastasis. Nature 446:765–770

96. Wolf MJ et al (2012) Endothelial CCR2 signaling induced by colon carcinoma cells enables extravasation via the JAK2-Stat5 and p38MAPK pathway. Cancer Cell 22:91–105
97. Weis S, Cui J, Barnes L, Cheresh D (2004) Endothelial barrier disruption by VEGF-mediated Src activity potentiates tumor cell extravasation and metastasis. J Cell Biol 167:223–229
98. Huang Y et al (2009) Pulmonary vascular destabilization in the premetastatic phase facilitates lung metastasis. Cancer Res 69:7529–7537
99. Schumacher D, Strilic B, Sivaraj KK, Wettschureck N, Offermanns S (2013) Platelet-derived nucleotides promote tumor-cell Transendothelial migration and metastasis via P2Y2 receptor. Cancer Cell 24:130–137
100. Lee E et al (2014) Breast cancer cells condition lymphatic endothelial cells within pre-metastatic niches to promote metastasis. Nat Commun 5:112
101. Cox TR et al (2015) The hypoxic cancer secretome induces pre-metastatic bone lesions through lysyl oxidase. Nature 522:106–110
102. Wong CCL et al (2014) Lysyl oxidase-like 2 is critical to tumor microenvironment and metastatic niche formation in hepatocellular carcinoma. Hepatology 60:1645–1658
103. Manisterski M, Golan M, Amir S, Weisman Y, Mabjeesh N (2014) Hypoxia induces PTHrP gene transcription in human cancer cells through the HIF-2α. Cell Cycle 9:3747–3753
104. Guise TA et al (1996) Evidence for a causal role of parathyroid hormone-related protein in the pathogenesis of human breast cancer-mediated osteolysis. J Clin Investig 98:1544–1549
105. Mazzieri R et al (2011) Targeting the ANG2/TIE2 Axis inhibits tumor growth and metastasis by impairing angiogenesis and disabling rebounds of proangiogenic myeloid cells. Cancer Cell 19:512–526
106. Oskarsson T et al (2011) Breast cancer cells produce tenascin C as a metastatic niche component to colonize the lungs. Nat Med 17:867–874
107. Loo JM et al (2015) Extracellular metabolic energetics can promote cancer progression. Cell 160:393–406
108. Rankin EB, Giaccia AJ (2016) Hypoxic control of metastasis. Science 352:175–180

5 Hypoxia-Dependent Angiogenesis and Lymphangiogenesis in Cancer

Luana Schito

Abstract

Hypoxia (low O_2) is a ubiquitous feature of solid cancers, arising as a mismatch between cellular O_2 supply and consumption. Hypoxia is associated to metastatic disease and mortality owing to its ability to stimulate the formation of blood (angiogenesis) and lymphatic vessels (lymphangiogenesis), thereby allowing cancer cells to escape the unfavorable tumor microenvironment and disseminate into secondary sites. This review outlines molecular mechanisms by which intratumoral hypoxia regulates the expression of motogenic and mitogenic factors that induce angiogenesis and lymphangiogenesis, whilst discussing their implications for metastatic cancers.

Keywords

Angiogenesis · HIF · Hypoxia · Lymphangiogenesis · Metastasis

5.1 Introduction

Failure to control metastatic dissemination is the most common cause of mortality in patients bearing solid cancers, a direct result of the development of chemo(radio)therapy resistance. Recent advances in our understanding of the molecular underpinnings of the metastatic sequence have revealed that low O_2 (hypoxia) is a key microenvironmental factor promoting therapy resistance and cancer cell (CC) dissemination. The main focus of this chapter is to address the molecular mechanisms whereby hypoxia promotes the development of new blood (angiogenesis) and lymphatic (lymphangiogenesis) vessels, both essential *phenomena* required for the escape of CCs into metastatic sites. In addition, I will outline emerging evidence suggesting that other non-canonical vascular mechanisms can support metastasis through hypoxia-sensitive signaling pathways, whilst summarizing the implications of these insights for cancer therapy.

5.2 Pathophysiology of Tumor Hypoxia

Sustained proliferation and evasion from cellular death signaling, two major traits of human cancer pathogenesis [1], enable CCs to expand into an avascular mass that is exquisitely dependent upon O_2 diffusion [2]. This process becomes inefficient once the tumor reaches a diameter > 1000–2000 μm unless the primary lesion encounters sufficient host vessel density to warrant further expansion; in addition, tumoral regions located more than 160–180 μm from the nearest vessel become poorly

L. Schito (✉)
Biological Sciences Platform, Sunnybrook Research Institute, University of Toronto, Toronto, ON, Canada
e-mail: luana.schito@ls2r.science

D. M. Gilkes (ed.), *Hypoxia and Cancer Metastasis*, Advances in Experimental Medicine and Biology 1136,
https://doi.org/10.1007/978-3-030-12734-3_5

oxygenated, typically encountering O_2 levels between ≈ 0.7% and 1.8% [3, 4]. These hypoxic microregions present steep O_2 gradients (up to 0.84% O_2/100 μm) [5, 6], and a heterogenous spatiotemporal distribution which is dynamically regulated by intrinsic and extrinsic cues such as cellular O_2 consumption, composition and hydrostatic pressure of the *interstitium*, vessel density and vascular tree geometry [3, 7]. Importantly, *circa* half of all locally advanced solid tumors contain regions of hypoxia [8] which have been recognized as major contributors of tumor progression and resistance to therapy [3, 9]. Indeed, direct assessment of oxygenation in human cancers identified tumor hypoxia as an adverse prognostic factor that is independent of tumor stage, histological grade, and lymph node *status* [10].

5.3 Hypoxia-Inducible Factors

Hypoxia is transduced to the nucleus of cancer and stromal cells as activation of a transcriptional program mediated by hypoxia-inducible factors (HIFs), a family of evolutionary conserved bHLH-PAS heterodimers, consisting of an O_2-regulated HIFα subunit (designated as HIF-1α, HIF-2α, or HIF-3α) and a constitutively expressed HIF-1β subunit [11]. HIF-1α is expressed in all nucleated cells, whereas HIF-2α exhibits a cell-type specific expression, with high levels in the vascular *endothelium* [12–15]; consistently, patients presenting HIF-2α stabilization under non-hypoxic conditions are predisposed to hypervascular tumors such as hemangioblastomas, hemangiomas, and renal cell carcinomas (RCCs) [16]. By contrast, HIF-3α is present only in mammalian cells where it acts as a dominant negative regulator of HIF-1α function [17]. For simplicity, I will henceforth refer to HIFα when responses are mediated by both HIF-1α and -2α. Under non-hypoxic conditions (>2% O_2), HIFα subunits undergo enzymatic hydroxylation by prolyl-4-hydroxylases (PHDs) that contain non-heme-reduced Fe^{2+} in their catalytic center and utilize O_2 and α-ketoglutarate as substrates [18–22]; this posttranslational modification creates binding sites for the von Hippel-Lindau (VHL) tumor suppressor which ubiquitinylates HIFα, thereby triggering its proteasomal degradation [23–26]. Substrate (O_2) limitation impairs the catalytic activity of PHDs and allows HIFα to accumulate, dimerize with HIF-1β and translocate to the nucleus, thereby promoting the expression of target genes involved in critical aspects of cancer pathobiology, including the formation of blood and lymphatic vessels [27, 28]. Likewise, clinical immunohistochemical analyses have identified significant correlation among HIF-1α expression, blood and lymphatic vessel density, as well as hematogenous and lymphogenous metastasis in breast cancer [29–34], pancreatic cancer [35], oral squamous cell carcinoma [36], and esophageal cancer [37].

5.4 Hypoxia and Tumoral Angiogenesis

Judah Folkman provided seminal evidence leading to the hypothesis that nutrient and O_2 availability critically limit tumor growth above volumes of ≈1–2 mm^3, thus predicting the existence of a molecular mechanism allowing tumors to co-opt vessel formation [2, 38]. This hypothesis was subsequently shown to be correct, since angiogenesis (the budding of new capillaries from existing vessels) is an adaptive pathobiological response hijacked by CCs to promote O_2 delivery, an essential process for hematogenous metastasis. Importantly, intratumoral hypoxia stimulates the expression of several angiogenic factors with motogenic and mitogenic function, primarily *via* HIFα-dependent transcriptional activity; additionally, hypoxia activates HIFα-independent pro-angiogenic pathways such as the mechanistic target of rapamycin (mTOR) and unfolded protein response (UPR) (reviewed in [39]). It is worth to mention that, whilst physiological angiogenesis leads to the generation of functional vessels that enhance perfusion, tumoral angiogenesis induces an unharmonious angiogenic profile, that typically gives rise to vessels that are structurally unsound (*i.e.*, dilated, tortuous, disorganized and often containing blind ends), and functionally abnormal (*i.e.*, excessively permeable, leading to an edematous *interstitium*) [40]. Consequently, the overall effect of tumoral angiogenesis is a paradoxical decrease in perfusion that aggravates hypoxia in a feed-forward loop promoting HIFα stabilization and transactivity; likewise, it follows that vascular disruption by antiangiogenic agents could contribute to therapy resistance and metastatic progression by upregulating alternative HIFα-dependent and -independent hypoxia-sensitive angiogenic signaling [41].

5.4.1 Hypoxia-Inducible Angiogenic Pathways

I will hereby delineate angiogenic mechanisms downstream of canonical pro-angiogenic HIFα gene targets including vascular endothelial growth factor-A (VEGF-A), placenta growth factor (PGF), angiopoietin-2 (ANGPT2), chemokine (C-X-C motif), ligand (CXCL) 12 (also known as SDF-1), hepatocyte growth factor (HGF), and platelet-derived growth factor-B (PDGF-B) [42].

5.4.1.1 Vascular Endothelial Growth Factor-A

VEGF-A, initially referred to as vascular permeability factor, is an endothelial cell (EC) mitogenic and angiogenic factor, and the prototypical molecule of the VEGF family that in mammals includes VEGF-B, -C, -D and PGF [43–45]. Characteristically, VEGF-A signal transduction occurs upon binding to its cognate receptor tyrosine kinases (RTKs), VEGFR1 and VEGFR2 expressed by ECs [46–48], leading to the activation of downstream cellular events enhancing proliferation, survival, migration, extracellular matrix (ECM) degradation, vascular permeability and vessel branching into the organ parenchyma (reviewed in [49]). Importantly, both VEGF-A and its receptors VEGFR1 and VEGFR2 are dependent upon HIFα transcriptional activity under hypoxia [50–53]. Likewise, VEGFR1 post-receptor signal transduction is potentiated by phosphodiesterase PDE4A and PDE4D, which increase cAMP levels, resulting in augmented tumoral VEGF-A levels [54].

5.4.1.2 Placenta Growth Factor

PGF is a member of the VEGF family and a hypoxia inducible HIFα target that upon binding to its cognate receptor VEGFR1, commonly expressed by ECs, stimulates endothelial and mural cell proliferation, whilst upregulating the expression of several angiogenic factors such as VEGF-A, PDGF-B and fibroblast growth factor-2 (FGF-2), or matrix metalloproteinases promoting ECM remodeling (reviewed in [55]). In addition, PGF induces the recruitment of hematopoietic progenitor cells into the tumor microenvironment, whilst promoting CC motility and stromal invasion [56]. Interestingly, PGF can promote dilation and structural reinforcement of angiogenic capillaries (referred to as 'vessel normalization'), a *phenomenon* that improves tumor perfusion and oxygenation, thereby resulting in sensitization to anti-VEGF or anti-VEGFR2 therapies; in this context, PGF targeting might counteract the beneficial effect of VEGF→VEGFR2 blockade [41, 57, 58]. Importantly, *in vitro* studies suggest that supraphysiological levels of PGF might promote desensitization to anti-PGF therapies, thus prompting careful evaluation of the angiogenic cytokine profile in patients as a necessary requirement for effective PGF targeting [59].

5.4.1.3 Angiopoietin-2

ANGPT2 is a HIFα target gene bearing an hypoxia response element (HRE) sequence within the first intron of the *Angpt2* gene [60]. ANGPT2 plays a critical role in tumor angiogenesis by disrupting the adhesive interactions between endothelial and perivascular mural cells [61]. Consequently, ANGPT2 → TIE2 signaling has been shown to promote VEGF-A-dependent EC proliferation within hypoxic tumor regions, thereby disrupting vascular wall integrity [62–65].

5.4.1.4 CXCL12 → CXCR4 Signaling

CXCL12 and its cognate receptor CXCR4 are two well-characterized HIFα targets activating an autocrine/paracrine loop that is essential for homing and migration of vascular and hematopoietic progenitor cells into the tumor microenvironment [66, 67]. More recently, significant cross-talk between VEGF-A and the CXCL12 → CXCR4 has been uncovered, since the former enhances CXCR4 expression in ECs, thereby synergizing with CXCL12 to induce angiogenesis [68]. Consequently, the simultaneous induction of CXCL12 and VEGF-A upstream of CXCL12 → CXCR4 provides significant pathobiological reinforcement of angiogenesis within the hypoxic tumor microenvironment.

5.4.1.5 HGF → MET Signaling

HGF is a potent regulator of angiogenesis enhancing the proliferation and motility of ECs expressing its cognate receptor MET [69, 70] under non-hypoxic conditions. In the malignant context, HGF → MET→PI3K → ERK signaling promotes IL-8 and VEGF secretion in glioma and head and neck carcinoma cell lines [71, 72]. In addition, HGF represses the antiangiogenic factor thrombospondin-1, known to induce EC apoptosis and inhibit EC proliferation [73, 74]. Importantly, these molecular mechanisms are thought to be further activated under hypoxia *via* HIF-1α-dependent transcription of MET, which in turn sustains hypoxic HIF-1α protein translation, thereby establishing a closed feed-forward signaling loop facilitating tumor angiogenesis and CC invasion [75, 76].

5.4.1.6 PDGF-B → PDGFRβ Signaling

PDGF-B is a dimeric factor and a potent mitogen for mesenchymal cells and fibroblasts, inducing a variety of angiogenic effects including recruitment of PDGFRβ^{+} pericytes to the wall of newly formed tumor vessels *via* the CXCL12 → CXCR4 axis, and VEGF-A induction [77, 78]. Of relevance for the hypoxic tumor microenvironment, PDGF-B expression was shown to be directly upregulated by HIF-1α, owing to the presence of an intronic HRE in the *Pdgfb* gene; moreover, PDGFRβ has been shown to be induced by hypoxia in lymphatic endothelial cells (LECs) [33].

5.4.1.7 Nitric Oxide

A critical mediator of angiogenesis is nitric oxide (NO), a gaseous molecule generated from L-arginine by three distinct nitric oxide synthase (NOS) isoforms (NOS-1, -2 and -3), which are regulated by HIF-1α in a cell-type specific manner [79, 80]. NO mediates multiple vascular processes including vasodilation, cell proliferation and migration; moreover, NO establishes a positive feedback loop that further promotes tumor vascularization by inducing VEGF-A and FGF-2 [81, 82]. These processes depend on NO gradients established within the tumor microenvironment [83].

5.4.2 Bone Marrow-Derived Angiogenic Cells

Bone marrow-derived angiogenic cells (BMDACs) are a heterogeneous population of myeloid-derived cells exerting paracrine pro-angiogenic functions, also proposed to fulfill an endothelial progenitor cell role equivalent to embryonic hemangioblasts [84–86]. Tumor angiogenesis relies upon the mobilization of BMDACs into the circulation that subsequently home into the tumor parenchyma; nonetheless, multiple lines of evidence show that adult BMDACs include myelomonocytic and mesenchymal stem cells exhibiting pro-angiogenic roles upon activation of the CXCL12 → CXCR4, ANGPT2 → TIE2 and SCF → c-KIT axes [86, 87], followed by their recruitment into the hypoxic tumor microenvironment [27, 88]. Additionally, the HIF-1α-dependent secretion of VEGF-A and CXCL12 by hypoxic CCs increases the abundance of circulating CD34^{+}, c-kit^{+} or Sca1^{+} angiogenic cells, a mobilized BMDAC subtype recruited into the tumor microenvironment [88].

5.4.3 Non-angiogenic Mechanisms

Notwithstanding the well-established role of angiogenesis in cancer progression, recent studies indicate that solid tumors can acquire a microcirculation through two distinct, non-angiogenic mechanisms: vessel co-option and vasculogenic mimicry.

5.4.3.1 Vessel Co-option

CC proliferation in richly vascularized organs such as the brain, liver and lungs, relies upon vessel co-option, a pathological *phenomenon* wherein expanding and invading CCs hijack vessels from the surrounding normal tissue to extract O_2 and nutrients (reviewed in [89]). Additionally, a mechanism of non-angiogenic tumor growth resembling vessel co-option has been recently identified in lymphatic metastases which were able to survive within lymph nodes by utilizing preexisting intranodal vasculature [90, 91]. Taken together, these studies suggest vessel co-option as a clinically relevant mechanism countering the oncolytic effect of anti-angiogenic therapies [92–94]. Importantly, a recent preclinical study utilizing a liver cancer model uncovered that actin-related protein (Arp)-2/3 complex was necessary for vessel co-option and metastasis [93]. Intriguingly, Arp2/3 complex is a key mediator of actin filament nucleation and branching that initiates *invadopodia*, a process known to be induced by hypoxia (reviewed in [95]), thereby providing a plausible mechanistic basis for the pathogenic role of hypoxia in vessel co-option.

5.4.3.2 Vasculogenic Mimicry

Malignant tumors arising in mesoderm-derived organs are thought to engage in vasculogenic mimicry (VM), the *de novo* formation of perfusable, CC-lined pseudovessels that promote hematogenous metastases [96].

VM channel-forming CCs express genes belonging to embryonic/stem cell, vascular, and hypoxia signaling pathways, the latter occurring *via* HIFα-dependent upregulation of TWIST, a critical mediator of epithelial-to-mesenchymal transition (EMT) [97–99]. Interestingly, hypoxia and HIFα-dependent induction of VM have been described in melanoma [100, 101], hepatocellular carcinoma cells [99], and *Ewing* sarcoma cells [102]; moreover, CCs engaged in VM recruit pericytes *via* PDGF-B → PDGFRβ signaling, thus facilitating sprouting as well as providing structural support to nascent VM pseudovascular networks [103]. Consequently, it seems likely that hypoxia/EMT-driven genetic programs, recapitulating a mesenchymal phenotype in CCs, might underlie VM-dependent metastatic dissemination.

5.5 Structural Features of the Lymphatic System

The lymphatic system consists of capillaries, collecting vessels, lymphoid organs, trunks and ducts, configuring a unidirectional network that drains fluid, proteins, pathogens and migrating cells (*i.e.*, lymph) from the interstitial space to the blood circulation. In contrast to the closed circulatory system, the lymphatic system is open, endowed with blind-ended and fenestrated vessels lacking a continuous basement membrane, pericytic/smooth muscle coverage, and tight cell junctions. These morphological features allow lymphatic vessels to balance osmotic and hydrostatic pressure, key factors driving the *Starling* forces governing fluid, nutrient and cellular movement in and out of tissues.

Compared to physiological lymphangiogenesis, tumoral lymphangiogenesis generates vessels with even wider *lumina* and further loosened intercellular spacing [104]. Consequently, CC intravasation into lymphatics is facilitated by the absence of a basement membrane and widened LEC↔LEC junctions. These morphological features occur in parallel with an hypoxic LEC microenvironment; indeed, O_2 measurements in the mesenteric lymphatic microvasculature show levels <3% O_2, reaching anoxic levels in parenchymal regions devoid of surrounding blood capillaries [105, 106].

5.5.1 Molecular LEC Markers

The identification of molecular markers specifically expressed by LECs allowed to distinguish lymphatic from blood vessels, and subsequently understand their role in health (*e.g.*, regulation of interstitial pressure, immune surveillance, and dietary fat absorption), and disease (*e.g.*, lymphedema, inflammatory diseases, and cancer metastasis) [107]. The markers characterizing LECs include: *first*, PROX-1, a homeobox transcription factor implicated in lymphovascular development, and a HIF-1α and -2α target; *second*, podoplanin-1, a transmembrane glycoprotein; and *third*, lymphatic vascular endothelial-cell hyaluronan receptor-1 (LYVE-1), a homolog of the CD44 hyaluronic acid receptor [108–113].

5.5.2 Specific Lymphangiogenic Signals: VEGF-C/–D → VEGFR3

A considerable body of data established a central role for VEGF-C and VEGF-D on lymphangiogenic signaling carried out through binding to their receptor VEGFR3 in LECs [114–118]; however, similar to other members of the VEGF family, VEGF-C and VEGF-D can bind VEGFR2, thereby also inducing EC proliferation and migration [116, 119]. Nonetheless, lymphangiogenic pathway specificity is achieved to the fact that VEGF-C and VEGF-D do not activate VEGFR1 and VEGFR2 in ECs. In addition to LEC proliferation and sprouting, tumor-derived VEGF-C and VEGF-D can cause structural changes within lymphatic vessels located at the tumor periphery, such as increased diameter and/or cell size [120–122], or induce dilation of draining lymphatics located proximal or distal to sentinel lymph nodes [115, 123]. Immunohistochemical studies indicate a significant association between VEGF-C/D and HIF-1α expression in various human cancers [32, 124]. At the molecular level, genome wide analysis uncovered that hypoxia (1% O_2) induced VEGF-C and VEGF-D mRNA expression in LECs [125], whereas recent data indicate that HIF-1α regulates VEGF-C expression *via* an HRE located within the *Vegfc* promoter in macrophages [126]. Moreover, it has been shown that hypoxia can induce VEGF-C mRNA translation *via* an internal ribosome entry site-dependent mechanism, irrespective of HIF-1α signaling [127]. In addition to CCs, VEGF-C and VEGF-D can be expressed by tumor-associated fibroblasts and immune cells, thereby contributing to tumor progression within the hypoxic tumor microenvironment [128, 129].

VEGFR3 is a transmembrane tyrosine kinase receptor for VEGF-C and VEGF-D, whose expression is restricted to adult LECs and induced in lymphovascular progenitors during embryonic development [111, 114, 116, 117, 130, 131]. Importantly, human cancers can mimic the ontogenic expression of VEGFR3 by inducing its expression within ECs and LECs during tumorigenesis [132, 133].

Following the characterization of VEGF-C and VEGF-D as LEC mitogens [116, 134, 135], several studies documented the expression of VEGF-C and VEGF-D, as well as the occurrence of lymphangiogenesis in a variety of solid cancers [32, 124, 136–139]. It is now well established that VEGF-C and VEGF-D are major drivers of tumor lymphangiogenesis and metastatic dissemination, predicting adverse clinical outcomes [140–146]. Consistently, VEGF-C and VEGF-D expression levels are positively correlated with metastatic dissemination [136], whereas inhibition of lymphangiogenesis *via* VEGF-C or VEGFR3 blockade prevents lymph node and distant metastases in preclinical models [144]. Of relevance, lymphangiogenesis can also occur in sentinel lymph nodes where it precedes and contributes to distant metastasis [137, 142, 147, 148].

5.5.3 Lymphatic↔Vascular Crosstalk: VEGF-A and PDGF-B

Similar to tumor angiogenesis, induction of tumor lymphangiogenesis relies on the interplay of several growth and motogenic factors; as a result, a significant overlap exists between the spectrum of cytokines exerting angiogenic and lymphangiogenic properties. In preclinical studies, tumor-derived VEGF-A was shown to stimulate lymphangiogenesis and lymphatic metastasis [149]; moreover, in a manner analogous to solid tumors, VEGF-A positively correlated with VEGF-C and lymphatic vessel density in malignant lymphomas [118, 150–152]. Of relevance, VEGFR3 neutralizing antibody was not able to inhibit VEGF-A-induced lymphangiogenesis, thus suggesting that VEGF-A-stimulated lymphatic vessel growth *via* VEGFR2 signaling is independent of VEGF-C → VEGFR3 and VEGF-D → VEGFR3 activation [150, 151]. These data were subsequently corroborated by the observation that tumor LECs expressed VEGFR2 in addition to VEGFR3 [153]. Notwithstanding, direct evidence linking VEGF-A with hypoxia/HIFα within the context of tumoral lymphangiogenesis is currently lacking, regardless of the well documented role of HIFα→VEGF-A signaling in tumor angiogenesis.

Pioneering studies showed that PDGF-B overexpression induced tumoral lymphangiogenesis and lymph node metastasis in a preclinical fibrosarcoma mouse model [154], whereas PDGF-B and PDGFRβ expression was associated with lymphatic metastasis in patients with gastric cancer [155]. Intriguingly, PDGF-B mRNA was upregulated by hypoxia and HIF-1α in a number of non-malignant and cancer cell lines [64, 65, 156, 157]. The molecular mechanism underlying PDGF-B-dependent lymphangiogenesis and lymphovascular metastasis was subsequently characterized through an exhaustive bioinformatic screening of the *Pdgfb* gene promoter region, thereby identifying a functional HRE located 9622 *bp* upstream of the translation start site. This non-canonical intronic HRE was necessary and sufficient to mediate hypoxic HIF-1α → PDGF-B signaling, whilst promoting the migration of hypoxic CCs towards PDGFRβ$^+$ LECs [33]. Consistently, HIF-1α or PDGF-B inhibition *via* digoxin or imatinib administration, respectively decreased intratumoral lymphangiogenesis and breast CC nodal invasion. Importantly, immunohistochemical analysis followed by multiple linear regression showed that HIF-1α, PDGF-B expression, and lymphatic vessel area were linearly correlated and predicted the Scarff-Bloom-Richardson histopathological prognosis index in a small cohort of breast cancer patients, thus illustrating the clinical importance of the HIFα→PDGF-B signaling [33].

5.5.4 Other Pleiotropic Lymphangiogenic Modulators

Recent data suggest that signaling mechanisms not traditionally considered to participate in vascular or lymphovascular homeostasis can support these processes in hypoxic cancers. The potent vasoconstrictor peptide endothelin (ET)-1, signaling through the endothelin-B receptor (ET_BR), represents one such pathway underlying lymphangiogenesis and lymphatic metastasis. Mechanistically, ET-1 → ET_BR promotes LEC growth and differentiation *via* HIF-1α-dependent expression of VEGF-A and VEGF-C [158]. Importantly, ET-1 is known to stabilize HIF-1α (reviewed in [159]), whereas HIF-1α directly controls ET-1 expression, owing to the presence of an HRE on the *Edn1* gene [160], therefore indicating the existence of a feedback mechanism synergistically promoting lymphangiogenesis and lymphatic metastasis. Likewise, ET-1-dependent HIF-1α activity augments the expression of CCR7, whose signal transduction (CCL21 → CCR7) contributes to lymphatic metastasis [161, 162]. Similarly, the hypoxia-dependent CCL21 → CXCR4 system reinforces tumor lymphangiogenesis and metastasis [163].

In addition, insulin-like growth factors-1 and -2 promote lymphangiogenesis by stimulating LEC proliferation and migration [150, 151], whereas hypoxia can induce the expression of PGF in LYVE1$^+$ tumor vessels [164], thus pointing to a role for the PGF → VEGR1 axis in lymphangiogenesis and/or lymphatic metastasis.

5.6 Therapeutic Implications and Concluding Remarks

Hypoxia-induced angiogenesis and lymphangiogenesis have recently emerged as essential mechanisms supporting the dissemination of CCs from the primary tumor to distant, secondary sites. Examination of hypoxic angiogenic and lymphangiogenic signaling pathways within the tumor microenvironment have revealed multiple auto/paracrine feedback mechanisms that represent potential therapeutic targets to halt malignant progression. Current data suggest that targeted therapies blocking angiogenic- and lymphangiogenic-dependent metastasis should be aimed, in principle, at two hypoxia-inducible molecular hubs: HIFα and VEGFRs; in particular, VEGFR3 emerges as a promising target in early, localized malignancies, since it holds the potential of impeding lymphovascular metastasis.

Conversely, PDGF-B and VEGF-A targeting, in combination with HIFα inhibition, might simultaneously impede hematogenous and lymphogenous CC dissemination, in advanced malignancies that rely upon both mechanisms. Importantly, uncovering the molecular nodes enabling the hypoxic crosstalk between angiogenesis and lymphangiogenesis can help in the design of targeted therapeutics blocking functional redundancies in said pathways, thereby preventing resistance, recurrence and metastatic-disease related mortality.

References

1. Hanahan D, Weinberg RA (2011) Hallmarks of cancer: the next generation. Cell 144:646–674. https://doi.org/10.1016/j.cell.2011.02.013
2. Folkman J (1971) Tumor angiogenesis: therapeutic implications. N Engl J Med 285:1182–1186. https://doi.org/10.1056/NEJM197111182852108
3. Rey S, Schito L, Koritzinsky M, Wouters BG (2017a) Molecular targeting of hypoxia in radiotherapy. Adv Drug Deliv Rev 109:45–62. https://doi.org/10.1016/j.addr.2016.10.002
4. Thomlinson RH, Gray LH (1955) The histological structure of some human lung cancers and the possible implications for radiotherapy. Br J Cancer 9:539–549
5. Höckel M, Vaupel P (2001) Tumor hypoxia: definitions and current clinical, biologic, and molecular aspects. J Natl Cancer Inst 93:266–276. https://doi.org/10.1093/jnci/93.4.266
6. Vaupel P, Mayer A (2017) Tumor oxygenation status: facts and fallacies. Adv Exp Med Biol 977:91–99. https://doi.org/10.1007/978-3-319-55231-6_13
7. Schito L, Rey S (2018) Cell-autonomous metabolic reprogramming in hypoxia. Trends Cell Biol 28:128–142. https://doi.org/10.1016/j.tcb.2017.10.006
8. Vaupel P, Mayer A, Höckel M (2004) Tumor hypoxia and malignant progression. Methods Enzymol 381:335–354. https://doi.org/10.1016/S0076-6879(04)81023-1
9. Schito L, Rey S, Konopleva M (2017) Integration of hypoxic HIF-α signaling in blood cancers. Oncogene 36:5331–5340. https://doi.org/10.1038/onc.2017.119
10. Vaupel P (2009) Prognostic potential of the pre-therapeutic tumor oxygenation status. Adv Exp Med Biol 645:241–246. https://doi.org/10.1007/978-0-387-85998-9_36
11. Wang GL, Jiang BH, Rue EA, Semenza GL (1995) Hypoxia-inducible factor 1 is a basic-helix-loop-helix-PAS heterodimer regulated by cellular O2 tension. Proc Natl Acad Sci USA 92:5510–5514
12. Ema M, Taya S, Yokotani N, Sogawa K, Matsuda Y, Fujii-Kuriyama Y (1997) A novel bHLH-PAS factor with close sequence similarity to hypoxia-inducible factor 1alpha regulates the VEGF expression and is potentially involved in lung and vascular development. Proc Natl Acad Sci USA 94:4273–4278
13. Flamme I, Fröhlich T, von Reutern M, Kappel A, Damert A, Risau W (1997) HRF, a putative basic helix-loop-helix-PAS-domain transcription factor is closely related to hypoxia-inducible factor-1 alpha and developmentally expressed in blood vessels. Mech Dev 63:51–60
14. Peng J, Zhang L, Drysdale L, Fong GH (2000) The transcription factor EPAS-1/hypoxia-inducible factor 2alpha plays an important role in vascular remodeling. Proc Natl Acad Sci USA 97:8386–8391. https://doi.org/10.1073/pnas.140087397
15. Tian H, McKnight SL, Russell DW (1997) Endothelial PAS domain protein 1 (EPAS1), a transcription factor selectively expressed in endothelial cells. Genes Dev 11:72–82
16. Shen C, Kaelin WG (2013) The VHL/HIF axis in clear cell renal carcinoma. Semin Cancer Biol 23:18–25. https://doi.org/10.1016/j.semcancer.2012.06.001
17. Rytkönen KT, Williams TA, Renshaw GM, Primmer CR, Nikinmaa M (2011) Molecular evolution of the metazoan PHD-HIF oxygen-sensing system. Mol Biol Evol 28:1913–1926. https://doi.org/10.1093/molbev/msr012
18. Bruick RK, McKnight SL (2001) A conserved family of prolyl-4-hydroxylases that modify HIF. Science 294:1337–1340. https://doi.org/10.1126/science.1066373
19. Epstein AC, Gleadle JM, McNeill LA, Hewitson KS, O'Rourke J, Mole DR, Mukherji M, Metzen E, Wilson MI, Dhanda A, Tian YM, Masson N, Hamilton DL, Jaakkola P, Barstead R, Hodgkin J, Maxwell PH, Pugh CW, Schofield CJ, Ratcliffe PJ (2001) C. elegans EGL-9 and mammalian homologs define a family of dioxygenases that regulate HIF by prolyl hydroxylation. Cell 107:43–54
20. Ivan M, Kondo K, Yang H, Kim W, Valiando J, Ohh M, Salic A, Asara JM, Lane WS, Kaelin WG (2001) HIFalpha targeted for VHL-mediated destruction by proline hydroxylation: implications for O2 sensing.

Science 292:464–468. https://doi.org/10.1126/science.1059817
21. Jaakkola P, Mole DR, Tian YM, Wilson MI, Gielbert J, Gaskell SJ, von Kriegsheim A, Hebestreit HF, Mukherji M, Schofield CJ, Maxwell PH, Pugh CW, Ratcliffe PJ (2001) Targeting of HIF-alpha to the von Hippel-Lindau ubiquitylation complex by O2-regulated prolyl hydroxylation. Science 292:468–472. https://doi.org/10.1126/science.1059796
22. Masson N, Willam C, Maxwell PH, Pugh CW, Ratcliffe PJ (2001) Independent function of two destruction domains in hypoxia-inducible factor-alpha chains activated by prolyl hydroxylation. EMBO J 20:5197–5206. https://doi.org/10.1093/emboj/20.18.5197
23. Cockman ME, Masson N, Mole DR, Jaakkola P, Chang GW, Clifford SC, Maher ER, Pugh CW, Ratcliffe PJ, Maxwell PH (2000) Hypoxia inducible factor-alpha binding and ubiquitylation by the von Hippel-Lindau tumor suppressor protein. J Biol Chem 275:25733–25741. https://doi.org/10.1074/jbc.M002740200
24. Huang LE, Gu J, Schau M, Bunn HF (1998) Regulation of hypoxia-inducible factor 1alpha is mediated by an O2-dependent degradation domain via the ubiquitin-proteasome pathway. Proc Natl Acad Sci USA 95:7987–7992
25. Kamura T, Sato S, Iwai K, Czyzyk-Krzeska M, Conaway RC, Conaway JW (2000) Activation of HIF1alpha ubiquitination by a reconstituted von Hippel-Lindau (VHL) tumor suppressor complex. Proc Natl Acad Sci USA 97:10430–10435. https://doi.org/10.1073/pnas.190332597
26. Maxwell PH, Wiesener MS, Chang GW, Clifford SC, Vaux EC, Cockman ME, Wykoff CC, Pugh CW, Maher ER, Ratcliffe PJ (1999) The tumour suppressor protein VHL targets hypoxia-inducible factors for oxygen-dependent proteolysis. Nature 399:271–275. https://doi.org/10.1038/20459
27. Rey S, Semenza GL (2010) Hypoxia-inducible factor-1-dependent mechanisms of vascularization and vascular remodelling. Cardiovasc Res 86:236–242. https://doi.org/10.1093/cvr/cvq045
28. Schito L, Rey S (2017) Hypoxic pathobiology of breast cancer metastasis. Biochim Biophys Acta 1868:239–245. https://doi.org/10.1016/j.bbcan.2017.05.004
29. Bos R, van der Groep P, Greijer AE, Shvarts A, Meijer S, Pinedo HM, Semenza GL, van Diest PJ, van der Wall E (2003) Levels of hypoxia-inducible factor-1alpha independently predict prognosis in patients with lymph node negative breast carcinoma. Cancer 97:1573–1581. https://doi.org/10.1002/cncr.11246
30. Bos R, van Diest PJ, de Jong JS, van der Groep P, van der Valk P, van der Wall E (2005) Hypoxia-inducible factor-1alpha is associated with angiogenesis, and expression of bFGF, PDGF-BB, and EGFR in invasive breast cancer. Histopathology 46:31–36. https://doi.org/10.1111/j.1365-2559.2005.02045.x
31. Nalwoga H, Ahmed L, Arnes JB, Wabinga H, Akslen LA (2016) Strong expression of hypoxia-inducible factor-1α (HIF-1α) is associated with Axl expression and features of aggressive tumors in African breast Cancer. PLoS One 11:e0146823. https://doi.org/10.1371/journal.pone.0146823
32. Okada K, Osaki M, Araki K, Ishiguro K, Ito H, Ohgi S (2005) Expression of hypoxia-inducible factor (HIF-1alpha), VEGF-C and VEGF-D in non-invasive and invasive breast ductal carcinomas. Anticancer Res 25:3003–3009
33. Schito L, Rey S, Tafani M, Zhang H, Wong CC-L, Russo A, Russo MA, Semenza GL (2012) Hypoxia-inducible factor 1-dependent expression of platelet-derived growth factor B promotes lymphatic metastasis of hypoxic breast cancer cells. Proc Natl Acad Sci USA 109:E2707–E2716. https://doi.org/10.1073/pnas.1214019109
34. Schoppmann SF, Fenzl A, Schindl M, Bachleitner-Hofmann T, Nagy K, Gnant M, Horvat R, Jakesz R, Birner P (2006) Hypoxia inducible factor-1alpha correlates with VEGF-C expression and lymphangiogenesis in breast cancer. Breast Cancer Res Treat 99:135–141. https://doi.org/10.1007/s10549-006-9190-3
35. Matsuo Y, Ding Q, Desaki R, Maemura K, Mataki Y, Shinchi H, Natsugoe S, Takao S (2014) Hypoxia inducible factor-1 alpha plays a pivotal role in hepatic metastasis of pancreatic cancer: an immunohistochemical study. J Hepato-Biliary-Pancreat Sci 21:105–112. https://doi.org/10.1002/jhbp.6
36. Liang X, Yang D, Hu J, Hao X, Gao J, Mao Z (2008) Hypoxia inducible factor-alpha expression correlates with vascular endothelial growth factor-C expression and lymphangiogenesis/angiogenesis in oral squamous cell carcinoma. Anticancer Res 28:1659–1666
37. Katsuta M, Miyashita M, Makino H, Nomura T, Shinji S, Yamashita K, Tajiri T, Kudo M, Ishiwata T, Naito Z (2005) Correlation of hypoxia inducible factor-1alpha with lymphatic metastasis via vascular endothelial growth factor-C in human esophageal cancer. Exp Mol Pathol 78:123–130. https://doi.org/10.1016/j.yexmp.2004.11.002
38. Folkman J (1995) Angiogenesis in cancer, vascular, rheumatoid and other disease. Nat Med 1:27–30. https://doi.org/10.1038/nm0195-27
39. Schito L (2018) Bridging angiogenesis and immune evasion in the hypoxic tumor microenvironment. Am J Physiol Regul Integr Comp Physiol. https://doi.org/10.1152/ajpregu.00209.2018
40. Rapisarda A, Melillo G (2012) Overcoming disappointing results with antiangiogenic therapy by targeting hypoxia. Nat Rev Clin Oncol 9:378–390. https://doi.org/10.1038/nrclinonc.2012.64
41. Rey S, Schito L, Wouters BG, Eliasof S, Kerbel RS (2017b) Targeting hypoxia-inducible factors for antiangiogenic Cancer therapy. Trends

Cancer 3:529–541. https://doi.org/10.1016/j.trecan.2017.05.002
42. Schito L, Semenza GL (2016) Hypoxia-inducible factors: master regulators of cancer progression. Trends Cancer 2:758–770. https://doi.org/10.1016/j.trecan.2016.10.016
43. Ferrara N, Henzel WJ (1989) Pituitary follicular cells secrete a novel heparin-binding growth factor specific for vascular endothelial cells. Biochem Biophys Res Commun 161:851–858
44. Keck PJ, Hauser SD, Krivi G, Sanzo K, Warren T, Feder J, Connolly DT (1989) Vascular permeability factor, an endothelial cell mitogen related to PDGF. Science 246:1309–1312
45. Leung DW, Cachianes G, Kuang WJ, Goeddel DV, Ferrara N (1989) Vascular endothelial growth factor is a secreted angiogenic mitogen. Science 246:1306–1309
46. de Vries C, Escobedo JA, Ueno H, Houck K, Ferrara N, Williams LT (1992) The fms-like tyrosine kinase, a receptor for vascular endothelial growth factor. Science 255:989–991
47. Jakeman LB, Winer J, Bennett GL, Altar CA, Ferrara N (1992) Binding sites for vascular endothelial growth factor are localized on endothelial cells in adult rat tissues. J Clin Invest 89:244–253. https://doi.org/10.1172/JCI115568
48. Terman BI, Dougher-Vermazen M, Carrion ME, Dimitrov D, Armellino DC, Gospodarowicz D, Böhlen P (1992) Identification of the KDR tyrosine kinase as a receptor for vascular endothelial cell growth factor. Biochem Biophys Res Commun 187:1579–1586
49. Claesson-Welsh L, Welsh M (2013) VEGFA and tumour angiogenesis. J Intern Med 273:114–127. https://doi.org/10.1111/joim.12019
50. Bougatef F, Menashi S, Khayati F, Naïmi B, Porcher R, Podgorniak M-P, Millot G, Janin A, Calvo F, Lebbé C, Mourah S (2010) EMMPRIN promotes melanoma cells malignant properties through a HIF-2alpha mediated up-regulation of VEGF-receptor-2. PLoS One 5:e12265. https://doi.org/10.1371/journal.pone.0012265
51. Elvert G, Kappel A, Heidenreich R, Englmeier U, Lanz S, Acker T, Rauter M, Plate K, Sieweke M, Breier G, Flamme I (2003) Cooperative interaction of hypoxia-inducible factor-2alpha (HIF-2alpha) and Ets-1 in the transcriptional activation of vascular endothelial growth factor receptor-2 (Flk-1). J Biol Chem 278:7520–7530. https://doi.org/10.1074/jbc.M211298200
52. Forsythe JA, Jiang BH, Iyer NV, Agani F, Leung SW, Koos RD, Semenza GL (1996) Activation of vascular endothelial growth factor gene transcription by hypoxia-inducible factor 1. Mol Cell Biol 16:4604–4613
53. Gerber HP, Condorelli F, Park J, Ferrara N (1997) Differential transcriptional regulation of the two vascular endothelial growth factor receptor genes. Flt-1, but not Flk-1/KDR, is up-regulated by hypoxia. J Biol Chem 272:23659–23667
54. Pullamsetti SS, Banat GA, Schmall A, Szibor M, Pomagruk D, Hänze J, Kolosionek E, Wilhelm J, Braun T, Grimminger F, Seeger W, Schermuly RT, Savai R (2013) Phosphodiesterase-4 promotes proliferation and angiogenesis of lung cancer by crosstalk with HIF. Oncogene 32:1121–1134. https://doi.org/10.1038/onc.2012.136
55. Fischer C, Mazzone M, Jonckx B, Carmeliet P (2008a) FLT1 and its ligands VEGFB and PlGF: drug targets for anti-angiogenic therapy? Nat Rev Cancer 8:942–956. https://doi.org/10.1038/nrc2524
56. Marcellini M, De Luca N, Riccioni T, Ciucci A, Orecchia A, Lacal PM, Ruffini F, Pesce M, Cianfarani F, Zambruno G, Orlandi A, Failla CM (2006) Increased melanoma growth and metastasis spreading in mice overexpressing placenta growth factor. Am J Pathol 169:643–654. https://doi.org/10.2353/ajpath.2006.051041
57. Hedlund E-M, Hosaka K, Zhong Z, Cao R, Cao Y (2009) Malignant cell-derived PlGF promotes normalization and remodeling of the tumor vasculature. Proc Natl Acad Sci USA 106:17505–17510. https://doi.org/10.1073/pnas.0908026106
58. Hedlund E-ME, Yang X, Zhang Y, Yang Y, Shibuya M, Zhong W, Sun B, Liu Y, Hosaka K, Cao Y (2013) Tumor cell-derived placental growth factor sensitizes antiangiogenic and antitumor effects of anti-VEGF drugs. Proc Natl Acad Sci USA 110:654–659. https://doi.org/10.1073/pnas.1209310110
59. Fischer C, Mazzone M, Jonckx B, Carmeliet P (2008b) FLT1 and its ligands VEGFB and PlGF: drug targets for anti-angiogenic therapy? Nat Rev Cancer 8:942–956. https://doi.org/10.1038/nrc2524
60. Simon M-P, Tournaire R, Pouyssegur J (2008) The angiopoietin-2 gene of endothelial cells is up-regulated in hypoxia by a HIF binding site located in its first intron and by the central factors GATA-2 and Ets-1. J Cell Physiol 217:809–818. https://doi.org/10.1002/jcp.21558
61. Thomas M, Augustin HG (2009) The role of the angiopoietins in vascular morphogenesis. Angiogenesis 12:125–137. https://doi.org/10.1007/s10456-009-9147-3
62. Holash J, Maisonpierre PC, Compton D, Boland P, Alexander CR, Zagzag D, Yancopoulos GD, Wiegand SJ (1999) Vessel cooption, regression, and growth in tumors mediated by angiopoietins and VEGF. Science 284:1994–1998
63. Schmittnaegel M, De Palma M (2017) Reprogramming tumor blood vessels for enhancing immunotherapy. Trends Cancer 3:809–812. https://doi.org/10.1016/j.trecan.2017.10.002
64. Zhang L, Yang N, Park J-W, Katsaros D, Fracchioli S, Cao G, O'Brien-Jenkins A, Randall TC, Rubin SC, Coukos G (2003a) Tumor-derived vascular endothelial growth factor up-regulates angiopoietin-2 in host endothelium and destabilizes host vasculature, sup-

porting angiogenesis in ovarian cancer. Cancer Res 63:3403–3412
65. Zhang SXL, Gozal D, Sachleben LR, Rane M, Klein JB, Gozal E (2003b) Hypoxia induces an autocrine-paracrine survival pathway via platelet-derived growth factor (PDGF)-B/PDGF-beta receptor/phosphatidylinositol 3-kinase/Akt signaling in RN46A neuronal cells. FASEB J 17:1709–1711. https://doi.org/10.1096/fj.02-1111fje
66. Ceradini DJ, Kulkarni AR, Callaghan MJ, Tepper OM, Bastidas N, Kleinman ME, Capla JM, Galiano RD, Levine JP, Gurtner GC (2004) Progenitor cell trafficking is regulated by hypoxic gradients through HIF-1 induction of SDF-1. Nat Med 10:858–864. https://doi.org/10.1038/nm1075
67. Staller P, Sulitkova J, Lisztwan J, Moch H, Oakeley EJ, Krek W (2003) Chemokine receptor CXCR4 downregulated by von Hippel–Lindau tumour suppressor pVHL. Nature 425:307–311. https://doi.org/10.1038/nature01874
68. Kryczek I, Lange A, Mottram P, Alvarez X, Cheng P, Hogan M, Moons L, Wei S, Zou L, Machelon V, Emilie D, Terrassa M, Lackner A, Curiel TJ, Carmeliet P, Zou W (2005) CXCL12 and vascular endothelial growth factor synergistically induce neoangiogenesis in human ovarian cancers. Cancer Res 65:465–472
69. Bussolino F, Di Renzo MF, Ziche M, Bocchietto E, Olivero M, Naldini L, Gaudino G, Tamagnone L, Coffer A, Comoglio PM (1992) Hepatocyte growth factor is a potent angiogenic factor which stimulates endothelial cell motility and growth. J Cell Biol 119:629–641
70. Grant DS, Kleinman HK, Goldberg ID, Bhargava MM, Nickoloff BJ, Kinsella JL, Polverini P, Rosen EM (1993) Scatter factor induces blood vessel formation in vivo. Proc Natl Acad Sci USA 90:1937–1941
71. Dong G, Chen Z, Li ZY, Yeh NT, Bancroft CC, Van Waes C (2001) Hepatocyte growth factor/scatter factor-induced activation of MEK and PI3K signal pathways contributes to expression of proangiogenic cytokines interleukin-8 and vascular endothelial growth factor in head and neck squamous cell carcinoma. Cancer Res 61:5911–5918
72. Moriyama T, Kataoka H, Hamasuna R, Yokogami K, Uehara H, Kawano H, Goya T, Tsubouchi H, Koono M, Wakisaka S (1998) Up-regulation of vascular endothelial growth factor induced by hepatocyte growth factor/scatter factor stimulation in human glioma cells. Biochem Biophys Res Commun 249:73–77. https://doi.org/10.1006/bbrc.1998.9078
73. Jiménez B, Volpert OV, Crawford SE, Febbraio M, Silverstein RL, Bouck N (2000) Signals leading to apoptosis-dependent inhibition of neovascularization by thrombospondin-1. Nat Med 6:41–48. https://doi.org/10.1038/71517
74. Sargiannidou I, Zhou J, Tuszynski GP (2001) The role of thrombospondin-1 in tumor progression. Exp Biol Med (Maywood) 226:726–733
75. Glück AA, Orlando E, Leiser D, Poliaková M, Nisa L, Quintin A, Gavini J, Stroka DM, Berezowska S, Bubendorf L, Blaukat A, Aebersold DM, Medová M, Zimmer Y (2018) Identification of a MET-eIF4G1 translational regulation axis that controls HIF-1α levels under hypoxia. Oncogene 37:4181–4196. https://doi.org/10.1038/s41388-018-0256-6
76. Pennacchietti S, Michieli P, Galluzzo M, Mazzone M, Giordano S, Comoglio PM (2003) Hypoxia promotes invasive growth by transcriptional activation of the met protooncogene. Cancer Cell 3:347–361
77. Guo P, Hu B, Gu W, Xu L, Wang D, Huang H-JS, Cavenee WK, Cheng S-Y (2003) Platelet-derived growth factor-B enhances glioma angiogenesis by stimulating vascular endothelial growth factor expression in tumor endothelia and by promoting pericyte recruitment. Am J Pathol 162:1083–1093. https://doi.org/10.1016/S0002-9440(10)63905-3
78. Song N, Huang Y, Shi H, Yuan S, Ding Y, Song X, Fu Y, Luo Y (2009) Overexpression of platelet-derived growth factor-BB increases tumor pericyte content via stromal-derived factor-1alpha/CXCR4 axis. Cancer Res 69:6057–6064. https://doi.org/10.1158/0008-5472.CAN-08-2007
79. Palmer LA, Semenza GL, Stoler MH, Johns RA (1998) Hypoxia induces type II NOS gene expression in pulmonary artery endothelial cells via HIF-1. Am J Phys 274:L212–L219
80. Ward ME, Toporsian M, Scott JA, Teoh H, Govindaraju V, Quan A, Wener AD, Wang G, Bevan SC, Newton DC, Marsden PA (2005) Hypoxia induces a functionally significant and translationally efficient neuronal NO synthase mRNA variant. J Clin Invest 115:3128–3139. https://doi.org/10.1172/JCI20806
81. Dulak J, Józkowicz A, Dembinska-Kiec A, Guevara I, Zdzienicka A, Zmudzinska-Grochot D, Florek I, Wójtowicz A, Szuba A, Cooke JP (2000) Nitric oxide induces the synthesis of vascular endothelial growth factor by rat vascular smooth muscle cells. Arterioscler Thromb Vasc Biol 20:659–666. https://doi.org/10.1161/01.ATV.20.3.659
82. Ziche M, Parenti A, Ledda F, Dell'Era P, Granger HJ, Maggi CA, Presta M (1997) Nitric oxide promotes proliferation and plasminogen activator production by coronary venular endothelium through endogenous bFGF. Circ Res 80:845–852
83. Kashiwagi S, Tsukada K, Xu L, Miyazaki J, Kozin SV, Tyrrell JA, Sessa WC, Gerweck LE, Jain RK, Fukumura D (2008) Perivascular nitric oxide gradients normalize tumor vasculature. Nat Med 14:255–257. https://doi.org/10.1038/nm1730
84. Asahara T, Murohara T, Sullivan A, Silver M, van der Zee R, Li T, Witzenbichler B, Schatteman G, Isner JM (1997) Isolation of putative progenitor endothelial cells for angiogenesis. Science 275:964–967
85. Rehman J, Li J, Orschell CM, March KL (2003) Peripheral blood "endothelial progenitor cells" are derived from monocyte/macrophages and

secrete angiogenic growth factors. Circulation 107:1164–1169
86. Rey S, Lee K, Wang CJ, Gupta K, Chen S, McMillan A, Bhise N, Levchenko A, Semenza GL (2009) Synergistic effect of HIF-1alpha gene therapy and HIF-1-activated bone marrow-derived angiogenic cells in a mouse model of limb ischemia. Proc Natl Acad Sci USA 106:20399–20404. https://doi.org/10.1073/pnas.0911921106
87. Du R, Lu KV, Petritsch C, Liu P, Ganss R, Passegué E, Song H, Vandenberg S, Johnson RS, Werb Z, Bergers G (2008) HIF1alpha induces the recruitment of bone marrow-derived vascular modulatory cells to regulate tumor angiogenesis and invasion. Cancer Cell 13:206–220. https://doi.org/10.1016/j.ccr.2008.01.034
88. Lee K, Qian DZ, Rey S, Wei H, Liu JO, Semenza GL (2009) Anthracycline chemotherapy inhibits HIF-1 transcriptional activity and tumor-induced mobilization of circulating angiogenic cells. Proc Natl Acad Sci U S A 106:2353–2358. https://doi.org/10.1073/pnas.0812801106
89. Donnem T, Hu J, Ferguson M, Adighibe O, Snell C, Harris AL, Gatter KC, Pezzella F (2013) Vessel co-option in primary human tumors and metastases: an obstacle to effective anti-angiogenic treatment? Cancer Med 2:427–436. https://doi.org/10.1002/cam4.105
90. Jeong H-S, Jones D, Liao S, Wattson DA, Cui CH, Duda DG, Willett CG, Jain RK, Padera TP (2015) Investigation of the lack of angiogenesis in the formation of lymph node metastases. J Natl Cancer Inst 107. https://doi.org/10.1093/jnci/djv155
91. Naresh KN, Nerurkar AY, Borges AM (2001) Angiogenesis is redundant for tumour growth in lymph node metastases. Histopathology 38:466–470
92. Bridgeman VL, Vermeulen PB, Foo S, Bilecz A, Daley F, Kostaras E, Nathan MR, Wan E, Frentzas S, Schweiger T, Hegedus B, Hoetzenecker K, Renyi-Vamos F, Kuczynski EA, Vasudev NS, Larkin J, Gore M, Dvorak HF, Paku S, Kerbel RS, Dome B, Reynolds AR (2016) Vessel co-option is common in human lung metastases and mediates resistance to anti-angiogenic therapy in preclinical lung metastasis models. J Pathol. n/a-n/a. https://doi.org/10.1002/path.4845
93. Frentzas S, Simoneau E, Bridgeman VL, Vermeulen PB, Foo S, Kostaras E, Nathan MR, Wotherspoon A, Gao Z, Shi Y, Van den Eynden G, Daley F, Peckitt C, Tan X, Salman A, Lazaris A, Gazinska P, Berg TJ, Eltahir Z, Ritsma L, van Rheenen J, Khashper A, Brown G, Nyström H, Sund M, Van Laere S, Loyer E, Dirix L, Cunningham D, Metrakos P, Reynolds AR (2016) Vessel co-option mediates resistance to anti-angiogenic therapy in liver metastases. Nat Med 22:1294–1302. https://doi.org/10.1038/nm.4197
94. Kuczynski EA, Yin M, Bar-Zion A, Lee CR, Butz H, Man S, Daley F, Vermeulen PB, Yousef GM, Foster FS, Reynolds AR, Kerbel RS (2016) Co-option of liver vessels and not sprouting angiogenesis drives acquired Sorafenib resistance in hepatocellular carcinoma. J Natl Cancer Inst 108. https://doi.org/10.1093/jnci/djw030
95. Gould CM, Courtneidge SA (2014) Regulation of invadopodia by the tumor microenvironment. Cell Adhes Migr 8:226–235
96. Seftor REB, Hess AR, Seftor EA, Kirschmann DA, Hardy KM, Margaryan NV, Hendrix MJC (2012) Tumor cell vasculogenic mimicry: from controversy to therapeutic promise. Am J Pathol 181:1115–1125. https://doi.org/10.1016/j.ajpath.2012.07.013
97. Kirschmann DA, Seftor EA, Hardy KM, Seftor REB, Hendrix MJC (2012) Molecular pathways: vasculogenic mimicry in tumor cells: diagnostic and therapeutic implications. Clin Cancer Res 18:2726–2732. https://doi.org/10.1158/1078-0432.CCR-11-3237
98. Li S, Meng W, Guan Z, Guo Y, Han X (2016) The hypoxia-related signaling pathways of vasculogenic mimicry in tumor treatment. Biomed Pharmacother 80:127–135. https://doi.org/10.1016/j.biopha.2016.03.010
99. Ma J, Han S, Zhu Q, Zhao J, Zhang D, Wang L, Lv Y (2011) Role of Twist in vasculogenic mimicry formation in hypoxic hepatocellular carcinoma cells in vitro. Biochem Biophys Res Commun 408:686–691. https://doi.org/10.1016/j.bbrc.2011.04.089
100. Sun B, Zhang D, Zhang S, Zhang W, Guo H, Zhao X (2007) Hypoxia influences vasculogenic mimicry channel formation and tumor invasion-related protein expression in melanoma. Cancer Lett 249:188–197. https://doi.org/10.1016/j.canlet.2006.08.016
101. Zhang S, Li M, Zhang D, Xu S, Wang X, Liu Z, Zhao X, Sun B (2009) Hypoxia influences linearly patterned programmed cell necrosis and tumor blood supply patterns formation in melanoma. Lab Investig J Tech Methods Pathol 89:575–586. https://doi.org/10.1038/labinvest.2009.20
102. van der Schaft DWJ, Hillen F, Pauwels P, Kirschmann DA, Castermans K, Egbrink MGAO, Tran MGB, Sciot R, Hauben E, Hogendoorn PCW, Delattre O, Maxwell PH, Hendrix MJC, Griffioen AW (2005) Tumor cell plasticity in Ewing sarcoma, an alternative circulatory system stimulated by hypoxia. Cancer Res 65:11520–11528. https://doi.org/10.1158/0008-5472.CAN-05-2468
103. Thijssen VL, Paulis YW, Nowak-Sliwinska P, Deumelandt KL, Hosaka K, Soetekouw PM, Cimpean AM, Raica M, Pauwels P, van den Oord JJ, Tjan-Heijnen VC, Hendrix MJ, Heldin C-H, Cao Y, Griffioen AW (2018) Targeting PDGF-mediated recruitment of pericytes blocks vascular mimicry and tumor growth. J Pathol. https://doi.org/10.1002/path.5152
104. Alitalo K (2011) The lymphatic vasculature in disease. Nat Med 17:1371–1380. https://doi.org/10.1038/nm.2545
105. Hangai-Hoger N, Cabrales P, Briceño JC, Tsai AG, Intaglietta M (2004) Microlymphatic and tissue oxygen tension in the rat mesentery. Am J Physiol

Heart Circ Physiol 286:H878–H883. https://doi.org/10.1152/ajpheart.00913.2003
106. Hangai-Hoger N, Tsai AG, Cabrales P, Intaglietta M (2007) Terminal lymphatics: the potential "lethal corner" in the distribution of tissue pO2. Lymphat Res Biol 5:159–168. https://doi.org/10.1089/lrb.2007.5303
107. Karaman S, Leppänen V-M, Alitalo K (2018) Vascular endothelial growth factor signaling in development and disease. Dev Camb Engl 145. https://doi.org/10.1242/dev.151019
108. Banerji S, Ni J, Wang SX, Clasper S, Su J, Tammi R, Jones M, Jackson DG (1999) LYVE-1, a new homologue of the CD44 glycoprotein, is a lymph-specific receptor for hyaluronan. J Cell Biol 144:789–801
109. Breiteneder-Geleff S, Soleiman A, Kowalski H, Horvat R, Amann G, Kriehuber E, Diem K, Weninger W, Tschachler E, Alitalo K, Kerjaschki D (1999) Angiosarcomas express mixed endothelial phenotypes of blood and lymphatic capillaries: podoplanin as a specific marker for lymphatic endothelium. Am J Pathol 154:385–394. https://doi.org/10.1016/S0002-9440(10)65285-6
110. Jussila L, Valtola R, Partanen TA, Salven P, Heikkilä P, Matikainen MT, Renkonen R, Kaipainen A, Detmar M, Tschachler E, Alitalo R, Alitalo K (1998) Lymphatic endothelium and Kaposi's sarcoma spindle cells detected by antibodies against the vascular endothelial growth factor receptor-3. Cancer Res 58:1599–1604
111. Kaipainen A, Korhonen J, Mustonen T, van Hinsbergh VW, Fang GH, Dumont D, Breitman M, Alitalo K (1995) Expression of the fms-like tyrosine kinase 4 gene becomes restricted to lymphatic endothelium during development. Proc Natl Acad Sci USA 92:3566–3570
112. Wigle JT, Oliver G (1999) Prox1 function is required for the development of the murine lymphatic system. Cell 98:769–778
113. Zhou B, Si W, Su Z, Deng W, Tu X, Wang Q (2013) Transcriptional activation of the Prox1 gene by HIF-1α and HIF-2α in response to hypoxia. FEBS Lett 587:724–731. https://doi.org/10.1016/j.febslet.2013.01.053
114. Achen MG, Jeltsch M, Kukk E, Mäkinen T, Vitali A, Wilks AF, Alitalo K, Stacker SA (1998) Vascular endothelial growth factor D (VEGF-D) is a ligand for the tyrosine kinases VEGF receptor 2 (Flk1) and VEGF receptor 3 (Flt4). Proc Natl Acad Sci USA 95:548–553
115. He Y, Rajantie I, Pajusola K, Jeltsch M, Holopainen T, Yla-Herttuala S, Harding T, Jooss K, Takahashi T, Alitalo K (2005) Vascular endothelial cell growth factor receptor 3-mediated activation of lymphatic endothelium is crucial for tumor cell entry and spread via lymphatic vessels. Cancer Res 65:4739–4746. https://doi.org/10.1158/0008-5472.CAN-04-4576
116. Joukov V, Pajusola K, Kaipainen A, Chilov D, Lahtinen I, Kukk E, Saksela O, Kalkkinen N, Alitalo K (1996) A novel vascular endothelial growth factor, VEGF-C, is a ligand for the Flt4 (VEGFR-3) and KDR (VEGFR-2) receptor tyrosine kinases. EMBO J 15:290–298
117. Kukk E, Lymboussaki A, Taira S, Kaipainen A, Jeltsch M, Joukov V, Alitalo K (1996) VEGF-C receptor binding and pattern of expression with VEGFR-3 suggests a role in lymphatic vascular development. Dev Camb Engl 122:3829–3837
118. Whitehurst B, Flister MJ, Bagaitkar J, Volk L, Bivens CM, Pickett B, Castro-Rivera E, Brekken RA, Gerard RD, Ran S (2007) Anti-VEGF-A therapy reduces lymphatic vessel density and expression of VEGFR-3 in an orthotopic breast tumor model. Int J Cancer 121:2181–2191. https://doi.org/10.1002/ijc.22937
119. Stacker SA, Williams SP, Karnezis T, Shayan R, Fox SB, Achen MG (2014) Lymphangiogenesis and lymphatic vessel remodelling in cancer. Nat Rev Cancer 14:159–172. https://doi.org/10.1038/nrc3677
120. Hoshida T, Isaka N, Hagendoorn J, di Tomaso E, Chen Y-L, Pytowski B, Fukumura D, Padera TP, Jain RK (2006) Imaging steps of lymphatic metastasis reveals that vascular endothelial growth factor-C increases metastasis by increasing delivery of cancer cells to lymph nodes: therapeutic implications. Cancer Res 66:8065–8075. https://doi.org/10.1158/0008-5472.CAN-06-1392
121. Leu AJ, Berk DA, Lymboussaki A, Alitalo K, Jain RK (2000) Absence of functional lymphatics within a murine sarcoma: a molecular and functional evaluation. Cancer Res 60:4324–4327
122. Padera TP, Kadambi A, di Tomaso E, Carreira CM, Brown EB, Boucher Y, Choi NC, Mathisen D, Wain J, Mark EJ, Munn LL, Jain RK (2002) Lymphatic metastasis in the absence of functional intratumor lymphatics. Science 296:1883–1886. https://doi.org/10.1126/science.1071420
123. Karnezis T, Shayan R, Caesar C, Roufail S, Harris NC, Ardipradja K, Zhang YF, Williams SP, Farnsworth RH, Chai MG, Rupasinghe TWT, Tull DL, Baldwin ME, Sloan EK, Fox SB, Achen MG, Stacker SA (2012) VEGF-D promotes tumor metastasis by regulating prostaglandins produced by the collecting lymphatic endothelium. Cancer Cell 21:181–195. https://doi.org/10.1016/j.ccr.2011.12.026
124. Currie MJ, Hanrahan V, Gunningham SP, Morrin HR, Frampton C, Han C, Robinson BA, Fox SB (2004) Expression of vascular endothelial growth factor D is associated with hypoxia inducible factor (HIF-1alpha) and the HIF-1alpha target gene DEC1, but not lymph node metastasis in primary human breast carcinomas. J Clin Pathol 57:829–834. https://doi.org/10.1136/jcp.2003.015644
125. Irigoyen M, Ansó E, Martínez E, Garayoa M, Martínez-Irujo JJ, Rouzaut A (2007) Hypoxia alters the adhesive properties of lymphatic endothelial cells. A transcriptional and functional study. Biochim Biophys Acta 1773:880–890. https://doi.org/10.1016/j.bbamcr.2007.03.001

126. Guo Y-C, Zhang M, Wang F-X, Pei G-C, Sun F, Zhang Y, He X, Wang Y, Song J, Zhu F-M, Pandupuspitasari NS, Liu J, Huang K, Yang P, Xiong F, Zhang S, Yu Q, Yao Y, Wang C-Y (2017) Macrophages regulate unilateral ureteral obstruction-induced renal Lymphangiogenesis through C-C motif chemokine receptor 2-dependent phosphatidylinositol 3-kinase-AKT-mechanistic target of rapamycin signaling and hypoxia-inducible factor-1α/vascular endothelial growth factor-C expression. Am J Pathol 187:1736–1749. https://doi.org/10.1016/j.ajpath.2017.04.007
127. Morfoisse F, Kuchnio A, Frainay C, Gomez-Brouchet A, Delisle M-B, Marzi S, Helfer A-C, Hantelys F, Pujol F, Guillermet-Guibert J, Bousquet C, Dewerchin M, Pyronnet S, Prats A-C, Carmeliet P, Garmy-Susini B (2014) Hypoxia induces VEGF-C expression in metastatic tumor cells via a HIF-1α-independent translation-mediated mechanism. Cell Rep 6:155–167. https://doi.org/10.1016/j.celrep.2013.12.011
128. Farnsworth RH, Achen MG, Stacker SA (2018) The evolving role of lymphatics in cancer metastasis. Curr Opin Immunol 53:64–73. https://doi.org/10.1016/j.coi.2018.04.008
129. Schoppmann SF, Birner P, Stöckl J, Kalt R, Ullrich R, Caucig C, Kriehuber E, Nagy K, Alitalo K, Kerjaschki D (2002) Tumor-associated macrophages express lymphatic endothelial growth factors and are related to peritumoral lymphangiogenesis. Am J Pathol 161:947–956. https://doi.org/10.1016/S0002-9440(10)64255-1
130. Dumont DJ, Jussila L, Taipale J, Lymboussaki A, Mustonen T, Pajusola K, Breitman M, Alitalo K (1998) Cardiovascular failure in mouse embryos deficient in VEGF receptor-3. Science 282:946–949
131. Tammela T, Zarkada G, Wallgard E, Murtomäki A, Suchting S, Wirzenius M, Waltari M, Hellström M, Schomber T, Peltonen R, Freitas C, Duarte A, Isoniemi H, Laakkonen P, Christofori G, Ylä-Herttuala S, Shibuya M, Pytowski B, Eichmann A, Betsholtz C, Alitalo K (2008) Blocking VEGFR-3 suppresses angiogenic sprouting and vascular network formation. Nature 454:656–660. https://doi.org/10.1038/nature07083
132. Laakkonen P, Waltari M, Holopainen T, Takahashi T, Pytowski B, Steiner P, Hicklin D, Persaud K, Tonra JR, Witte L, Alitalo K (2007) Vascular endothelial growth factor receptor 3 is involved in tumor angiogenesis and growth. Cancer Res 67:593–599. https://doi.org/10.1158/0008-5472.CAN-06-3567
133. Petrova TV, Bono P, Holnthoner W, Chesnes J, Pytowski B, Sihto H, Laakkonen P, Heikkilä P, Joensuu H, Alitalo K (2008) VEGFR-3 expression is restricted to blood and lymphatic vessels in solid tumors. Cancer Cell 13:554–556. https://doi.org/10.1016/j.ccr.2008.04.022
134. Jeltsch M, Kaipainen A, Joukov V, Meng X, Lakso M, Rauvala H, Swartz M, Fukumura D, Jain RK, Alitalo K (1997) Hyperplasia of lymphatic vessels in VEGF-C transgenic mice. Science 276:1423–1425
135. Oh SJ, Jeltsch MM, Birkenhäger R, McCarthy JE, Weich HA, Christ B, Alitalo K, Wilting J (1997) VEGF and VEGF-C: specific induction of angiogenesis and lymphangiogenesis in the differentiated avian chorioallantoic membrane. Dev Biol 188:96–109. https://doi.org/10.1006/dbio.1997.8639
136. He Y, Karpanen T, Alitalo K (2004) Role of lymphangiogenic factors in tumor metastasis. Biochim Biophys Acta 1654:3–12. https://doi.org/10.1016/j.bbcan.2003.07.003
137. Hirakawa S, Brown LF, Kodama S, Paavonen K, Alitalo K, Detmar M (2007) VEGF-C-induced lymphangiogenesis in sentinel lymph nodes promotes tumor metastasis to distant sites. Blood 109:1010–1017. https://doi.org/10.1182/blood-2006-05-021758
138. Karpanen T, Egeblad M, Karkkainen MJ, Kubo H, Ylä-Herttuala S, Jäättelä M, Alitalo K (2001) Vascular endothelial growth factor C promotes tumor lymphangiogenesis and intralymphatic tumor growth. Cancer Res 61:1786–1790
139. Kopfstein L, Veikkola T, Djonov VG, Baeriswyl V, Schomber T, Strittmatter K, Stacker SA, Achen MG, Alitalo K, Christofori G (2007) Distinct roles of vascular endothelial growth factor-D in lymphangiogenesis and metastasis. Am J Pathol 170:1348–1361. https://doi.org/10.2353/ajpath.2007.060835
140. Brakenhielm E, Burton JB, Johnson M, Chavarria N, Morizono K, Chen I, Alitalo K, Wu L (2007) Modulating metastasis by a lymphangiogenic switch in prostate cancer. Int J Cancer 121:2153–2161. https://doi.org/10.1002/ijc.22900
141. Dadras SS, Paul T, Bertoncini J, Brown LF, Muzikansky A, Jackson DG, Ellwanger U, Garbe C, Mihm MC, Detmar M (2003) Tumor lymphangiogenesis: a novel prognostic indicator for cutaneous melanoma metastasis and survival. Am J Pathol 162:1951–1960. https://doi.org/10.1016/S0002-9440(10)64328-3
142. Mandriota SJ, Jussila L, Jeltsch M, Compagni A, Baetens D, Prevo R, Banerji S, Huarte J, Montesano R, Jackson DG, Orci L, Alitalo K, Christofori G, Pepper MS (2001) Vascular endothelial growth factor-C-mediated lymphangiogenesis promotes tumour metastasis. EMBO J 20:672–682. https://doi.org/10.1093/emboj/20.4.672
143. Mattila MM-T, Ruohola JK, Karpanen T, Jackson DG, Alitalo K, Härkönen PL (2002) VEGF-C induced lymphangiogenesis is associated with lymph node metastasis in orthotopic MCF-7 tumors. Int J Cancer 98:946–951
144. Podgrabinska S, Skobe M (2014) Role of lymphatic vasculature in regional and distant metastases. Microvasc Res 95:46–52. https://doi.org/10.1016/j.mvr.2014.07.004
145. Skobe M, Hawighorst T, Jackson DG, Prevo R, Janes L, Velasco P, Riccardi L, Alitalo K, Claffey K, Detmar M (2001) Induction of tumor lymphangiogenesis by VEGF-C promotes breast can-

cer metastasis. Nat Med 7:192–198. https://doi.org/10.1038/84643
146. Stacker SA, Caesar C, Baldwin ME, Thornton GE, Williams RA, Prevo R, Jackson DG, Nishikawa S, Kubo H, Achen MG (2001) VEGF-D promotes the metastatic spread of tumor cells via the lymphatics. Nat Med 7:186–191. https://doi.org/10.1038/84635
147. Harrell MI, Iritani BM, Ruddell A (2007) Tumor-induced sentinel lymph node lymphangiogenesis and increased lymph flow precede melanoma metastasis. Am J Pathol 170:774–786. https://doi.org/10.2353/ajpath.2007.060761
148. Hirakawa S (2009) From tumor lymphangiogenesis to lymphvascular niche. Cancer Sci 100:983–989. https://doi.org/10.1111/j.1349-7006.2009.01142.x
149. Hirakawa S, Kodama S, Kunstfeld R, Kajiya K, Brown LF, Detmar M (2005) VEGF-A induces tumor and sentinel lymph node lymphangiogenesis and promotes lymphatic metastasis. J Exp Med 201:1089–1099. https://doi.org/10.1084/jem.20041896
150. Björndahl M, Cao R, Nissen LJ, Clasper S, Johnson LA, Xue Y, Zhou Z, Jackson D, Hansen AJ, Cao Y (2005a) Insulin-like growth factors 1 and 2 induce lymphangiogenesis in vivo. Proc Natl Acad Sci USA 102:15593–15598. https://doi.org/10.1073/pnas.0507865102
151. Björndahl MA, Cao R, Burton JB, Brakenhielm E, Religa P, Galter D, Wu L, Cao Y (2005b) Vascular endothelial growth factor-a promotes peritumoral lymphangiogenesis and lymphatic metastasis. Cancer Res 65:9261–9268. https://doi.org/10.1158/0008-5472.CAN-04-2345
152. Kadowaki I, Ichinohasama R, Harigae H, Ishizawa K, Okitsu Y, Kameoka J, Sasaki T (2005) Accelerated lymphangiogenesis in malignant lymphoma: possible role of VEGF-A and VEGF-C. Br J Haematol 130:869–877. https://doi.org/10.1111/j.1365-2141.2005.05695.x
153. Zeng Y, Opeskin K, Goad J, Williams ED (2006) Tumor-induced activation of lymphatic endothelial cells via vascular endothelial growth factor receptor-2 is critical for prostate cancer lymphatic metastasis. Cancer Res 66:9566–9575. https://doi.org/10.1158/0008-5472.CAN-06-1488
154. Cao R, Björndahl MA, Religa P, Clasper S, Garvin S, Galter D, Meister B, Ikomi F, Tritsaris K, Dissing S, Ohhashi T, Jackson DG, Cao Y (2004) PDGF-BB induces intratumoral lymphangiogenesis and promotes lymphatic metastasis. Cancer Cell 6:333–345. https://doi.org/10.1016/j.ccr.2004.08.034
155. Kodama M, Kitadai Y, Sumida T, Ohnishi M, Ohara E, Tanaka M, Shinagawa K, Tanaka S, Yasui W, Chayama K (2010) Expression of platelet-derived growth factor (PDGF)-B and PDGF-receptor β is associated with lymphatic metastasis in human gastric carcinoma. Cancer Sci 101:1984–1989. https://doi.org/10.1111/j.1349-7006.2010.01639.x
156. Kelly BD, Hackett SF, Hirota K, Oshima Y, Cai Z, Berg-Dixon S, Rowan A, Yan Z, Campochiaro PA, Semenza GL (2003) Cell type-specific regulation of angiogenic growth factor gene expression and induction of angiogenesis in nonischemic tissue by a constitutively active form of hypoxia-inducible factor 1. Circ Res 93:1074–1081. https://doi.org/10.1161/01.RES.0000102937.50486.1B
157. Ullerås E, Wilcock A, Miller SJ, Franklin GC (2001) The sequential activation and repression of the human PDGF-B gene during chronic hypoxia reveals antagonistic roles for the depletion of oxygen and glucose. Growth Factors Chur Switz 19:233–245
158. Spinella F, Garrafa E, Di Castro V, Rosanò L, Nicotra MR, Caruso A, Natali PG, Bagnato A (2009) Endothelin-1 stimulates lymphatic endothelial cells and lymphatic vessels to grow and invade. Cancer Res 69:2669–2676. https://doi.org/10.1158/0008-5472.CAN-08-1879
159. Grimshaw MJ (2007) Endothelins and hypoxia-inducible factor in cancer. Endocr Relat Cancer 14:233–244. https://doi.org/10.1677/ERC-07-0057
160. Camenisch G, Stroka DM, Gassmann M, Wenger RH (2001) Attenuation of HIF-1 DNA-binding activity limits hypoxia-inducible endothelin-1 expression. Pflugers Arch 443:240–249. https://doi.org/10.1007/s004240100679
161. Shields JD, Fleury ME, Yong C, Tomei AA, Randolph GJ, Swartz MA (2007) Autologous chemotaxis as a mechanism of tumor cell homing to lymphatics via interstitial flow and autocrine CCR7 signaling. Cancer Cell 11:526–538. https://doi.org/10.1016/j.ccr.2007.04.020
162. Wilson JL, Burchell J, Grimshaw MJ (2006) Endothelins induce CCR7 expression by breast tumor cells via endothelin receptor a and hypoxia-inducible factor-1. Cancer Res 66:11802–11807. https://doi.org/10.1158/0008-5472.CAN-06-1222
163. Zhuo W, Jia L, Song N, Lu X-A, Ding Y, Wang X, Song X, Fu Y, Luo Y (2012) The CXCL12-CXCR4 chemokine pathway: a novel axis regulates lymphangiogenesis. Clin Cancer Res 18:5387–5398. https://doi.org/10.1158/1078-0432.CCR-12-0708
164. Tudisco L, Orlandi A, Tarallo V, De Falco S (2017) Hypoxia activates placental growth factor expression in lymphatic endothelial cells. Oncotarget 8:32873–32883. https://doi.org/10.18632/oncotarget.15861

Hypoxia and Metabolism in Metastasis

6

Tong Zhang, Caixia Suo, Chenyang Zheng, and Huafeng Zhang

Abstract

The hypoxic microenvironment is one of the major features of solid tumors, which regulates cell malignancy in multiple ways. As a response to hypoxia, a large number of target genes involved in cell growth, metabolism, metastasis and immunity are activated in cancer cells. Hypoxia-inducible factor 1 (HIF-1), as a heterodimeric DNA-binding complex, is comprised of a constitutively expressed HIF-1β subunit and an oxygen sensitive HIF-1α subunit, thus, adapts to decreased oxygen availability as a transcriptional factor. HIF-1 regulates many genes involved in tumorigenesis. Here, we focus on cancer cell metabolism and metastasis regulated by hypoxia.

T. Zhang · C. Zheng · H. Zhang (✉)
Division of Molecular Medicine, Hefei National Laboratory for Physical Sciences at Microscale, the CAS Key Laboratory of Innate Immunity and Chronic Disease, School of Life Sciences, University of Science and Technology of China, Hefei, China
e-mail: hzhang22@ustc.edu.cn

C. Suo
Guangzhou First People's Hospital, School of Medicine, Institutes for Life Sciences, South China University of Technology of China, Guangzhou, China

Keywords

Hypoxia · HIF1 · Glycolysis · Glycogen synthesis · Lipid metabolism · Metastasis · EMT · Metastatic niche · Mitochondria · Metabolic enzymes

6.1 Hypoxia and Metabolism

Hypoxia has profound effects on cellular metabolism, including glycolysis, glycogen synthesis, oxidative phosphorylation and lipid metabolism. Tumor cells utilize glucose more often than normal cells. Under hypoxic conditions, HIF-1 directly promotes the expression of most genes involved in glycolysis, including glucose transporters (GLUT1 and GLUT3) and glycolytic enzymes (HK1/2, GPI, PFKL, ALDOA, TPI, GAPDH, PGK1, PGAM1, ENO1, PKM2 and LDHA) [1]. Thus, HIF-1 pathway increases glucose uptake and lactate secretion by upregulating GLUT1, GLUT3 and LDHA, respectively. Enhanced LDHA mediates the conversion of pyruvate to lactate, accompanied by the recycle of cytosolic NAD^+ that is necessary for further glycolysis. Consequently, accumulated lactate in hypoxic cells together with H^+ are exported via monocarboxylate transporter 1/4 (MCT-1/4), carbonic anhydrase (CA9), and Na^+/H^+ exchanger (NHE-1) which are all activated by HIF-1 under hypoxia [2]. In addition, cancer cells reserve glucose by activating glycogen syn-

D. M. Gilkes (ed.), *Hypoxia and Cancer Metastasis*, Advances in Experimental Medicine and Biology 1136,
https://doi.org/10.1007/978-3-030-12734-3_6

thesis pathway. HIF-1 upregulates the expression of glycogen synthase (GYS1), UDP-glucose pyrophosphorylase (UGP2) and 1.4α glucan branching enzyme (GBE1), resulting in hypoxic tolerance via enhancing glycogen synthesis. Furthermore, HIF-1 targeted PP1-complex phosphatase PPP1R3C activates GYS1 while decreasing glycogen breakdown by inhibiting glycogen phosphorylase (PYGL) [3]. Accumulated glycogen can be utilized as a source of glucose under nutrient starvation.

As is known, cancer cells often exhibit reduced mitochondrial respiration to avoid the cell damage induced by the accumulated mitochondrial ROS. Under hypoxic conditions, HIF-1 regulates the mitochondrial respiration through different ways. First, HIF-1 directly promotes the expression of PDK1 that phosphorylates and inactivates PDH, which results in decreased acetyl-CoA into the TCA cycle [4]. HIF-1 reduces PGC-1β-mediated mitochondrial biogenesis by upregulating MXI-1, which decreases mitochondrial respiration [5]. HIF-1 also modulates respiratory chain function by switching from COX4-1 (high activity, high ROS) to COX4-2 (low activity, low ROS) [6]. HIF-1 activates BNIP3 and BNIP3Lwhich trigger mitochondrial autophagy, leading to down-regulated mitochondrial mass and function [7, 8]. Furthermore, hypoxia also affects mitochondrial complex activity. On one hand, HIF-1 suppresses mitochondrial complex I activity by activating NADH dehydrogenase 1α sub-complex subunit 4-like 2 (NDUFA4L2) [9]. On the other hand, hypoxia inhibits mitochondrial ROS generation through inducing miR-210 which represses the expression of the Fe-S cluster assembly proteins ISCU-1 and ISCU-2 [10]. Thus, HIF-1 mediates adaptive responses of cancer cells to hypoxia by regulating glucose metabolism as well as mitochondrial activity.

Besides the important regulation of glucose metabolism, hypoxia plays important roles in lipid metabolism, such as, fatty acids uptake, *de novo* fatty acids synthesis and fatty acids oxidation. FABP3 and FABP7 mediate fatty acids uptake and ADRP is required for the formation of lipid droplet (LD). It has been shown that FABP3, FABP7 and ADRP are induced by HIF-1α, thus leading to LD accumulation under hypoxia. Accumulation of lipid droplets protects against ROS and maintains survival of cells under hypoxia-reoxygenation conditions [11]. Regarding the *de novo* fatty acid synthesis, hypoxic cells exhibit increased rates of glutamine uptake to recover the decreased entry of glucose-derived carbons into mitochondria. Specifically, HIF-1α-dependent induction of the ubiquitin ligase SIAH2 triggers proteasomal degradation of 2-oxoglutarate dehydrogenase (OGDH), which blocks Suc-CoA generation and promotes *de novo* fatty acid synthesis by increasing citrate substrate [12]. Recently, several interesting studies have shown that hypoxic cancer cells absorb acetate from plasma to generate acetyl-CoA by acetyl-CoA synthetase 2 (ACSS2), then acetyl-CoA is utilized for anabolic metabolism including the *de novo* lipid synthesis. On the other hand, HIF-1 suppresses fatty acids oxidation (FAO) by inhibiting the medium- and long-chain acyl-CoA dehydrogenases (MCAD and LCAD) expression, resulting in reduced ROS and suppression of PTEN pathway, thus promoting cancer cell proliferation [13]. To eliminate cellular ROS, cancer cells also generate more NADPH from serine biosynthesis pathway (SSP) under hypoxia by activating phosphoglycerate dehydrogenase (PHGDH) and serine hydroxymethyl transferase 2 (SHMT2) [14, 15]. In summary, lipid metabolism and serine biosynthesis pathways not only eliminate ROS but also provide lipids and nucleotides for cancer cell proliferation (Fig. 6.1).

There are many other kinds of cells, such as T cells, NK cells, cancer-associated fibroblasts (CAFs), tumor-associated macrophages (TAMs) and mesenchymal stem cells (MSCs), which surround cancer cells in the tumor microenvironment. Therefore, it is unavoidable for cancer cells to compete with other cells for nutrients and metabolites. Cancer cells have stronger competition ability for glucose, while T cells are restricted in glucose uptake. The restricted glucose uptake leads to a drop in the production of glycolytic intermediate phosphoenolpyruvate in T cells, which shuts down T cell receptor-mediated calcium signaling, eventually dimin-

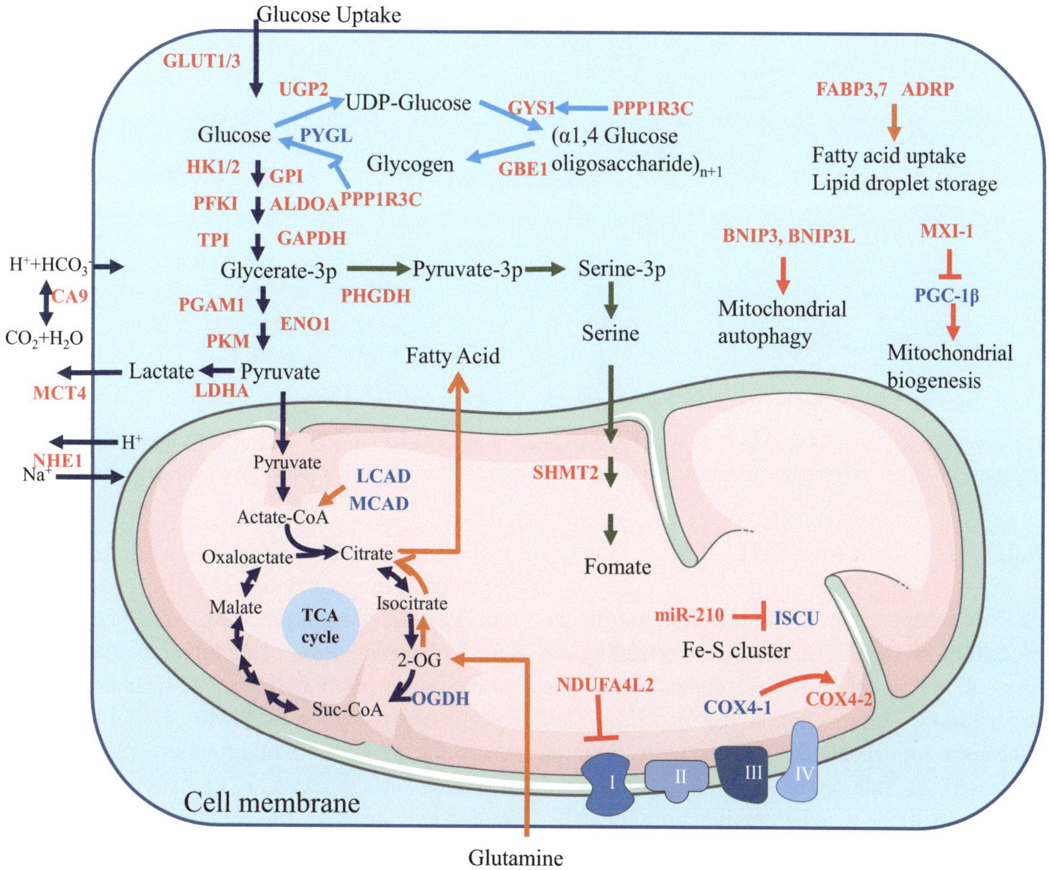

Fig. 6.1 Hypoxia regulated metabolic reprogramming. HIF-1 plays important roles in glucose metabolism, regulation of mitochondrial respiration, lipid metabolism and pentose phosphate pathway. Genes marked with red color are upregulated by HIF-1, while genes marked with blue color are downregulated by HIF-1

ishing the anti-tumor responses of T cells [16]. On one hand, as one of the major metabolite in cancer cells, lactate is used for bioenergetics and also serves as a signaling metabolite. LDHA-associated lactic acid production from human melanoma cells leads to immune evasion by impairing cytokine production, in particular IFN-γ in tumor-infiltrating T cells and NK cells, which promotes tumor growth [17]. This study suggests that lactic acid is a potential signaling molecule in regulation of tumor immune escape. On the other hand, lactic acid generated by tumor cells induces the expression of vascular endothelial growth factor (VEGF) and arginase 1 (Arg1), which polarizes macrophages to an M2-like state that is critical for tumor growth [18]. Further, lactate produced from hypoxic cancer cells and cancer-associated fibroblasts (CAFs) feeds the cancer cell populations that are within the same tumor but are proximal to the vasculature [19]. All in all, the interaction between tumor cells and surrounding cells in the microenvironment emerges as critical events in cancer progression.

6.2 Hypoxia and Metastasis

Tumor metastasis leads to more than 90% of cancer patients' death. Hypoxia affects multiple steps of metastasis, including epithelial-mesenchymal transition (EMT), intravasation, extravasation and pre-metastatic niche formation (Fig. 6.2).

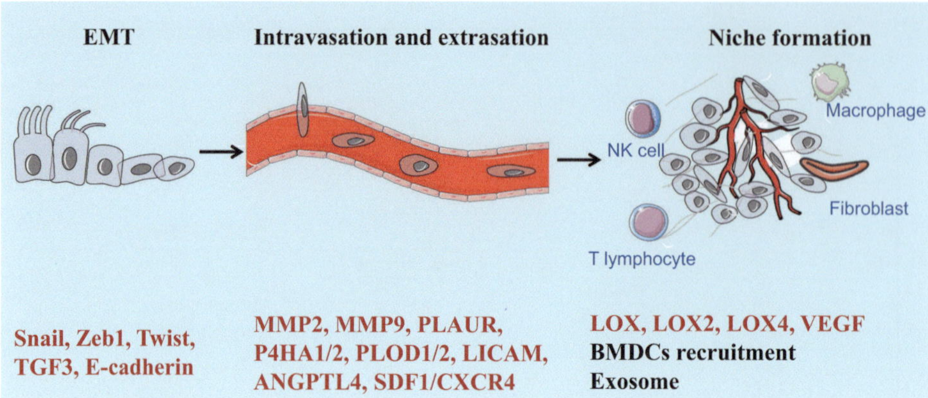

Fig. 6.2 Hypoxic regulation of cancer metastasis. Hypoxia influences multiple steps within the metastatic cascade, including EMT, intravasation, extravasation and metastatic niche formation. HIF-1 target genes are shown in red

6.2.1 Hypoxia and EMT

EMT has been regarded as the first step in the process of metastasis and is characterized by the loss of epithelial cell-cell contact and the acquisition of mesenchymal features [20]. Cell adhesions are mediated by cadherin proteins expressed at intercellular junctions. Loss of E-cadherin reduces adherent junctions of cells. On the other hand, mesenchymal protein vimentin reorganizes the cytoskeleton to promote cell motility. Hypoxia is sufficient to induce EMT and invasion through both direct and indirect mechanisms. For instance, HIF-1 directly induces the transcription of Snail, ZEB1, TWIST and TCF3 [21]. In breast cancer cells, HIF-1 activates EMT through ZEB1-MYB-E-cadherin signaling axis [22]. The HIF-1 pathway also indirectly promotes EMT via other signaling pathways, including TGF-β, Notch, tyrosine kinase receptors, Wnt signaling pathway and so on.

6.2.2 Hypoxia and Intravasation

In order to invade the surrounding tissues and infiltrate into blood vessels, cancer cells firstly need to degrade the surrounding basement membrane (BM). Proteinases contribute to collagen degradation, mainly including the matrix metalloproteinases (MMPs), which are a family of zinc-dependent enzymes. Hypoxia is associated with an increase of the expression of type IV collagen-degrading enzymes (MMP2 and MMP9) via HIF-1 dependent mechanism [23]. HIF-2 promotes the expression of membrane-bound membrane-type 1 MMP (MMP14) [24]. In addition to collagen degradation by MMPs, HIF-1 also activates the expression of urokinase plasminogen activator surface receptor (PLAUR), which promotes cell invasion by altering the interactions between integrins and the extracellular matrix (ECM) [25].

On the other hand, collagen deposition and biogenesis are very important for cell migration. P4HA1 and P4HA2 are required for collagen deposition and tumor fibrosis and stiffness, while PLOD1 and PLOD2 are important for collagen fiber crosslinking. Recent studies showed that HIF-1 plays a critical role in regulating collagen biogenesis by activating prolyl 4-hydroxylase α-subunit isoform 1 and 2 (P4HA1 and P4HA2) and procollagen-lysine 2-oxyglutarate 5-dioxygenase 1 and 2 (PLOD1 and PLOD2) [26].

6.2.3 Hypoxia and Extravasation

Extravasation is a limiting step in the metastatic process, only 0.1% of cancer cells entering the vascular system establish a metastatic lesion. To extravasate, cancer cells must first adhere to endothelial cells (ECs), and disrupt the interaction between vascular ECs. Hypoxia facilitates the extravasation of cancer cells to promote the metastasis of breast cancer cells to the lung tissues via HIF1-regulated LICAM and ANGPTL4 [27]. LICAM promotes the adhesion of cancer cells to EC monolayers via hemophilic interactions or heterophilic interactions with neuropilin 1 (NRP1), integrins and CD24. Meanwhile, ANGPTL4 inhibits EC-EC interaction, facilitating the extravasation of the breast cancer cells into the lung parenchyma. Moreover, HIF-1-regulated SDF-1/CXCR4 also contributes to the adhesion of cancer cells to ECs [28]. Thus, hypoxia promotes the extravasation of cancer cells through inhibiting EC-EC interaction and accelerating the adhesion of cancer cells to EC monolayers.

6.2.4 Hypoxia and Metastatic Niche Formation

The late stages of metastasis comprise of the colonization and growth of cancer cells within the distant tissue microenvironment. During metastasis, bone marrow-derived cells (BMDCs) are recruited to metastatic sites that are colonized by metastatic cancer cells. Furthermore, cancer cells produce lysyl oxidase (LOX), an enzyme that crosslinks extracellular matrix proteins such as collagen, which promotes breast cancer metastasis. In breast cancer cells, HIF-1 induces the expression of multiple members of the lysyl oxidase (LOX), including LOX, LOXL2 and LOXL4. Therefore, HIF-1 also contributes to premetastatic niche formation. A clinical study revealed that LOX was associated with metastases of breast cancer patients and it demonstrates that a monoclonal antibody directed against LOX might be useful for breast cancer therapy [29, 30].

Recent studies demonstrate that hypoxia promotes the formation of the premetastatic niche by exosomes. Exosomes carry and transfer molecules including proteins, lipids, microRNAs and mRNAs, and are involved in the premetastatic niche formation that promotes vascular permeability and BMDC recruitment [31]. Recent studies showed that exosomes from the glioblastoma multiforme (GBM) patients contain hypoxia-regulated proteins, including MMP9, MMP8, platelet-derived growth factor (PDGF), and insulin-like growth factor-binding protein 3 (IGFBP3), suggesting that hypoxia regulates metastasis via altering exosome cargo contents [32]. Moreover, hypoxia promotes microvesicle shedding by up-regulating guanosine triphosphatase RAB22A to accelerate invasion and metastasis [33].Thus, hypoxia plays multiple roles in regulation of cancer metastasis.

6.3 Metabolic Reprogramming During Metastasis Under Hypoxia

Although metabolic reprogramming is a hallmark of cancer cells, little is known about the relationship between metabolic alterations and metastasis. Energy supply is a pivotal factor for cancer metastasis. The liver microenvironment is hypoxic and glucose imbalanced, especially in the interstitial space due to competition from neighboring hepatocytes. For instance, to overcome the shortage of oxygen and glucose, colon cancer cells that metastasize to the liver must use alternative energy sources other than glucose. Loo et al. [34] demonstrates that colon cancer cells secrete a kinase that acts extracellularly to generate phosphocreatinine, a metabolite that directly fuels colon cancer growth and facilitates their metastasis to the liver. Mechanistically, in metastatic colon cancer cells, reduced miR-483 and miR-551a levels elevated the expression of their target protein, creatine kinase (CKB), which is secreted and phosphorylates extracellular creatine produced by hepatocytes to generate phosphocreatine. Extracellular phosphocreatine is imported into colon cancer cells by the transporter SLC6A8 and is used to generate ATP for metastasized colon cancer cells. In this way, metastatic colon cancer cells obtain sufficient energy to overcome metabolic stress for further liver colonization.

As is known, ovarian cancer cells preferentially metastasize to adipose tissue, but the molecular mechanisms are still elusive. Nieman et al. found that adipocytes provide an energy source for ovarian cancer cells by inducing lipolysis in adipocytes and β-oxidation in cancer cells [35]. Fatty acid-binding protein 4 (FABP4) transports fatty acids from the adipocyte to ovarian cancer cells and accelerates the β-oxidation of fatty acids in cancer cells. FABP4 deficiency impaired metastatic tumor growth in mice, indicating that FABP4 has an important role in metastasis of ovarian cancer cells. Hypoxia promotes FABP4 expression via suppression of miR-409-3p, suggesting the important role of hypoxia in ovarian cancer metastasis by regulating lipid metabolism [36]. Another study has revealed that knockdown of the respiratory enzyme citrate synthetase (CS) promotes cervical cancer cell metastasis and proliferation by inhibiting p53/TIGAR and SCO2 pathway due to decreased ATP production [37].

Hypoxia regulates tumor metastasis by altering the expression of metabolic enzymes. Dupuy et al. [38] found that liver metastatic breast cancer cells are highly dependent on glycolysis for survival and metastasis in comparison to cells metastasized to bone or lung tissues. Liver metastatic cells are with lower oxidative glutamine metabolism, however, increased glutamine uptake and glutamine-derived TCA cycle intermediates are found in bone and lung metastatic cells. The metabolic reprogramming of liver-metastatic breast cancer cells is attributed to activated pyruvate dehydrogenase kinase (PDK1), a known target of HIF-1. PDK1 blocks the conversion of pyruvate to acetyl-CoA and inhibits the TCA cycle by inactivating pyruvate dehydrogenase (PDH), leading to glycolysis accompanied with repressed mitochondrial function. Therefore, liver metastases of breast cancer cells rely on a HIF-1/PDK1 axis for their metabolic reprogramming to accelerate their efficient colonization and growth in the liver. Mechanistically, Wnt/β-catenin signaling induces vessel growth by facilitating the delivery of oxygen and nutrient upon PDK1-driven glycolytic metabolism. Similar to breast cancer cells, colorectal cancer cells also undergo metabolic reprogramming when they metastasize to the liver. Colon cancer cells metastasized to the liver tissue possess high levels of aldolase B (ALDOB), one enzyme that enhances fructose metabolism to fuel glycolysis, gluconeogenesis and pentose phosphate pathway. Targeting ALDOB significantly reduces liver metastases of the colorectal cancer but shows no effect on primary tumors, highlighting the importance of fructose metabolism and microenvironment during cancer metastasis [39] (Fig. 6.3).

On the other hand, some metabolic enzymes or metabolites regulate metastasis via HIF-1 pathway. Mutations of genes encoding succinate dehydrogenase (SDH), fumarate hydratase (FH) and isocitratedehydrogenase (IDH1 and IDH2) lead to the accumulation of succinate, fumarate and 2 hydroxyglutarate (2-HG), respectively. These metabolites stabilize HIF-1α by inhibiting prolyl-hydroxylases (PHDs), which accelerates multiple steps of metastasis [40]. Similarly, proinflammatory protein transglutaminase 2 (TG2) activated nuclear factor NF-κB binds to the HIF-1α promoter and induces its expression in mammary epithelial cells [41]. Monoamine oxidase A (MAOA), a mitochondrial enzyme that degrades monoamine neurotransmitters and amines, induces EMT by generating ROS that inhibits PHD3 activity and stabilizes HIF-1α. As a result, MAOA-mediated HIF-1α/VEGF/FOXO1/TWIST1 pathway was activated, which drives prostate cancer tumorigenesis and metastasis [42]. Luo et al. found that pyruvate kinase 2 (PKM2), a glycolytic enzyme, interacts with HIF-1α and promotes the expression of HIF-1 target genes by recruiting p300 to hypoxia response elements [43]. Further study demonstrates that PKM2 is hydroxylated on proline-403/408 by prolyl hydroxylase 3 (PHD3), which enhances its binding activity to HIF-1α. Therefore, some metabolic enzymes promote metastasis by stabilizing HIF-1α and may serve as new biomarkers and therapeutic targets for metastatic disease.

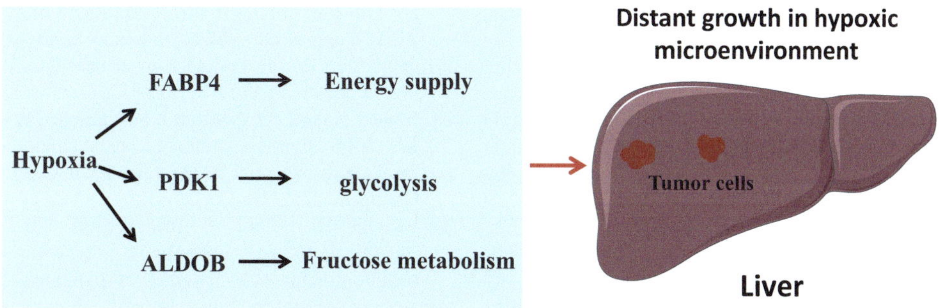

Fig. 6.3 Metabolic reprogramming during metastasis under hypoxia. HIF-1 supplies energy by upregulating FABP4. Concurrently, HIF-1 promotes PDK1 and ALDOB to regulate glycolysis and fructose metabolism, respectively. All of these are beneficial to cancer cells for their distant growth in hypoxic microenvironment (in the liver, for example, as shown)

6.4 Perspectives and Summary

Hypoxia plays important roles in metabolism and metastasis. Hypoxia regulates glycolysis, glycogen synthesis, lipid metabolism and oxidative phosphorylation to promote cell survival and proliferation. Moreover, hypoxia regulates both the early and late stages of metastasis. Herein, we summarized the recent progress of hypoxia and metabolism in metastasis. On one hand, current studies demonstrate that hypoxia reprograms energy supply or nutrient metabolism by regulating metabolic enzymes, which are essential for efficient metastatic colonization. For example, colon cancer cells release CKB into the extracellular space to generate ATP to adapt to the metabolic stress within the hypoxic liver, while breast cancer cells adapt to hypoxic liver by increasing the expression of PDK1. On the other hand, some metabolic enzymes regulate metastasis via HIF-1 pathway, such as SDH, FH, IDH and PKM2. More detailed mechanisms between metabolism and metastasis are still elusive and are inviting more in-depth studies in this field. The identification of new biomarkers from the perspective of metabolism and hypoxia would be valuable for the treatment of tumor metastasis.

Acknowledgments This work was supported in part by grants from the National Natural Science Foundation of China (81525022), the strategic priority research program(Pilot study)of the Chinese Academy of Sciences (XDPB10), and the major innovative program of development foundation of Hefei center for physical science and technology (2017FXZY004).

References

1. Semenza GL (2012) Hypoxia-inducible factors: mediators of cancer progression and targets for cancer therapy. Trends Pharmacol Sci 33:207–214
2. Schito L, Semenza GL (2016) Hypoxia-inducible factors: master regulators of cancer progression. Trends Cancer 2:758–770
3. Favaro E, Bensaad K, Chong MG, Tennant DA, Ferguson DJ, Snell C, Steers G, Turley H, Li JL, Gunther UL et al (2012) Glucose utilization via glycogen phosphorylase sustains proliferation and prevents premature senescence in cancer cells. Cell Metab 16:751–764
4. Kim JW, Tchernyshyov I, Semenza GL, Dang CV (2006) HIF-1-mediated expression of pyruvate dehydrogenase kinase: a metabolic switch required for cellular adaptation to hypoxia. Cell Metab 3:177–185
5. Zhang H, Gao P, Fukuda R, Kumar G, Krishnamachary B, Zeller KI, Dang CV, Semenza GL (2007) HIF-1 inhibits mitochondrial biogenesis and cellular respiration in VHL-deficient renal cell carcinoma by repression of C-MYC activity. Cancer Cell 11:407–420
6. Fukuda R, Zhang H, Kim JW, Shimoda L, Dang CV, Semenza GL (2007) HIF-1 regulates cytochrome oxidase subunits to optimize efficiency of respiration in hypoxic cells. Cell 129:111–122
7. Bellot G, Garcia-Medina R, Gounon P, Chiche J, Roux D, Pouyssegur J, Mazure NM (2009) Hypoxia-induced autophagy is mediated through hypoxia-inducible factor induction of BNIP3 and BNIP3L via their BH3 domains. Mol Cell Biol 29:2570–2581
8. Zhang H, Bosch-Marce M, Shimoda LA, Tan YS, Baek JH, Wesley JB, Gonzalez FJ, Semenza GL (2008) Mitochondrial autophagy is an HIF-1-

dependent adaptive metabolic response to hypoxia. J Biol Chem 283:10892–10903
9. Tello D, Balsa E, Acosta-Iborra B, Fuertes-Yebra E, Elorza A, Ordonez A, Corral-Escariz M, Soro I, Lopez-Bernardo E, Perales-Clemente E et al (2011) Induction of the mitochondrial NDUFA4L2 protein by HIF-1alpha decreases oxygen consumption by inhibiting Complex I activity. Cell Metab 14:768–779
10. Chan SY, Zhang YY, Hemann C, Mahoney CE, Zweier JL, Loscalzo J (2009) MicroRNA-210 controls mitochondrial metabolism during hypoxia by repressing the iron-sulfur cluster assembly proteins ISCU1/2. Cell Metab 10:273–284
11. Bensaad K et al (2014) Fatty acid uptake and lipid storageinduced by HIF-1a contribute to cell growth and survival afterhypoxia-reoxygenation. Cell Rep 9:349–365
12. Sun RC, Denko NC (2014) Hypoxic regulation of glutamine metabolism through HIF1 and SIAH2 supports lipid synthesis that is necessary for tumor growth. Cell Metab 19:285–292
13. Huang, Li T, Li X, Zhang L, Sun L, He X, Zhong X, Jia D, Song L, Semenza GL et al (2014) HIF-1-mediated suppression of acyl-CoA dehydrogenases and fatty acid oxidation is critical for cancer progression. Cell Rep 8:1930–1942
14. Samanta D, Park Y, Andrabi SA, Shelton LM, Gilkes DM, Semenza GL (2016) PHGDH expression is required for mitochondrial redox homeostasis, breast cancer stem cell maintenance, and lung metastasis. Cancer Res 76:4430–4442
15. Ye J, Fan J, Venneti S, Wan YW, Pawel BR, Zhang J, Finley LW, Lu C, Lindsten T, Cross JR et al (2014) Serine catabolism regulates mitochondrial redox control during hypoxia. Cancer Discov 4:1406–1417
16. Ho PC, Bihuniak JD, Macintyre AN, Staron M, Liu X, Amezquita R, Tsui YC, Cui G, Micevic G, Perales JC et al (2015) Phosphoenolpyruvate Is a metabolic checkpoint of anti-tumor T cell responses. Cell 162:1217–1228
17. Brand A, Singer K, Koehl GE, Kolitzus M, Schoenhammer G, Thiel A, Matos C, Bruss C, Klobuch S, Peter K et al (2016) LDHA-associated lactic acid production blunts tumor immunosurveillance by T and NK cells. Cell Metab 24:657–671
18. Colegio OR, Chu NQ, Szabo AL, Chu T, Rhebergen AM, Jairam V, Cyrus N, Brokowski CE, Eisenbarth SC, Phillips GM et al (2014) Functional polarization of tumour-associated macrophages by tumour-derived lactic acid. Nature 513:559–563
19. Lyssiotis CA, Kimmelman AC (2017) Metabolic interactions in the tumor microenvironment. Trends Cell Biol 27:863–875
20. Chaffer CL, Weinberg RA (2011) A perspective on cancer cell metastasis. Science 331:1559–1564
21. Jiang J, Tang Y-l, Liang X-h (2014) EMT: a new vision of hypoxia promoting cancer progression. Cancer Biol Ther 11:714–723
22. Hugo HJ, Pereira L, Suryadinata R, Drabsch Y, Gonda TJ, Gunasinghe NP, Pinto C, Soo ET, van Denderen BJ, Hill P et al (2013) Direct repression of MYB by ZEB1 suppresses proliferation and epithelial gene expression during epithelial-to-mesenchymal transition of breast cancer cells. Breast Cancer Res 15:R113
23. Munoz-Najar UM, Neurath KM, Vumbaca F, Claffey KP (2006) Hypoxia stimulates breast carcinoma cell invasion through MT1-MMP and MMP-2 activation. Oncogene 25:2379–2392
24. Petrella BL, Lohi J, Brinckerhoff CE (2005) Identification of membrane type-1 matrix metalloproteinase as a target of hypoxia-inducible factor-2 alpha in von Hippel-Lindau renal cell carcinoma. Oncogene 24:1043–1052
25. Graham CH, Forsdike J, Fitzgerald CJ, Macdonald-Goodfellow S (1999) Hypoxia-mediated stimulation of carcinoma cell invasiveness via upregulation of urokinase receptor expression. Int J Cancer 80:617–623
26. Gilkes DM, Bajpai S, Chaturvedi P, Wirtz D, Semenza GL (2013) Hypoxia-inducible factor 1 (HIF-1) promotes extracellular matrix remodeling under hypoxic conditions by inducing P4HA1, P4HA2, and PLOD2 expression in fibroblasts. J Biol Chem 288:10819–10829
27. Zhang H, Wong CC, Wei H, Gilkes DM, Korangath P, Chaturvedi P, Schito L, Chen J, Krishnamachary B, Winnard PT Jr et al (2012) HIF-1-dependent expression of angiopoietin-like 4 and L1CAM mediates vascular metastasis of hypoxic breast cancer cells to the lungs. Oncogene 31:1757–1770
28. Jin F, Brockmeier U, Otterbach F, Metzen E (2012) New insight into the SDF-1/CXCR4 axis in a breast carcinoma model: hypoxia-induced endothelial SDF-1 and tumor cell CXCR4 are required for tumor cell intravasation. Mol Cancer Res 10:1021–1031
29. Erler JT, Bennewith KL, Nicolau M, Dornhofer N, Kong C, Le QT, Chi JT, Jeffrey SS, Giaccia AJ (2006) Lysyl oxidase is essential for hypoxia-induced metastasis. Nature 440:1222–1226
30. Wong CC, Gilkes DM, Zhang H, Chen J, Wei H, Chaturvedi P, Fraley SI, Wong CM, Khoo US, Ng IO et al (2011) Hypoxia-inducible factor 1 is a master regulator of breast cancer metastatic niche formation. Proc Natl Acad Sci USA 108:16369–16374
31. Peinado H, Lavotshkin S, Lyden D (2011) The secreted factors responsible for pre-metastatic niche formation: old sayings and new thoughts. Semin Cancer Biol 21:139–146
32. Kucharzewska P, Christianson HC, Welch JE, Svensson KJ, Fredlund E, Ringner M, Morgelin M, Bourseau-Guilmain E, Bengzon J, Belting M (2013) Exosomes reflect the hypoxic status of glioma cells and mediate hypoxia-dependent activation of vascular cells during tumor development. Proc Natl Acad Sci USA 110:7312–7317
33. Wang T, Gilkes DM, Takano N, Xiang L, Luo W, Bishop CJ, Chaturvedi P, Green JJ, Semenza GL (2014) Hypoxia-inducible factors and RAB22A mediate formation of microvesicles that stimulate

breast cancer invasion and metastasis. Proc Natl Acad Sci USA 111:E3234–E3242
34. Loo JM, Scherl A, Nguyen A, Man FY, Weinberg E, Zeng Z, Saltz L, Paty PB, Tavazoie SF (2015) Extracellular metabolic energetics can promote cancer progression. Cell 160:393–406
35. Nieman KM, Kenny HA, Penicka CV, Ladanyi A, Buell-Gutbrod R, Zillhardt MR, Romero IL, Carey MS, Mills GB, Hotamisligil GS et al (2011) Adipocytes promote ovarian cancer metastasis and provide energy for rapid tumor growth. Nat Med 17:1498–1503
36. Gharpure KM, Pradeep S, Sans M, Rupaimoole R, Ivan C, Wu SY, Bayraktar E, Nagaraja AS, Mangala LS, Zhang X et al (2018) FABP4 as a key determinant of metastatic potential of ovarian cancer. Nat Commun 9:2923
37. Lin CC, Cheng TL, Tsai WH, Tsai HJ, Hu KH, Chang HC, Yeh CW, Chen YC, Liao CC, Chang WT (2012) Loss of the respiratory enzyme citrate synthase directly links the Warburg effect to tumor malignancy. Sci Rep 2:785
38. Dupuy F, Tabaries S, Andrzejewski S, Dong Z, Blagih J, Annis MG, Omeroglu A, Gao D, Leung S, Amir E et al (2015) PDK1-dependent metabolic reprogramming dictates metastatic potential in breast cancer. Cell Metab 22:577–589
39. Bu P, Chen KY, Xiang K, Johnson C, Crown SB, Rakhilin N, Ai Y, Wang L, Xi R, Astapova I et al (2018) Aldolase B-mediated fructose metabolism drives metabolic reprogramming of colon cancer liver metastasis. Cell Metab 27:1249–1262 e1244
40. Morin A, Letouze E, Gimenez-Roqueplo AP, Favier J (2014) Oncometabolites-driven tumorigenesis: from genetics to targeted therapy. Int J Cancer 135:2237–2248
41. Kumar S, Donti TR, Agnihotri N, Mehta K (2014) Transglutaminase 2 reprogramming of glucose metabolism in mammary epithelial cells via activation of inflammatory signaling pathways. Int J Cancer 134:2798–2807
42. Wu JB, Shao C, Li X, Li Q, Hu P, Shi C, Li Y, Chen YT, Yin F, Liao CP et al (2014) Monoamine oxidase A mediates prostate tumorigenesis and cancer metastasis. J Clin Invest 124:2891–2908
43. Luo W, Hu H, Chang R, Zhong J, Knabel M, O'Meally R, Cole RN, Pandey A, Semenza GL (2011) Pyruvate kinase M2 is a PHD3-stimulated coactivator for hypoxia-inducible factor 1. Cell 145:732–744

Hypoxia and the Metastatic Niche

7

Cerise Yuen-Ki Chan, Vincent Wai-Hin Yuen, and Carmen Chak-Lui Wong

Abstract

Metastasis is considered the latest stage of cancer development; however, metastasis occurs earlier than it can be detected. Metastatic sites are actively remodeled by secretory factors including growth factors, chemokines and cytokines, extracellular matrix (ECM) enzymes, and exosomes produced by the primary cancer tissues. Many of the associated-secretory factors are abundantly induced by inflammation and hypoxia. These secretory factors modify the ECM, immune composition, and blood vessel permeability of the future metastatic sites, a process termed 'metastatic niche formation.' In general, ECM is modified to enhance the attachment of other cell types or cancer cells to establish a growth-factor rich metastatic niche. Immune-suppressive cells such as tumor-associated macrophages (TAMs) and regulatory T cells (Tregs) dominate the metastatic niche to allow metastatic cancer cells to bypass immune surveillance and propagate. Endothelial cell-to-cell junctions of blood vessels are loosened to enhance the penetrance of metastatic cancer cells to the metastatic sites. Different metastatic tissues have unique ECM constituents, resident immune cells, and anatomical positions linked with the circulatory system; therefore, many cancer types have their own metastatic pattern, and they favor metastasis to specific organs. Some of the remodeling events represent the earliest step of metastasis, even preceding the detachment of cancer cells from the primary tumor site. Understanding how the metastatic niche is formed is important for the development of drugs to prevent the earliest step of metastasis and advance our understanding of organotrophic metastasis. This review summarizes the major findings in the field of metastatic niche highlighting the role of hypoxia.

Keywords

PMN (Premetastatic niche) · BMDC (bone marrow-derived cell) · TAM (tumor-associated macrophage) · Treg (regulatory T cell) · MDSC (myeloid-derived suppressor cell) · Tumor microenvironment · ECM · Cytokine · Chemokine · Exosome

Authors Cerise Yuen-Ki Chan and Vincent Wai-Hin Yuen have equally contributed to this chapter.

C. Y.-K. Chan · V. W.-H. Yuen · C. C.-L. Wong (✉)
Department of Pathology, The University of Hong Kong, Pokfulam, Hong Kong
e-mail: carmencl@pathology.hku.hk

D. M. Gilkes (ed.), *Hypoxia and Cancer Metastasis*, Advances in Experimental Medicine and Biology 1136,
https://doi.org/10.1007/978-3-030-12734-3_7

7.1 Seed and Soil Theory

The concept of the metastatic niche was first proposed by Stephen Paget in 1889. He proposed that successful metastatic colonization of cancer cells (seeds) is determined by the permissive microenvironment of the metastatic organ (soil). However, in the 1930s, James Ewing argued that metastatic colonization is determined by the physiologies of the circulatory system including the vascular and the lymphatic systems. Isaiah Fidler later challenged that, in addition to the blood flow, metastasis could be completed only in specific organs for many cancer types. Each type of cancer has its non-random and particular patterns of metastasis, highlighting the concept of organotropism.

7.2 Metastasis Pattern

Autopsies revealed that lymph nodes and lungs were the most frequent metastatic sites for melanoma while bone, lung, liver are the most common metastatic organs for breast cancer [1, 2]. Esophageal cancer tends to metastasize to gastric organs, lungs, and livers [3, 4]. Gastric metastases are poor prognostic indicators for esophageal cancer patients and are also associated with the aggressiveness of the tumors including size and differentiation status of the tumors [5]. Upper esophageal cancer tends to metastasize to lungs while lower esophageal cancer tends to metastasize to the liver [6]. In colorectal cancer, 50% of patients are diagnosed with metastasis in the lymph nodes, lungs, livers, and the peritoneal cavities [7]. Hepatocellular carcinoma tends to metastasize to portal veins, bones, the diaphragm, and peritoneal cavities. Ovarian cancer tends to metastasize to the omentum and peritoneum. Cervical cancer often metastasizes to the vagina, parametria, bladder, and pelvic lymph nodes.

7.3 Metastatic Niche and Pre-metastatic Niche (PMN)

'Metastatic niche' refers to the soil, or the microenvironment, of the secondary site of the same organ or distant organs where metastasis occurs. Pre-metastatic niche (PMN) however refers to the formation of a microenvironment in the future metastatic site which favors the attachment, survival, colonization, and propagation of cancer cells. PMN is primarily formed by four factors: (1) inflammation, (2) immune suppressive cells, (3) extracellular matrix (ECM), (4) exosomes (Fig. 7.1). The formation of PMN in lung, bone, and liver has been described more extensively than other organs. The lung is a frequent metastatic site. PMN formation in lung tissues starts with the modification of ECM by secretory factors produced by primary cancer cells, accumulation of bone marrow-derived cells (BMDCs) which create a growth factor-rich environment for subsequent colonization of cancer cells (Figs. 7.1 and 7.2). The modification of the ECM and the arrival of BMDCs precedes the arrival of cancer cells. BMDCs are indispensable components of the PMN in lung tissues [8–13]. BMDCs at PMN potentiate metastasis in several ways. First, BMDCs at the metastatic sites are associated with increased expression of chemokines such as SDF-1, which attracts $CXCR4^+$ tumor cells to the metastatic sites [10]. Second, BMDC cells, particularly $CD11b^+$ BMDCs (myeloid-derived suppressor cells, MDSCs), are essential for the development of neovessels in the metastatic lesion supporting cancer cell growth [9, 12, 13]. Third, $CD11b^+$ BMDCs suppress natural killer cell-dependent immune responses in tumor-bearing animals, supporting cancer cell maintenance [11]. Bone is another common metastatic site, especially for breast cancer. Bone is actively remodeled and regenerated through bone formation and resorption carried out by osteoblasts and osteoclasts, respectively. Osteoblasts reside on the surface of the bone and produce a variety of growth factors and enzymes such as TGFβ, IGFs, osteocalcin, collagen, and collagenases. After osteoblasts generate bone matrix, they produce RANKL which stimulates osteoclasts through RANK-RANKL signaling. Osteoclasts are bone-resorbing cells which dissolve minerals and degrade the matrix of the bone. The homeostatic balance of osteoblast-mediated bone formation and osteoclast-mediated bone degradation is disrupted by cancer cells. Breast cancer cells stimulate osteoclasts, bone resorption, and osteolysis

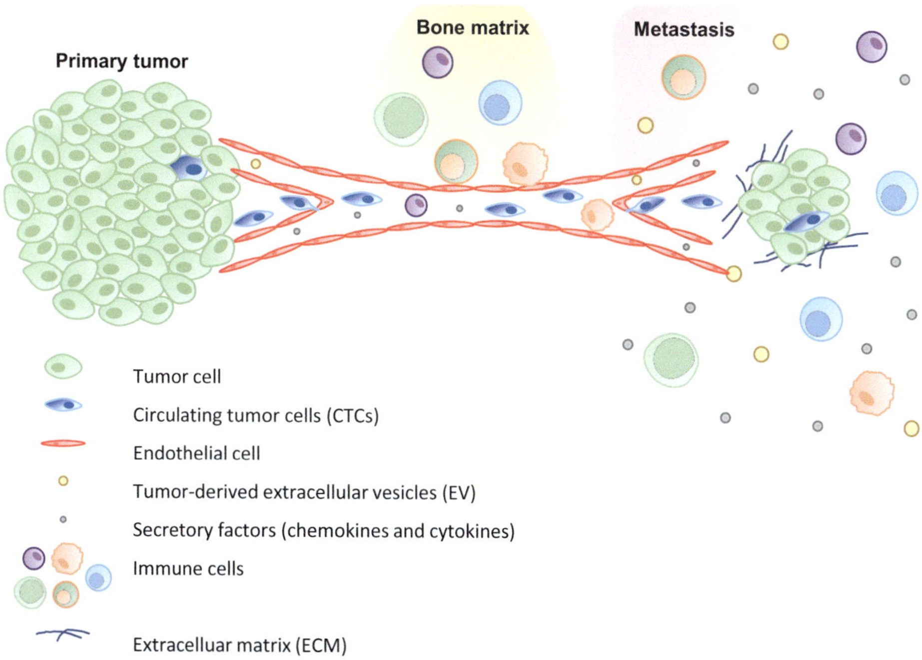

Fig. 7.1 Metastatic niches are shaped by tumor-derived extracellular vesicles (EV), tumor-derived secretory factors, endothelial cells, immune cells, and extracellular matrix (ECM)

(Fig. 7.3). Cancer cells secrete factors to induce bone marrow cells to create a permissive PMN which favors bone resorption and, subsequently, cancer cell colonization. The liver is a common metastatic site for pancreatic and colorectal cancers. The liver is a highly vascularized organ. Many pathogens in the blood pass through the liver. The liver is protected from pathogens by the resident macrophages known as Kupffer cells. Exosomes secreted by primary cancer cells activate Kupffer cells which in turn secrete TGFβ to activate fibronectin secretion in hepatic stellate cells, favoring the recruitment of tumor-associated macrophages (TAM) to the liver.

7.4 Macrophages

Macrophages are derived from myeloid progenitor cells. In mice, macrophages are defined as CD11b, F4/80, and colony-stimulating factor-1 receptor (CSF-1R/CD115) positive and Gr1 negative cells. In humans, macrophages are defined as CD68, CD163, CD16, CD132, and CD115 positive cells. Macrophages are heterogeneous and consist of many subpopulations. Although the re-classification of macrophages is underway, macrophages are still generally classified as M1 macrophages, or classically activated macrophages, and M2 macrophages, or alternatively activated macrophages [14]. M1 macrophages are pro-inflammatory. They are responsible for the eradication of pathogens and foreign cells through direct phagocytosis followed by presenting antigens to type I helper T (Th1) cells. M2 macrophages are anti-inflammatory and are responsible for wound healing and tissue repair by 'switching off' the immune system. Clinical evidence has demonstrated that macrophages can promote cancer development [15]. In cancer, macrophages are found in primary tumors and at metastatic sites [15]. A majority of publications have shown that an increased number of macrophages found in the primary tumor is associated with cancer aggressiveness [15]. Tumor-associated macrophages (TAMs) are generally thought to be M2 macrophages. They have been consistently found in thyroid, lung, and liver cancers. Patients with an increased number of macrophage have poor survival rates [16–18]. A

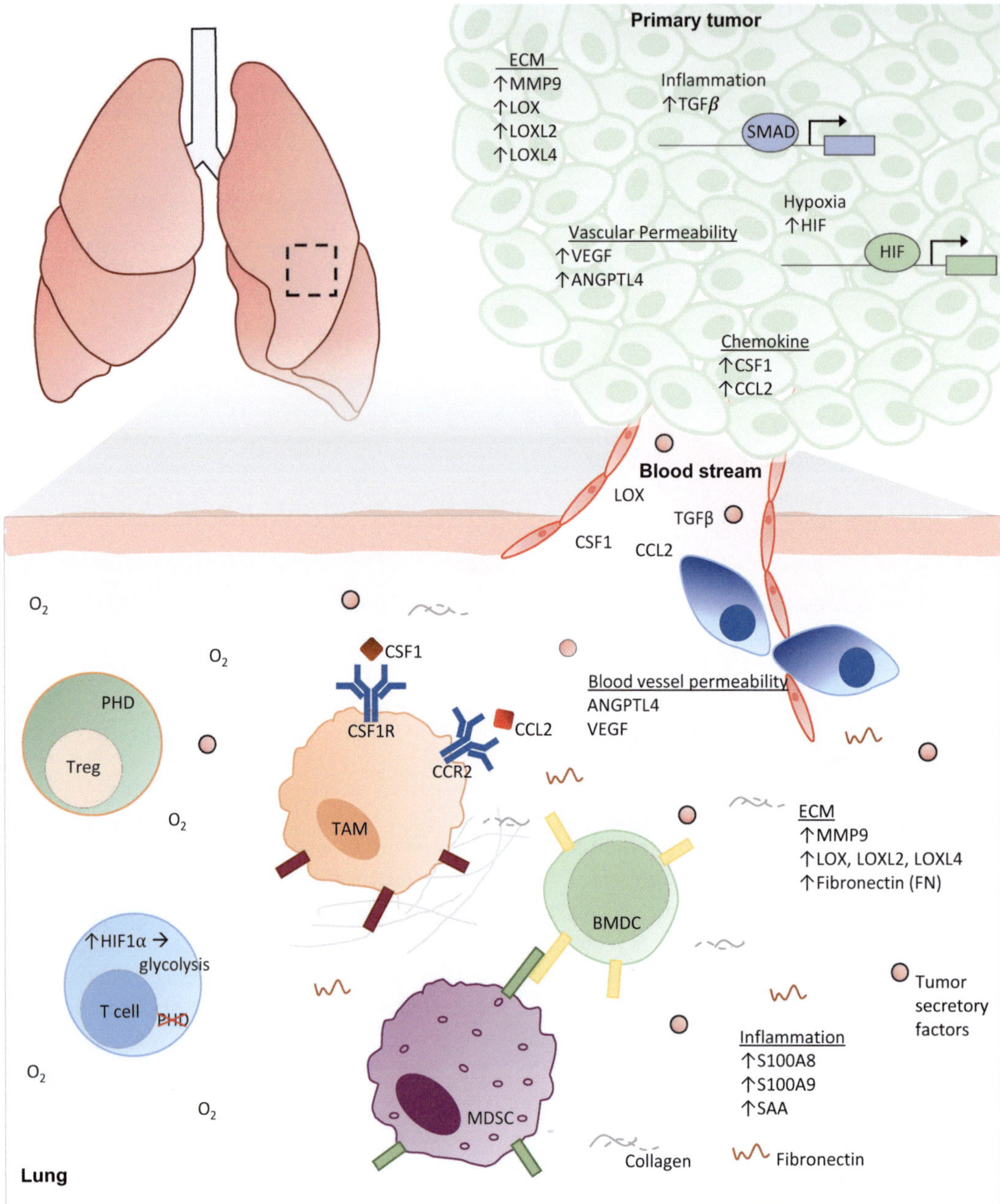

Fig. 7.2 Formation of metastatic lung niche. Inflammation and hypoxia drive cancer cells to produce secretory factors. Inflammation-mediated TGFβ/SMAD and hypoxia-mediated HIF activations which drive the transcriptions of genes involved in ECM modification, vascular permeability, and chemokine-driven immune cell recruitment. Blood stream carries these secretory factors (MMP9, LOX, LOXL2, LOXL4, VEGF, ANGPTL4, CSF1, CCL2) and immune cells to the lung tissues. LOX, LOXL2, LOXL4 cross-linked collagen at lung tissues to enhance attachment of BMDCs/MDSCs, creating a growth factor rich and immune suppressive microenvironments. CCL2 and CSF1 attracted TAMs to the metastatic sites through CSF1R and CCR2 receptors. VEGF and ANGPTL4 increased the permeability of blood vessels, facilitating the circulating cancer cells to extravasate. Lung metastatic niche is rich in oxygen (O_2). However, deletion of O_2 sensor PHD switched off the HIF-1-mediated glycolytic mode of T cells, favoring the development of Treg, an immune suppressive population of T cells. Other inflammatory factors that were enriched in lung tissues included S100A8, S100A9, and SAA. Increased MMP9 was hypothesized to facilitate the liberation of growth factors from the ECM of lung tissues. Fibronectin (FN) was shown to increase in lung tissues which also facilitated the adhesion of BMDC and cancer cells

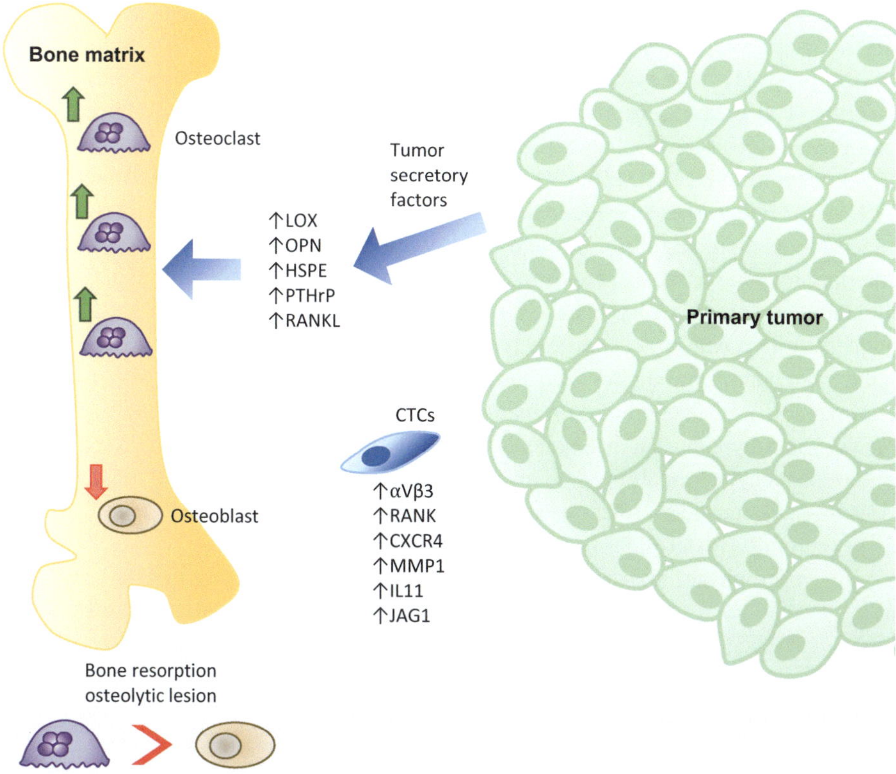

Fig. 7.3 Formation of metastatic bone niche. Cancer cells disrupt the homeostatic balance of osteoclast and osteoblast through tumor secretory factors such as LOX, OPN, HSPE, PTHrP, RANKL. Genes expressed on cancer cells which facilitate their attachment to the bone matrix include JAG1, integrin αVβ3, RANK, CXCR4, MMP1, IL11. Cancer cells favor osteoclast development which drives bone resorption and osteolytic lesion

transcriptomic analysis demonstrated that a macrophage gene signature was prognostic in follicular lymphoma [19]. Whereas, the TAM gene signature was not associated with survival in human breast cancer [20, 21].

Macrophages not only support tumor growth but are also an important component of the PMN that facilitates the attachment, colonization, and growth of metastatic cancer cells [15]. The differentiation, expansion, and motility of macrophages are generally driven by some key growth factors including Colony Stimulating Factor 1 (CSF-1), granulocyte-macrophage (GM-CSF), and (C-C Motif Chemokine Ligand 2 (CCL2) which are often abundantly produced by cancer cells [15]. Macrophages respond to CSF-1 and GM-CSF through CSF-1 receptor (CSF-1R) and respond to CCL2 through the CCR2 receptor [15]. CSF-1 up-regulation was a prognostic indicator for more aggressive breast, endometrial, ovarian, prostate, liver, and colorectal cancers [18, 22–27]. CCL2 up-regulation was also a prognostic indicator for more aggressive breast, colorectal, thyroid, and cervical cancers [24, 28–32]. The knockout of CSF-1 in mice reduced macrophage infiltration into tumors and suppressed cancer growth and metastasis in breast, colorectal cancer, and osteosarcoma mouse models [33, 34]. Therapeutically, CSF-1 or CSF-1R blockade by antibodies prevented macrophage homing to the tumor and suppressed growth and metastasis [35, 36]. Systemic depletion of macrophages by clodronate also suppressed growth and metastasis of different types of mouse cancers [15]. Macrophages promote cancer development mainly through two major mechanisms. TAMs are mostly immune suppressive. TAMs also secrete a large panel of growth and angiogenic factors as well as ECM enzymes to support the seeding and growth of cancer cells at the PMN [15]. Hypoxia produced tumor-necrosis factor (TNF), TNF receptor, and interleukin one expression in macrophages [37–39]. Hypoxia is known to induce CD11b (αM integrin) and CD18 (β2 integrin) expression in macrophages, facilitating their migration through the vasculature into hypoxic tissues and contributing to an M1 phenotype [40]. CCL2 has been reported to be induced by hypoxia in other tissue types [41]. On the other hand, hypoxia has been shown to be an unfavorable growth condition for macrophages *in vitro* and hypoxia has been shown to mitigate the phagocytic capability of macrophages [39, 42]. Therefore, further research is required to carefully determine the role of hypoxia in macrophage polarization and function.

7.5 Immunologically Permissive Metastatic Niche

In order to propagate at the metastatic site, cancer cells need to evade immune surveillance. A study showed that proly-hydroxylase (PHD) protein suppressed differentiation and development of T cells (Th1) and induced immune suppressive regulatory T cells (Treg) at lung tissues, creating an immunologically favorable metastatic niche (Fig. 7.2). PHD is an oxygen sensing protein which hydroxylates hypoxia-inducible factor (HIF)-1α leading to degradation of the HIF complexes [43]. T cell-specific PHD KO mice have hyper-activated T cells in lungs and prevented colonization of melanoma cells in the lungs as compared to wildtype mice [43]. Adoptive transfer of T cells treated with PHD inhibitor, DMOG, was able to exert a greater suppressive effect in lung metastasis [43]. T cells in lung tissues receive a high level of O_2, an environment that favors PHD-mediated enrichment of Tregs and suppression of T cells, therefore, favoring the propagation of melanoma cells [43]. Interestingly, it was shown that PHD-mediated immune tolerance in the metastatic niche was HIF-1 dependent [43]. Activated T cells are more dependent on glycolysis as compared to Tregs because glycolysis supports the nutrients required for the rapid expansion of effector T cells upon stimulation [43]. Tregs are quiescent and are relatively less metabolically active. HIF-1 activates glycolysis while PHD counteracts glycolysis through HIF-1 inactivation [43]. Therefore, it is proposed that PHD restricted T-cell expansion by reducing glycolytic activity of T cells through HIF-1 inhibition (Fig. 7.2). In other words, hypoxia, or low oxygen (O_2), and HIF-1 expression promoted T cell proliferation and suppressed metastasis in the metastatic lung niche in this model.

7.6 The Permeability of Metastatic Tissues

Cancer cells tend to colonize and form metastases at regions of secondary organs with higher permeability [44]. Conditioned medium from breast and lung cancer cells were able to induce vascular hyperpermeability in lung tissues [44]. Injection of Evan's blue (EB) dye through the tail vein followed by injection of metastatic cancer cells demonstrated that cancer cells tend to localize at lung regions with higher permeability [44]. Lodging of cancer cells in lung tissues with hyper-permeability was blocked by anti-VEGF antibody treatment, suggesting that VEGF might be one of the tumor-secreting factors responsible for making endothelial cells in the lung tissues more permeable [44]. It was later demonstrated that VEGF activated the Src-FAK complex in the endothelial cells to induce E-selectin. Enriched E-selectin expression in endothelial cells allows cancer cells to adhere to the lung endothelium [44]. A comparison of lung regions with high and low permeability based on EB intensity in mice injected with EB through the tail vein revealed that CCR2 was highly induced in hyper-permeable lung tissues, suggesting that CCR2/CCL2 signaling might play a role in vascular permeability in metastatic niche [45]. Knockout of

CCL2 or CCR2 impaired lung permeability in mice inoculated with Lewis lung tumors [45]. It was found that CCR2 induction in lung tissues was induced by MD2, a coreceptor for Toll-like receptor four which is important for innate immune signaling [45]. It was further shown that CCR2-CCL2 signaling induced the secretion of permeability factors including amyloid A3 and S100A2 [45]. Knockout of MD2 or CCR2 blocked cancer cell attachment in the metastatic niche. A recent study showed that immune cells from a third organ apart from the organ where the primary tumor is formed and the organ where metastases are formed might be involved in metastatic niche formation [46]. In the study, the authors showed that breast cancer, lung cancer, and melanoma induced hepatic $B220^{+}CD11c^{+}NK1.1^{+}$ (Natural Killer, NK) cells to increase coagulation factor which blocked fibrinogen depositions at the metastatic site (lung), thereby preventing cancer cell attachment in lung tissues. VEGF is one of the most well-described transcriptional targets of HIFs while CCL2 has also been reported to be induced by hypoxia [47]. Hypoxia is therefore speculated to increase the permeability of metastatic sites.

7.7 Secretory Factors from Primary Cancer

7.7.1 TGFβ-ANPGPTL4

The inflammatory factor, TGFβ, was found to be associated with lung metastasis in a large cohort of 368 patients. Breast cancer cells treated with TGFβ had an increased ability to colonize lung tissue in mice (Fig. 7.2). TGFβ induced SMAD-mediated transcription of ANGPTL4 which functioned to disrupt the endothelial junction of lung tissues enabling extravasation [48]. This explains why the TGFβ gene signature in breast cancer patients is only correlated with the presence of lung metastasis but not bone metastasis. ANGPTL4 is also a downstream transcriptional target of hypoxia-inducible factors (HIFs) which are stabilized during hypoxia [49]. HIF regulation of ANGPTL4 also increased extravasation by disrupting endothelial junctions in lung tissues [49].

7.7.2 VEGF

A critical component of the PMN is bone marrow-derived hematopoietic progenitor cells or BMDCs. In melanoma and Lewis lung cancer mouse models, $VEGFR1^{+}$ BMDCs arrived at metastatic sites before the arrival of cancer cells [10]. The $VEGFR1^{+}$ BMDCs also expressed integrin $\alpha_4\beta_1$ which facilitate their binding to fibronectin in the metastatic niche. Since cancer cells specifically secrete growth factors that induced fibronectin, cancer cells remotely facilitated the anchoring of BMDCs to the metastatic sites. These $VEGFR1^{+}$ BMDCs also expressed ID3 and it was shown that reconstitution of $VEGFR^{+}ID3^{+}$ BMDCs in mice was able to restore PMN and metastasis in Id3 knockout mice. Another important experiment was carried out using conditioned media from melanoma and Lewis lung cancer cells demonstrated that the conditioned media from these cancer cells could control the metastatic profile regardless of the type of primary cancers formed in mice. Specifically, the conditioned medium of melanoma cells was pre-injected into mice followed by implantation of Lewis lung cancer cell line. Reversely, the conditioned medium of Lewis lung cancer cells was pre-injected into mice followed by implantation of melanoma cells. Injection of conditioned medium of melanoma cells redirected Lewis lung cancer cells to metastasize to common metastatic sites of melanoma including kidney, spleen, intestine, and oviduct in mice and vice versa [10]. These interesting results highlighted that the secretory profiles of primary cancers determine where the metastases are formed. Therapeutically, inhibition of VEGFR1 by VEGFR1 antibody or depletion of $VEGFR1^{+}$ blocked PMN formation and subsequent metastasis [10]. VEGF and VEGFR1 are known to be induced by hypoxia and HIF-1, suggesting that hypoxia might be as well involved in the VEGF/VEGFR1-associated PMN formation.

7.7.3 S100A8 and S100A9

Gene expression profiles of lung tissue from normal, benign, malignant, and metastatic cancer-bearing mice demonstrated that S100A8 and S100A9 were significantly induced in pre-metastatic lungs [50]. S100A8 and S100A9 are pro-inflammatory factors which could simultaneously induce migration and recruitment of macrophages and tumor cells. S100A8 and S100A9 attract Mac1$^+$ (macrophage antigen 1) myeloid cells to be recruited to premetastatic lungs [50]. Meanwhile, S100A8 and S100A9 induced migration, invasion, and invadopodia formation in tumor cells through activation of the MAPK pathway [50]. Depletion of tumor-secreted factors such as VEGF-A, TNFα, and TGFβ blunted the induction of S100A8 and S100A9 in the PMN in vitro and in vivo, suggesting that S100A8 and S100A9-mediated PMN was initiated by the primary tumor [50]. Blockade of S100A8 and S100A9 by antibodies prevented the Mac1$^+$ myeloid cell recruitment and metastatic colonization in the lungs of mice [50]. It was later reported that S100A8 and S100A9 induced Serum amyloid A (SAA) 3 in the pre-metastatic lungs to induce NFκB pathway in macrophages through Toll-like receptor (TLR)4 which responds to SAA3 ligands [51]. Neutralizing SAA3 antibody suppressed Mac1$^+$ myeloid cell recruitment and cancer cell colonization in metastatic lung tissues of wildtype but not TLR4 knockout mice. Interestingly, this phenotype could not be observed in liver tissues, again suggesting organotropism. Interestingly, S100A8 and S100A9 were also found to be secreted by hypoxic normal prostate epithelial cells and prostate cancer cells. Ectopic expression of HIF-1α increased S100A8 and S100A9 expression [52]. Chromatin immunoprecipitation assay identified the binding locations of HIF-1 on *S100A8* and *S100A9* while luciferase reported assay further demonstrated these hypoxia response elements are functional in prostate cancer cell lines [52]. These findings suggested that S100A8 and S100A9 are very likely to be induced in hypoxic tumors and metastatic niches.

7.7.4 MMP9

MMP9 in the pre-metastatic lung tissues (endothelial and macrophages) was induced in the wildtype but not VEGFR1 knockout mice [53]. Tumor secretory factors, the VEGF and PIGF ligands, stimulated MMP9 expression in pre-metastatic lung tissues through VEGFR1 [53]. Increased MMP9 expression in lung tissues facilitated the invasion of Lewis lung cancer and melanoma cells [53]. Although the mechanisms have not been clarified, it is reasonable to hypothesize that MMP9 cleaves the ECM of the lung tissues to remove the physical barrier during cancer cell invasion and to liberate growth factors and other soluble molecules that were trapped in the ECM to stimulate colonization and propagation of cancer cells at the metastatic site. Clinically, lung tissues from patients carrying hepatocellular carcinoma, pancreatic cancer, esophageal cancer, colon cancer, cholangiocarcinoma, gastric cancer, melanoma, lymphoma, and ovarian cancer expressed a higher level of MMP9 than non-tumor-bearing patients [53]. MMP13 expression induced by HIF-1α-dependent manner [54] was reported to activate MMP9 by cleaving pro-MMP9 that promotes osteolytic breast cancer bone metastasis [55].

7.7.5 Lysyl Oxidase (LOX)

7.7.5.1 LOX and Lung Metastasis

The lysyl oxidase (LOX) family consists of LOX and LOXL1-4. LOX enzymes oxidatively deaminate lysine residues of collagen causing cross-linking of collagens which allow collagen fiber formation. This modification of collagen molecules increases stiffness and promotes adhesion and invasion of cells on the ECM. Cross-linked collagen provides an invasive track for cancer cells. The LOX family has been shown to be mainly induced by HIFs, in a hypoxia-dependent manner, and inflammation associated TGFβ-SMAD pathways [56]. It was shown that miR-26a and miR-29 also negatively regulated LOXL2 in liver cancer model [56]. Studies showed that LOX and LOXL2 cross-linked and stiffened the ECM of primary cancer tissue, thereby activating the focal adhesion formation and activation of the phosphatidylinositol 3-kinase signaling pathway, pathways that promote invasion of cancer cell [57]. Apart from influencing the microenvironments of primary cancer sites, secreted LOX enzymes could reach distant metastatic sites through the bloodstream to create a permissive metastatic niche. LOX, LOXL2, LOXL4 secreted by breast cancer cells increased cross-linked collagens in lung tissues [8, 58]. Cross-linked collagens facilitated the attachment of $CD11b^+$ bone marrow-derived cells (BMDCs) in the lung tissues prior to the arrival of cancer cells [8]. BMDCs acted as the source of angiogenic factors and MMP2 in the metastatic niche to promote cancer invasion and growth [8]. LOXL2 antibody also inhibited tetrachloride-induced liver fibrosis and bleomycin-induced lung fibrosis in mice [59]. LOX inhibitor (β-aminopropionitrile (βAPN) or LOX antibody prevented fibrosis caused by collagen cross-linking in a mouse breast cancer model [57]. HIF inhibitors, digoxin and acriflavine, also suppressed PMN formation and metastasis, through LOX inhibition, in mice with breast and liver cancers [56, 60].

7.7.5.2 LOX and Bone Metastasis

A proteomics study using SILAC demonstrated that LOX was also the most induced secreted proteins in hypoxic ER-breast cancer cells [61]. ER-negative breast cancer patients with bone relapse have higher LOX expression as compared to those patients without bone relapse [61]. LOX secreted by the tumors was found to disrupt the balance of bone remodeling causing osteolytic lesion formation [61]. Bone remodeling is performed by osteoblasts which generate bones tissues and osteoblasts which degrade bone tissues. LOX secreted by the tumors entered bone tissues and inhibited osteoblasts and activated osteoclasts causing extensive bone resorption creating a favorable microenvironment for breast cancer cell colonization [61]. LOX-mediated bone resorption through activating NFATc1 to promote osteoclastogenesis [61] (Fig. 7.3). Neutralizing LOX antibody and LOX inhibitor, β-aminopropionitrile (βAPN) were able to alleviate osteolytic bone lesions to repress bone metastasis [61].

7.7.6 CCL2

Macrophages are often found in microenvironments of both primary tumors and metastatic sites. Intriguingly, primary tumors and metastases recruited distinctive macrophage populations. An interesting study has reported that inflammatory monocytes were recruited to the metastatic niche via the CCL2-CCR2 axis [62]. Metastatic sites attracted inflammatory monocytes but not other types of immune cells such as T and B cells. Mouse monocytes were defined as CD11b and CD115 positive cells. These monocytes were further subdivided into resident monocytes lacking Gr1 and Ly6C and inflammatory monocytes expressing Gr1 and Ly6C. Adoptive transfer of both populations of monocytes in breast cancer-bearing mice revealed that resident monocytes were specifically recruited by primary tumors whereas inflammatory monocytes were specifically recruited by lung metastases [62]. Lung metastases particularly recruited CCR2-expressing monocytes which were attracted by chemokines CCL2 expressed by metastatic cells and other stromal tissues in the lungs [62]. CCR2 is highly expressed in inflammatory monocytes but not resident monocytes [62]. CCL2 neutralizing antibody was able to suppress inflammatory monocyte recruitment and lung metastasis and extend the survival of tumor-bearing mice [62]. CCR2-expressing cells were found to secrete VEGF which increased the permeability of blood vessels to promote the extravasation of breast cancer cells [62]. VEGF-knockout bone marrow cells were not able to increase permeability [62]. Interestingly, other CCR2-expressing immune cells such as T cells were not influenced by CCL2 antibody treatment [62]. Why CCL2 only recruited $CCR2^+$ monocytes to the metastatic niche but not other $CCR2^+$ immune cells remain to the elucidated. CCL2 has been shown to be regulated by HIF-1 in lung epithelial cells [47]. CCL2 is one of the senescence-associated secretory phenotypes induced by senescent cells [41]. Hypoxia causes cell cycle arrest and senescence. Whether CCL2 is induced in hypoxic tumors directly through HIFs or indirectly through senescence remains to be determined.

7.7.7 Secretory Factors that Support Bone Metastasis

Cancer cells secreted a soluble form of RANKL to activate osteoclasts. Cancer cells secreted osteopontin (OPN) to mediate bone marrow cell migration and cancer cell growth [63]. Cancer cells also secreted heparanase (HPSE) to reduce the length of the heparin sulfate chain in the ECM to expedite bone resorption [64]. Cancer cells also secrete a parathyroid hormone-related protein (PTHrP) which also activated bone resorption and increased CCL2 expression in the bone [65]. Interestingly, bone degradation caused the release of growth factors that are stored in bone matrix, such as TGFβ, which can in turn promoted tumor growth [66]. Cancer cells expressed αvβ3 integrin which facilitated cancer cell adhesion to bone niche [67]. Osteoblasts secreted CXCL12 to attract CXCR4-expressing cancer cells to the bone [68]. Cancer cells expressing MMP1, interleukin (IL)-11, and connective tissue growth factor (CTGF) also tend to metastasize to bone [69]. MMP1 and IL-11 enhanced RANKL production while CTGF drove osteoclast activation [69]. Together, these cancer-secreted factors drive bone resorption and formation of an osteolytic lesion (Fig. 7.3). Notably, some of the above factors were reported to be induced by hypoxia. For example, hypoxia-induced RANKL and its cognate receptor RANK in breast cancer cells in HIF-1α dependent manner [70]. PTHrP transcription was enhanced by HIF-2α in multiple human cancer cells [71]. Breast cancer cell lines with high propensity to metastasize to bone expressed high level of Notch ligand, JAG1 [72]. When breast cancer cells entered the bone niche, JAG1 on breast cancer cells interacted with NOTCH receptors on osteoblasts and osteoclasts. JAG1-NOTCH binding activates γ-secretase-mediated cleavage of NOTCH receptor, allowing the intracellular domain of the NOTCH (NCID) to be translocated to the nucleus to activate transcription of genes such as HEY1/HES [72]. Co-culturing of JAG1-over-expressing breast cancer cells with osteoblasts induced NOTCH/HEY1-mediated transcription and secretion of interleukin 6 (IL6) by osteoblasts [72]. IL6 promoted proliferation of breast cancer cells. In addition, JAG1 on breast cancer cells bound to osteoclasts in the bone metastatic niche to drive the differentiation of osteoclasts, causing osteolytic lesions and growth of bone metastases [72]. Meanwhile, it was found that JAG1 upregulation in breast cancer cells was mediated by SMAD4 which responded to TGFβ secreted by osteoclasts [72]. Therefore, JAG1-NOTCH interaction between breast cancer cells and bone cells formed a paracrine feed-forward signaling through IL6 and TGFβ [72]. MRK003, a γ-secretase inhibitor which prevents the cleavage and activation of the NOTCH pathway, inhibited osteolytic lesion formation and bone marrow metastasis in JAG1-over-expressing breast cancers [72]. JAG1 antibody prevented osteolytic lesion and bone metastasis in mouse breast cancer model. JAG1 antibody also sensitized breast cancer to chemotherapy treatment [73].

7.8 Exosomes

Exosomes are defined as extracellular vesicles (EVs) budded from the membranes of cells with size range from 30 to 150 nm. Exosomes could be detected in biological fluids including blood and urine. Exosomes are packed with a wide variety of molecules including cytokines, growth factors, extracellular matrix components, DNA, mRNA, non-coding RNA, proteins, and lipids. Given their circulating nature, exosomes have been shown by multiple studies for the roles in long-range cell-cell communications by transferring the above biomolecules (Fig. 7.1). Interestingly, recent studies showed that hypoxia-induced the release of exosomes from tumor cells that modulates tumor microenvironment to facilitate angiogenesis and metastasis [74]; therefore, hypoxia might be involved in the exosome-related pro-metastatic roles below. Exosomes also serve as diagnostic and prognostic indicators for cancer stages and metastasis in different cancers.

7.8.1 Pro-metastatic Roles of Exosomes

In stage 4 melanoma patients, it was shown that that high exosome content in plasma was correlated with shorter survival in patients. Exosomal proteins including TYRP2, VLA-4, HSP70 showed a step-wise increase in patients with advancing stages of melanoma from stage 1 to 4 [75]. In patients with stage 3 melanoma, TYRP2 was significantly increased in patients with metastasis [75]. Similarly, total exosomes from highly metastatic melanoma cells were more abundant as compared to non-metastatic melanoma cells [75]. Injection of exosomes from metastatic melanoma cells could induce S100A8/S100A9 in lung tissues of mice, suggesting that the metastatic site has been altered and primed to be pro-inflammatory [75]. Intravenous injection of exosomes from B16-F10 (highly metastatic melanoma cells) followed by injection of the same cell line increased formation of metastasis burden as compared to mice injected with control exosome particles or with exosomes derived from B16-F1 (lowly metastatic melanoma cells) [75]. Adoptive transfer of bone marrow cells that were pre-treated with B16-F10-derived exosomes increased homing of bone marrow cells to metastatic sites and increased metastatic burden [75]. Unlike $CD11b^+$, $CD11b^+Gr1^+$, $F4/80^+$ myeloid cells or $VEGFR1^+$ bone marrow cells which responded to and were mobilized by tumor-secreted growth factors and chemotactic molecules, exosomes mainly mobilized pro-angiogenic $c\text{-}Kit^+Tie2^+$ bone marrow cells to PMN [75]. Intriguingly, exosomes transferred oncoprotein MET horizontally from melanoma cells to bone marrow cells to enhance their homing ability to PMN [75]. Exosomes from melanoma cells with MET knockdown failed to educate bone marrow cells and promote metastasis [75]. Ablation of proteins involved in the exosome biogenesis cascades (RAb27a) was able to block exosome production and bone marrow cell-mediated PMN formation [75]. Hypoxic stress is able to change the contents and functions of exosomes. Hypoxic exosomes derived from glioma cells were enriched with hypoxia-related proteins including MMPs and LOX, which were associated with tumor progression and metastasis [76].

Pancreatic ductal adenocarcinoma (PDAC) cells frequently metastasize to the liver. Injection of exosomes isolated from metastatic PDAC cells prior to the injection of PDAC in mice was able to increase liver metastases [77]. Flow cytometry analysis coupled with imaging demonstrated that PDAC-derived exosomes were mainly localized in liver and were specifically taken up by Kupffer cells instead of other resident cells such as fibroblasts, endothelial, and epithelial cells [77]. Transcriptome sequencing of Kupffer cells further revealed that PDAC exosomes triggered the liver fibrosis gene signature with significant induction of CTGF, EDN, IGF, PDGF, and TGFβ [77]. TGFβ would subsequently activate hepatic stellate cells to produce fibronection [77]. Immunofluorescent analysis demonstrated that fibronection expression was markedly induced by PDAC exosomes in the liver [77]. Fibronectin-rich microenvironment favored the recruitment of F4/80 macrophage in the liver [77]. Depletion of macrophages by diphtheria toxin abrogated the effects induced by PDAC-derived exosomes and liver metastatic burden, suggesting these macrophages are probably tumor-associated macrophages (anti-inflammatory) [77]. PDAC-derived exosomes are rich in macrophage migration inhibitory factor (MIF). Exosomal MIF expression was significantly higher in mice with more advanced stages of PDAC [77]. Inhibition of MIF in PDAC cells could abolish the PMN-shaping effects of PDAC exosomes, suggesting that exosomal MIF played an essential role in pancreatic cancer metastasis to the liver [77].

7.8.2 Exosomes and Organotropisms

Recent evidence has highlighted that exosomes secreted by cancer cells determine which metastatic organs where cancer cells eventually colonize [78]. This relationship of exosomes and organotropism was simply and elegantly demonstrated in breast cancer cell lines which incline to form metastases in lungs and pancreatic cell lines which incline to form metastases in liver [78]. Exosomes from breast cancer cells were able to promote pancreatic cancer cells to metastasize to lungs while exosomes from pancreatic cell lines were able to promote breast cancer cells to metastasize to livers of mice [78]. A large scale of exosomal proteomic profiling revealed that the exosomal integrin profiles of cancer cell lines could predict the metastatic organ specificity [78]. ITGα6, ITGβ4, ITGβ1 were highly expressed in exosomes that favor lung metastasis [78]. ITGβ5, ITGαV were highly expressed in exosomes that favor liver metastasis [78]. ITGβ3 was highly expressed in exosomes that favor brain metastasis [78]. ITGα2β1, however, was generally expressed by all metastatic cancer cells. Fluorescence-labeled exosomes from different cancer cell lines were injected into mice and their metastatic organs were analyzed to identify the recipient cells [78]. This experiment has consolidated two important hypotheses. First, it was found that exosomes have the propensity to localize back to their favorable metastatic organs, suggesting that different integrin members might interact with the specific extracellular matrix substrates in their favorable metastatic niches. Second, it was found that different types of resident cells were responsible for the uptake of specific integrin-expressing exosomes. Lung metastasis-specific exosomes were taken up by S100A4-expressing fibroblasts and surfactant protein C (SPC)-expressing epithelial cells in lung tissues while pancreatic metastasis-specific exosomes were taken up by Kupffer cells in liver tissues [78]. Brain metastasis-specific exosomes were taken up by CD31-positive endothelial cells in the brain. ITGα6 and ITGβ4 facilitated exosomal attachment to laminin-rich lung tissues whereas ITGβ5 and ITGαV facilitated exosomal attachment to fibronectin-rich liver tissues [78]. Gene expression studies further identified the genes that were altered by exosomes in the resident cells. Apart from genes in cell motility and migration, S100A4, S100A6, S100A10, S100A11, S100A13, and S100A16 were induced in PMN of lungs while S100A8 and S100P were induced in PMN of livers [78]. Clinically, exosomal ITGαV was associated with lung metastasis in breast cancer patients while exosomal ITGβ4 was associated with liver metastasis in pancreatic cancer patients [78]. Blockade of specific integrins using specific decoy peptides abolished exosome attachment to their integrin and organ-specific metastatic organs. An interesting piece of data from this study showed that the expressions of integrins in cancer cells and their exosomes do not necessarily correlate, indicating that the exosome packaging may influence the final levels of exosomal integrins which later play an important role in metastatic organotropism [78].

7.8.3 Exosomes Elicited Immune Surveillance

While most studies have shown that highly metastatic cancer cells secrete exosomes to establish niche favorable to metastasis, a recent study controversially reported that non-metastatic cancer cells secreted specific exosomes which suppressed metastasis through igniting innate immune cells. Exosomes from melanoma cells with different degrees of metastatic capability were injected into mice followed by inoculation of syngeneic melanoma cells [79]. Interestingly, less extensive metastasis was found in mice which were pre-injected with exosomes from non-metastatic melanoma cells (Exo^{NM}) as compared to exosomes from metastatic melanoma cells (Exo^{M}) [79]. Exosomes injected into the animals were primarily located in lymph nodes, lungs, livers, and bone marrows [79]. PEDF on the surfaces of Exo^{NM} enriched the $CX3CR1^{+}Ly6C^{low}$ patrolling monocytes and increased their Nr4a1 and ITGB2 expression in the bone marrow [79]. PEDF also induced TRAIL and IL12 in $CX3CR1^{+}Ly6C^{low}$ patrolling monocytes to promote their differentiation into macrophages which later migrate to the metastatic site with enhanced cancer cell-killing capability [79]. Although less mechanistic details were provided, it was shown in the same study that ExoNM also promoted NK cell activity in tumor cell clearance [79]. Clinically, exosomal PEDF level in blood of melanoma patients was positively correlated with was associated with the survival length of patients [79]. Injection of sera of non-metastatic melanoma patients into nude mice followed by injection of melanoma cells resulted in less lung metastasis as compared to injection of sera of metastatic melanoma patients [79]. These data demonstrated that exosomes from cancer cells with different metastatic capabilities might have diverse effects on the metastatic niches by influencing the extracellular matrix and eliciting immune cell surveillance.

References

1. Leong SP, Cady B, Jablons DM, Garcia-Aguilar J, Reintgen D, Jakub J, Pendas S, Duhaime L, Cassell R, Gardner M, Giuliano R, Archie V, Calvin D, Mensha L, Shivers S, Cox C, Werner JA, Kitagawa Y, Kitajima M (2006) Clinical patterns of metastasis. Cancer Metastasis Rev 25:221–232
2. Patel JK, Didolkar MS, Pickren JW, Moore RH (1978) Metastatic pattern of malignant melanoma. A study of 216 autopsy cases. Am J Surg 135:807–810
3. Bhansali MS, Fujita H, Kakegawa T, Yamana H, Ono T, Hikita S, Toh Y, Fujii T, Tou U, Shirouzu K (1997) Pattern of recurrence after extended radical esophagectomy with three-field lymph node dissection for squamous cell carcinoma in the thoracic esophagus. World J Surg 21:275–281
4. Mafune KI, Tanaka Y, Takubo K (2000) Autopsy findings in patients with esophageal carcinoma: comparison between resection and nonresection groups. J Surg Oncol 74:196–200
5. Saito T, Iizuka T, Kato H, Watanabe H (1985) Esophageal carcinoma metastatic to the stomach. A clinicopathologic study of 35 cases. Cancer 56:2235–2241
6. Kato H, Miyazaki T, Nakajima M, Sohda M, Fukai Y, Masuda N, Fukuchi M, Manda R, Tsukada K, Kuwano H (2003) Prediction of hematogenous recurrence in patients with esophageal carcinoma. Jpn J Thorac Cardiovasc Surg 51:599–608
7. Niederhuber JE (1993) Colon and rectum cancer. Patterns of spread and implications for workup. Cancer 71:4187–4192
8. Erler JT, Bennewith KL, Cox TR, Lang G, Bird D, Koong A, Le QT, Giaccia AJ (2009) Hypoxia-induced lysyl oxidase is a critical mediator of bone marrow cell recruitment to form the premetastatic niche. Cancer Cell 15:35–44
9. Gao D, Nolan DJ, Mellick AS, Bambino K, McDonnell K, Mittal V (2008) Endothelial progenitor cells control the angiogenic switch in mouse lung metastasis. Science 319:195–198
10. Kaplan RN, Riba RD, Zacharoulis S, Bramley AH, Vincent L, Costa C, MacDonald DD, Jin DK, Shido K, Kerns SA, Zhu Z, Hicklin D, Wu Y, Port JL, Altorki N, Port ER, Ruggero D, Shmelkov SV, Jensen KK, Rafii S, Lyden D (2005) VEGFR1-positive haematopoietic bone marrow progenitors initiate the pre-metastatic niche. Nature 438:820–827
11. Liu C, Yu S, Kappes J, Wang J, Grizzle WE, Zinn KR, Zhang HG (2007) Expansion of spleen myeloid suppressor cells represses NK cell cytotoxicity in tumor-bearing host. Blood 109:4336–4342
12. Lyden D, Hattori K, Dias S, Costa C, Blaikie P, Butros L, Chadburn A, Heissig B, Marks W, Witte L, Wu Y, Hicklin D, Zhu Z, Hackett NR, Crystal RG, Moore MA, Hajjar KA, Manova K, Benezra R, Rafii S (2001) Impaired recruitment of bone-marrow-derived endothelial and hematopoietic precursor cells blocks tumor angiogenesis and growth. Nat Med 7:1194–1201
13. Yang L, DeBusk LM, Fukuda K, Fingleton B, Green-Jarvis B, Shyr Y, Matrisian LM, Carbone DP, Lin PC (2004) Expansion of myeloid immune suppressor Gr+CD11b+ cells in tumor-bearing host directly promotes tumor angiogenesis. Cancer Cell 6:409–421
14. Pollard JW (2009) Trophic macrophages in development and disease. Nat Rev Immunol 9:259–270
15. Qian BZ, Pollard JW (2010) Macrophage diversity enhances tumor progression and metastasis. Cell 141:39–51
16. Chen JJ, Lin YC, Yao PL, Yuan A, Chen HY, Shun CT, Tsai MF, Chen CH, Yang PC (2005) Tumor-associated macrophages: the double-edged sword in cancer progression. J Clin Oncol 23:953–964
17. Ryder M, Ghossein RA, Ricarte-Filho JC, Knauf JA, Fagin JA (2008) Increased density of tumor-associated macrophages is associated with decreased survival in advanced thyroid cancer. Endocr Relat Cancer 15:1069–1074
18. Zhu XD, Zhang JB, Zhuang PY, Zhu HG, Zhang W, Xiong YQ, Wu WZ, Wang L, Tang ZY, Sun HC (2008) High expression of macrophage colony-stimulating factor in peritumoral liver tissue is associated with poor survival after curative resection of hepatocellular carcinoma. J Clin Oncol 26:2707–2716
19. Farinha P, Masoudi H, Skinnider BF, Shumansky K, Spinelli JJ, Gill K, Klasa R, Voss N, Connors JM, Gascoyne RD (2005) Analysis of multiple biomarkers shows that lymphoma-associated macrophage (LAM) content is an independent predictor of survival in follicular lymphoma (FL). Blood 106:2169–2174

20. Ojalvo LS, King W, Cox D, Pollard JW (2009) High-density gene expression analysis of tumor-associated macrophages from mouse mammary tumors. Am J Pathol 174:1048–1064
21. Zabuawala T, Taffany DA, Sharma SM, Merchant A, Adair B, Srinivasan R, Rosol TJ, Fernandez S, Huang K, Leone G, Ostrowski MC (2010) An ets2-driven transcriptional program in tumor-associated macrophages promotes tumor metastasis. Cancer Res 70:1323–1333
22. Groblewska M, Mroczko B, Wereszczynska-Siemiatkowska U, Mysliwiec P, Kedra B, Szmitkowski M (2007) Serum levels of granulocyte colony-stimulating factor (G-CSF) and macrophage colony-stimulating factor (M-CSF) in pancreatic cancer patients. Clin Chem Lab Med 45:30–34
23. Lin EY, Gouon-Evans V, Nguyen AV, Pollard JW (2002) The macrophage growth factor CSF-1 in mammary gland development and tumor progression. J Mammary Gland Biol Neoplasia 7:147–162
24. Mantovani A, Sica A (2010) Macrophages, innate immunity and cancer: balance, tolerance, and diversity. Curr Opin Immunol 22:231–237
25. Mroczko B, Groblewska M, Wereszczynska-Siemiatkowska U, Okulczyk B, Kedra B, Laszewicz W, Dabrowski A, Szmitkowski M (2007) Serum macrophage-colony stimulating factor levels in colorectal cancer patients correlate with lymph node metastasis and poor prognosis. Clin Chim Acta 380:208–212
26. Sapi E, Kacinski BM (1999) The role of CSF-1 in normal and neoplastic breast physiology. Proc Soc Exp Biol Med 220:1–8
27. Smith HO, Anderson PS, Kuo DY, Goldberg GL, DeVictoria CL, Boocock CA, Jones JG, Runowicz CD, Stanley ER, Pollard JW (1995) The role of colony-stimulating factor 1 and its receptor in the etiopathogenesis of endometrial adenocarcinoma. Clin Cancer Res 1:313–325
28. Bailey C, Negus R, Morris A, Ziprin P, Goldin R, Allavena P, Peck D, Darzi A (2007) Chemokine expression is associated with the accumulation of tumour associated macrophages (TAMs) and progression in human colorectal cancer. Clin Exp Metastasis 24:121–130
29. Saji H, Koike M, Yamori T, Saji S, Seiki M, Matsushima K, Toi M (2001) Significant correlation of monocyte chemoattractant protein-1 expression with neovascularization and progression of breast carcinoma. Cancer 92:1085–1091
30. Tanaka K, Kurebayashi J, Sohda M, Nomura T, Prabhakar U, Yan L, Sonoo H (2009) The expression of monocyte chemotactic protein-1 in papillary thyroid carcinoma is correlated with lymph node metastasis and tumor recurrence. Thyroid 19:21–25
31. Yoshidome H, Kohno H, Shida T, Kimura F, Shimizu H, Ohtsuka M, Nakatani Y, Miyazaki M (2009) Significance of monocyte chemoattractant protein-1 in angiogenesis and survival in colorectal liver metastases. Int J Oncol 34:923–930
32. Zijlmans HJ, Fleuren GJ, Baelde HJ, Eilers PH, Kenter GG, Gorter A (2006) The absence of CCL2 expression in cervical carcinoma is associated with increased survival and loss of heterozygosity at 17q11.2. J Pathol 208:507–517
33. Kubota Y, Takubo K, Shimizu T, Ohno H, Kishi K, Shibuya M, Saya H, Suda T (2009) M-CSF inhibition selectively targets pathological angiogenesis and lymphangiogenesis. J Exp Med 206:1089–1102
34. Oguma K, Oshima H, Aoki M, Uchio R, Naka K, Nakamura S, Hirao A, Saya H, Taketo MM, Oshima M (2008) Activated macrophages promote Wnt signalling through tumour necrosis factor-alpha in gastric tumour cells. EMBO J 27:1671–1681
35. Abraham D, Zins K, Sioud M, Lucas T, Schafer R, Stanley ER, Aharinejad S (2010) Stromal cell-derived CSF-1 blockade prolongs xenograft survival of CSF-1-negative neuroblastoma. Int J Cancer 126:1339–1352
36. Aharinejad S, Sioud M, Lucas T, Abraham D (2009) Targeting stromal-cancer cell interactions with siRNAs. Methods Mol Biol 487:243–266
37. Ghezzi P et al (1991) Hypoxia increases production of interleukin-1 and tumor necrosis factor by human mononuclear cells. Cytokine 3:189–194
38. Scannell G et al (1993) Hypoxia induces a human macrophage cell line to release tumor necrosis factor-alpha and its soluble receptors in vitro. J Surg Res 54:281–285. https://doi.org/10.1006/jsre.1993.1044
39. Yun JK et al (1997) Inflammatory mediators are perpetuated in macrophages resistant to apoptosis induced by hypoxia. Proc Natl Acad Sci USA 94:13903–13908
40. Scannell G et al (1995) Effects of trauma on leukocyte intercellular adhesion molecule-1, CD11b, and CD18 expressions. J Trauma 39:641–644
41. Acosta JC et al (2013) A complex secretory program orchestrated by the inflammasome controls paracrine senescence. Nat Cell Biol 15:978–990. https://doi.org/10.1038/ncb2784
42. Leeper-Woodford SK, Mills JW (1992) Phagocytosis and ATP levels in alveolar macrophages during acute hypoxia. Am J Respir Cell Mol Biol 6:326–334. https://doi.org/10.1165/ajrcmb/6.3.326
43. Clever D, Roychoudhuri R, Constantinides MG, Askenase MH, Sukumar M, Klebanoff CA, Eil RL, Hickman HD, Yu Z, Pan JH, Palmer DC, Phan AT, Goulding J, Gattinoni L, Goldrath AW, Belkaid Y, Restifo NP (2016) Oxygen sensing by T cells establishes an immunologically tolerant metastatic niche. Cell 166:1117–1131. e1114
44. Hiratsuka S, Goel S, Kamoun WS, Maru Y, Fukumura D, Duda DG, Jain RK (2011) Endothelial focal adhesion kinase mediates cancer cell homing to discrete regions of the lungs via E-selectin up-regulation. Proc Natl Acad Sci USA 108:3725–3730
45. Hiratsuka S, Ishibashi S, Tomita T, Watanabe A, Akashi-Takamura S, Murakami M, Kijima H, Miyake K, Aburatani H, Maru Y (2013) Primary tumours modulate innate immune signalling to create pre-

metastatic vascular hyperpermeability foci. Nat Commun 4:1853
46. Hiratsuka S, Tomita T, Mishima T, Matsunaga Y, Omori T, Ishibashi S, Yamaguchi S, Hosogane T, Watarai H, Omori-Miyake M, Yamamoto T, Shibata N, Watanabe A, Aburatani H, Tomura M, High KA, Maru Y (2018) Hepato-entrained B220(+)CD11c(+) NK1.1(+) cells regulate pre-metastatic niche formation in the lung. EMBO Mol Med 10
47. Baay-Guzman GJ et al (2012) HIF-1 expression is associated with CCL2 chemokine expression in airway inflammatory cells: implications in allergic airway inflammation. Respir Res 13:60. https://doi.org/10.1186/1465-9921-13-60
48. Padua D, Zhang XH, Wang Q, Nadal C, Gerald WL, Gomis RR, Massague J (2008) TGFbeta primes breast tumors for lung metastasis seeding through angiopoietin-like 4. Cell 133:66–77
49. Zhang H, Wong CCL, Wei H, Gilkes DM, Korangath P, Chaturvedi P, Schito L, Chen J, Krishnamachary B, Winnard PT, Raman V, Zhen L, Mitzner WA, Sukumar S, Semenza GL (2012) HIF-1-dependent expression of angiopoietin-like 4 and L1CAM mediates vascular metastasis of hypoxic breast cancer cells to the lungs. Oncogene 31:1757–1770
50. Hiratsuka S, Watanabe A, Aburatani H, Maru Y (2006) Tumour-mediated upregulation of chemoattractants and recruitment of myeloid cells predetermines lung metastasis. Nat Cell Biol 8:1369–1375
51. Hiratsuka S, Watanabe A, Sakurai Y, Akashi-Takamura S, Ishibashi S, Miyake K, Shibuya M, Akira S, Aburatani H, Maru Y (2008) The S100A8-serum amyloid A3-TLR4 paracrine cascade establishes a pre-metastatic phase. Nat Cell Biol 10:1349–1355
52. Grebhardt S, Veltkamp C, Strobel P, Mayer D (2012) Hypoxia and HIF-1 increase S100A8 and S100A9 expression in prostate cancer. Int J Cancer 131:2785–2794. https://doi.org/10.1002/ijc.27591
53. Hiratsuka S, Nakamura K, Iwai S, Murakami M, Itoh T, Kijima H, Shipley JM, Senior RM, Shibuya M (2002) MMP9 induction by vascular endothelial growth factor receptor-1 is involved in lung-specific metastasis. Cancer Cell 2:289–300
54. Shan Y et al (2018) Hypoxia-induced matrix metalloproteinase-13 expression in exosomes from nasopharyngeal carcinoma enhances metastases. Cell Death Dis 9:382. https://doi.org/10.1038/s41419-018-0425-0
55. Nannuru KC et al (2010) Matrix metalloproteinase (MMP)-13 regulates mammary tumor-induced osteolysis by activating MMP9 and transforming growth factor-beta signaling at the tumor-bone interface. Cancer Res 70:3494–3504. https://doi.org/10.1158/0008-5472.CAN-09-3251
56. Wong CC, Tse AP, Huang YP, Zhu YT, Chiu DK, Lai RK, Au SL, Kai AK, Lee JM, Wei LL, Tsang FH, Lo RC, Shi J, Zheng YP, Wong CM, Ng IO (2014) Lysyl oxidase-like 2 is critical to tumor microenvironment and metastatic niche formation in hepatocellular carcinoma. Hepatology 60:1645–1658
57. Levental KR, Yu H, Kass L, Lakins JN, Egeblad M, Erler JT, Fong SF, Csiszar K, Giaccia A, Weninger W, Yamauchi M, Gasser DL, Weaver VM (2009) Matrix crosslinking forces tumor progression by enhancing integrin signaling. Cell 139:891–906
58. Wong CC, Gilkes DM, Zhang H, Chen J, Wei H, Chaturvedi P, Fraley SI, Wong CM, Khoo US, Ng IO, Wirtz D, Semenza GL (2011) Hypoxia-inducible factor 1 is a master regulator of breast cancer metastatic niche formation. Proc Natl Acad Sci USA 108:16369–16374
59. Barry-Hamilton V, Spangler R, Marshall D, McCauley S, Rodriguez HM, Oyasu M, Mikels A, Vaysberg M, Ghermazien H, Wai C, Garcia CA, Velayo AC, Jorgensen B, Biermann D, Tsai D, Green J, Zaffryar-Eilot S, Holzer A, Ogg S, Thai D, Neufeld G, Van Vlasselaer P, Smith V (2010) Allosteric inhibition of lysyl oxidase-like-2 impedes the development of a pathologic microenvironment. Nat Med 16:1009–1017
60. Wong CC, Zhang H, Gilkes DM, Chen J, Wei H, Chaturvedi P, Hubbi ME, Semenza GL (2012) Inhibitors of hypoxia-inducible factor 1 block breast cancer metastatic niche formation and lung metastasis. J Mol Med (Berl) 90:803–815
61. Cox TR, Rumney RMH, Schoof EM, Perryman L, Hoye AM, Agrawal A, Bird D, Latif NA, Forrest H, Evans HR, Huggins ID, Lang G, Linding R, Gartland A, Erler JT (2015) The hypoxic cancer secretome induces pre-metastatic bone lesions through lysyl oxidase. Nature 522:106–110
62. Qian BZ, Li J, Zhang H, Kitamura T, Zhang J, Campion LR, Kaiser EA, Snyder LA, Pollard JW (2011) CCL2 recruits inflammatory monocytes to facilitate breast-tumour metastasis. Nature 475:222–225
63. McAllister SS, Gifford AM, Greiner AL, Kelleher SP, Saelzler MP, Ince TA, Reinhardt F, Harris LN, Hylander BL, Repasky EA, Weinberg RA (2008) Systemic endocrine instigation of indolent tumor growth requires osteopontin. Cell 133:994–1005
64. Kelly T, Suva LJ, Huang Y, Macleod V, Miao HQ, Walker RC, Sanderson RD (2005) Expression of heparanase by primary breast tumors promotes bone resorption in the absence of detectable bone metastases. Cancer Res 65:5778–5784
65. Guise TA, Yin JJ, Taylor SD, Kumagai Y, Dallas M, Boyce BF, Yoneda T, Mundy GR (1996) Evidence for a causal role of parathyroid hormone-related protein in the pathogenesis of human breast cancer-mediated osteolysis. J Clin Invest 98:1544–1549
66. Korpal M, Yan J, Lu X, Xu S, Lerit DA, Kang Y (2009) Imaging transforming growth factor-beta signaling dynamics and therapeutic response in breast cancer bone metastasis. Nat Med 15:960–966
67. Kang Y, Siegel PM, Shu W, Drobnjak M, Kakonen SM, Cordon-Cardo C, Guise TA, Massague J (2003) A multigenic program mediating breast cancer metastasis to bone. Cancer Cell 3:537–549
68. Liang Z, Wu T, Lou H, Yu X, Taichman RS, Lau SK, Nie S, Umbreit J, Shim H (2004) Inhibition of breast

cancer metastasis by selective synthetic polypeptide against CXCR4. Cancer Res 64:4302–4308
69. Lu X, Wang Q, Hu G, Van Poznak C, Fleisher M, Reiss M, Massague J, Kang Y (2009) ADAMTS1 and MMP1 proteolytically engage EGF-like ligands in an osteolytic signaling cascade for bone metastasis. Genes Dev 23:1882–1894
70. Tang ZN, Zhang F, Tang P, Qi XW, Jiang J (2011) Hypoxia induces RANK and RANKL expression by activating HIF-1alpha in breast cancer cells. Biochem Biophys Res Commun 408:411–416. https://doi.org/10.1016/j.bbrc.2011.04.035
71. Manisterski M, Golan M, Amir S, Weisman Y, Mabjeesh NJ (2010) Hypoxia induces PTHrP gene transcription in human cancer cells through the HIF-2alpha. Cell Cycle 9:3723–3729
72. Sethi N, Dai X, Winter CG, Kang Y (2011) Tumor-derived JAGGED1 promotes osteolytic bone metastasis of breast cancer by engaging notch signaling in bone cells. Cancer Cell 19:192–205
73. Zheng H, Bae Y, Kasimir-Bauer S, Tang R, Chen J, Ren G, Yuan M, Esposito M, Li W, Wei Y, Shen M, Zhang L, Tupitsyn N, Pantel K, King C, Sun J, Moriguchi J, Jun HT, Coxon A, Lee B, Kang Y (2017) Therapeutic antibody targeting tumor- and osteoblastic niche-derived Jagged1 sensitizes bone metastasis to chemotherapy. Cancer Cell 32(731–747):e736
74. Park JE et al (2010) Hypoxic tumor cell modulates its microenvironment to enhance angiogenic and metastatic potential by secretion of proteins and exosomes. Mol Cell Proteomics 9:1085–1099. https://doi.org/10.1074/mcp.M900381-MCP200
75. Peinado H, Aleckovic M, Lavotshkin S, Matei I, Costa-Silva B, Moreno-Bueno G, Hergueta-Redondo M, Williams C, Garcia-Santos G, Ghajar C, Nitadori-Hoshino A, Hoffman C, Badal K, Garcia BA, Callahan MK, Yuan J, Martins VR, Skog J, Kaplan RN, Brady MS, Wolchok JD, Chapman PB, Kang Y, Bromberg J, Lyden D (2012) Melanoma exosomes educate bone marrow progenitor cells toward a pro-metastatic phenotype through MET. Nat Med 18:883–891
76. Kucharzewska P et al (2013) Exosomes reflect the hypoxic status of glioma cells and mediate hypoxia-dependent activation of vascular cells during tumor development. Proc Natl Acad Sci USA 110:7312–7317. https://doi.org/10.1073/pnas.1220998110
77. Costa-Silva B, Aiello NM, Ocean AJ, Singh S, Zhang H, Thakur BK, Becker A, Hoshino A, Mark MT, Molina H, Xiang J, Zhang T, Theilen TM, Garcia-Santos G, Williams C, Ararso Y, Huang Y, Rodrigues G, Shen TL, Labori KJ, Lothe IM, Kure EH, Hernandez J, Doussot A, Ebbesen SH, Grandgenett PM, Hollingsworth MA, Jain M, Mallya K, Batra SK, Jarnagin WR, Schwartz RE, Matei I, Peinado H, Stanger BZ, Bromberg J, Lyden D (2015) Pancreatic cancer exosomes initiate pre-metastatic niche formation in the liver. Nat Cell Biol 17:816–826
78. Hoshino A, Costa-Silva B, Shen TL, Rodrigues G, Hashimoto A, Tesic Mark M, Molina H, Kohsaka S, Di Giannatale A, Ceder S, Singh S, Williams C, Soplop N, Uryu K, Pharmer L, King T, Bojmar L, Davies AE, Ararso Y, Zhang T, Zhang H, Hernandez J, Weiss JM, Dumont-Cole VD, Kramer K, Wexler LH, Narendran A, Schwartz GK, Healey JH, Sandstrom P, Labori KJ, Kure EH, Grandgenett PM, Hollingsworth MA, de Sousa M, Kaur S, Jain M, Mallya K, Batra SK, Jarnagin WR, Brady MS, Fodstad O, Muller V, Pantel K, Minn AJ, Bissell MJ, Garcia BA, Kang Y, Rajasekhar VK, Ghajar CM, Matei I, Peinado H, Bromberg J, Lyden D (2015) Tumour exosome integrins determine organotropic metastasis. Nature 527:329–335
79. Plebanek MP, Angeloni NL, Vinokour E, Li J, Henkin A, Martinez-Marin D, Filleur S, Bhowmick R, Henkin J, Miller SD, Ifergan I, Lee Y, Osman I, Thaxton CS, Volpert OV (2017) Pre-metastatic cancer exosomes induce immune surveillance by patrolling monocytes at the metastatic niche. Nat Commun 8:1319

8 Mechanistic Justifications of Systemic Therapeutic Oxygenation of Tumors to Weaken the Hypoxia Inducible Factor 1α-Mediated Immunosuppression

Stephen Hatfield, Katarina Veszeleiova, Joe Steingold, Jyothi Sethuraman, and Michail Sitkovsky

Abstract

Long-term studies of anti-pathogen and anti-tumor immunity have provided complementary genetic and pharmacological evidence for the immunosuppressive and immunomodulatory effects of Hypoxia-HIF-1α and adenosine-mediated suppression via the A2A adenosine receptor signaling pathway (Hypoxia-A2A-adenosinergic). This pathway is life saving when it protects inflamed tissues of vital organs from collateral damage by overactive anti-pathogen immune cells or enables the differentiation of cells of adaptive immunity. However, the Hypoxia-A2A-adenosinergic immunosuppression can also prevent tumor rejection by inhibiting the anti-tumor effects of T and NK cells. In addition, this suppressive pathway has been shown to mask tumors due to the hypoxia-HIF-α-mediated loss of MHC Class I molecules on tumor cells. It is suggested that it will be impossible to realize the full anti-tumor capacities of current cancer immunotherapies without simultaneous administration of anti-Hypoxia-A2A-Adenosinergic drugs that inactivate this tumor-protecting mechanism in hypoxic and adenosine-rich tumors.

Here, we overview the supporting evidence for the conceptually novel immunotherapeutic motivation to breathe supplemental oxygen (40–60%) or to repurpose already available oxygenation agents in combination with current immunotherapies. Preclinical studies provide strong support for oxygen immunotherapy to enable much stronger tumor regression by weakening immunosuppression by A2A adenosine receptors and by the Hypoxia→HIF-1α axis. The results of these studies emphasize the value of systemic oxygenation as clinically feasible, promising, and as a valuable tool for mechanistic investigations of tumor biology and cancer immunology. Perhaps the most effective and feasible among individual members of this novel class of anti-tumor drugs are oxygenation agents.

S. Hatfield · K. Veszeleiova · J. Steingold
J. Sethuraman · M. Sitkovsky (✉)
New England Inflammation and Tissue Protection Institute, Northeastern University, Boston, MA, USA
e-mail: m.sitkovsky@northeastern.edu

D. M. Gilkes (ed.), *Hypoxia and Cancer Metastasis*, Advances in Experimental Medicine and Biology 1136,
https://doi.org/10.1007/978-3-030-12734-3_8

Keywords
Hypoxia · Hypoxia-inducible factor-1α · HIF-1α · Hypoxia reduction · Oxygen · Oxygenation agents · Supplemental oxygenation · Cancer immunotherapy · Tumor microenvironment · Adenosine · A2A adenosine receptor (A2AR) · Cyclic AMP (cAMP) · T cells · Natural killer cells

Narrative
The uncovering of the Hypoxia-HIF-1α-mediated mechanism of immunosuppression highlighted the clinical need to eliminate hypoxia in tumors as the root of major problems that negate therapeutic efficacy of current cancer immunotherapies. This, in turn, pointed toward the immunotherapeutic use of systemic oxygenation that has been shown to be effective in the elimination of solid tumors and metastases in preclinical studies.

8.1 Hypoxia-A2-Adenosinergic Suppression

The exquisite balance between the immunostimulatory and immunosuppressive mechanisms weakens the excessive collateral damage to non-infected tissues, preventing autoimmunity while allowing for the elimination of pathogens. However, this balance often needs to be therapeutically disturbed in patients undergoing cancer therapies to allow for immunotherapeutic destruction of cancerous tissues, even at the risk of manageable autoimmunity. The studies described here were motivated by the need to identify pathophysiological negative regulators of immunity in order to devise therapeutic approaches that prevent the inhibition of the anti-tumor immune response and develop therapeutic strategies that achieve complete tumor regression.

We summarize the reasoning behind our approach and conclusions reached as a result of parallel biochemical, immunological and genetic studies of the role of hypoxia-HIF-1α and cAMP-elevating adenosine receptors. While there are four subtypes of adenosine receptors, the high affinity A2A receptor has been reported as the most critical in the regulation of immune cells. It is now established that tissue hypoxia is associated with the accumulation of extracellular adenosine, which signals through intracellular cAMP-elevating A2A adenosine receptors to inhibit effector functions of immune cells including cytotoxicity and the secretion of important proinflammatory cytokines (e.g. IFNγ, TNFα) [1, 3, 8, 9].

Initial studies provided experimental support for the immunosuppressive role of hypoxia-HIF-1α and A2A adenosine receptor-mediated (hypoxia-adenosinergic) signaling pathway in inflamed and cancerous tissues [1–10], reviewed in [11–13]. Hypoxia-A2A-adenosinergic immunoregulation is also implicated in the control of autoimmunity, B cell differentiation and antibody affinity maturation in germinal centers [14–16]. Our observations were confirmed and greatly extended by others investigating the role of adenosine-generating enzymes [17–20] such as Smyth, Stagg, Zhang and co-authors in studies of immunotherapeutic targeting of CD73 [21, 22] and Robson and others investigating CD39 [23]. This, in turn, has accelerated the interest of the pharmaceutical industry and has led to ongoing clinical trials of anti-hypoxia-adenosinergic treatments to prevent the inhibition of anti-tumor T cells and NK cells.

The Hypoxia-HIF-1α and A2A adenosine receptor-cAMP axis has been implicated in tumor-protection by experimental approaches based on the inhibition of hypoxia-adenosinergic immunosuppression [1, 3–5, 12]. These studies provided conclusive evidence for the important role of this pathway and a feasible approach to weaken immunosuppression in order to increase anti-pathogen or anti-tumor immune responses. The observations of enhanced tumor rejection after genetic or pharmacological blockade of Hypoxia-HIF-1α or Adenosine→A2A-mediated signaling provided conclusive evidence for the power of this suppressive pathway in hypoxic and adenosine-rich tumors.

8.2 Direct In vivo Evidence that Hypoxia Is Pathophysiologically Immunosuppressive in Inflamed Tissues

Evidence that pathophysiological hypoxia is immunosuppressive due to the accumulation of extracellular adenosine and A2A adenosine receptor signaling were initially obtained in a model of T cell-mediated viral and autoimmune fulminant hepatitis [24]. In mice with ongoing hepatitis, tissue hypoxia was increased by placing mice in hypoxic (10% oxygen) or normoxic (21% oxygen) conditions for 8 h followed by evaluation of inflammation and liver damage. Systemic hypoxic intervention resulted in significant reduction of liver damage and decreased levels of proinflammatory cytokines. These data are in agreement with earlier demonstrations that genetic inactivation of the A2A adenosine receptor resulted in a dramatic increase of inflammatory damage in the same autoimmune and viral hepatitis model [1]. Mechanistically, these data demonstrated that systemic hypoxia-triggered immunosuppression protects inflamed tissues and that hypoxia (upstream) and A2AR (downstream) function in the same immunosuppressive and tissue-protecting pathway.

Similar conclusions were reached in studies of lung damage during acute respiratory distress syndrome (ARDS) [25]. These experiments examined whether the weakening of hypoxia by systemic oxygenation (breathing 60% oxygen) in mice with acute lung inflammation would enhance inflammatory lung damage by disengaging this tissue-protecting mechanism. Indeed, oxygenation decreased hypoxia and led to the strong exacerbation of lung injury. The immunosuppressive effects of hypoxia in lungs were also shown to be mediated by A2AR since no additional oxygenation-induced exacerbation of inflammation was observed in mice with A2AR blockade [25]. In agreement with this molecular understanding, the treatment of mice with an A2AR agonist prevented the oxygenation-induced exacerbation of lung injury [25].

8.3 Direct In vivo Evidence that Hypoxia Is Pathophysiologically Immunosuppressive in Cancerous Tissues

It has been established that hypoxic areas of tumors are immunosuppressive and tumor-protecting. This was confirmed in studies demonstrating the anti-immunosuppressive effects of systemic oxygenation [8], where tumor-bearing mice were housed in hyperoxia chambers infused with 60% oxygen. This resulted in the i) dramatic reduction of areas of low oxygen tension (hypoxic) in the TME (tumor microenvironment) and ii) prevention of immunosuppression in the previously hypoxic TMEs. Indeed, supplemental oxygenation was accompanied by a significant decrease in the intratumoral levels of hypoxia, despite original doubts that abnormal and leaky tumor microvasculature would not be able to provide sufficient oxygenation in chaotically grown tumor tissues.

Parallel proteomics studies revealed that breathing 60% oxygen reduced the protein expression of known HIF-1α targets and resulted in the downregulation of hypoxia- and HIF-1α-related gene products, confirming previous data obtained on the mRNA level [26]. These studies also point to the use of systemic oxygenation protocols as a tool in proteomics screens to identify relationships between molecules of the hypoxia-adenosinergic pathway and tumor progression. The confidence in using such a technique was strengthened by findings of the reduction in other hypoxia-dependent pathways including carbohydrate catabolic and metabolic processes and gluconeogenesis pathway, which are known to be regulated by oxygen tension. This decrease in the levels of HIF-1α in tumors after breathing 60% oxygen justifies follow up careful investigation of other novel proteins with similar hypoxia-dependency. These studies are consistent with demonstrations of the immunosuppressive and immunoregulatory effects of HIF-1α. The pathophysiologically immunosuppressive role of hypoxia is supported by observations of decreased protein levels of other tested immunosuppressive molecules in mice where intatrumoral hypoxia was reduced or eliminated by oxygenation [8, 9].

8.4 Physiologically Immunoregulatory and Pathophysiologically Immunosuppressive HIF-1α

HIF-1α-mediated signaling has now been firmly implicated in the upstream events that trigger the anti-inflammatory CD39/CD73➔[Adenosine]^{High}➔A2AR suppressive pathway [12, 27]. However, the first direct and conclusive evidence that HIF-1α is immunosuppressive and immunoregulatory was provided in studies of Rag-2-deficient blastocyst complementation $Hif1\alpha^{-/-} \rightarrow Rag2^{-/-}$ chimeric mice. These mice are characterized by T cell and B cell targeted knockout of HIF-1α [2]. The absence of HIF-1α resulted in the loss of inflammatory control and dramatic autoimmunity in these mice. This suggested that when HIF-1α is present, it functions to prevent autoimmunity by recruiting a powerful immunosuppressive mechanism that is likely to control endogenous immune cells even in the absence of immunostimulatory signals. Signs of autoimmunity included the accumulation of anti-dsDNA antibodies and rheumatoid factor in serum and deposits of IgG and IgM in kidney and proteinuria. These studies also revealed that HIF-1α deficiency in T and B cells results in dramatic and cell lineage-specific defects, which include appearance of abnormal peritoneal B-1-like lymphocytes, with high expression of CD45 receptor-associated protein tyrosine phosphatase as well as abnormal maturation of B-2 lymphocytes in bone marrow.

Subsequent biochemical studies indicated that HIF-1α-enabled glycolysis during B cell development is required in a developmental stage-specific manner since the glycolytic pathway in HIF-1α-deficient B220(+) bone marrow cells is much less functionally effective than in wild-type control cells [14]. Increased expression levels of respiratory chain-related genes and TCA cycle-related genes detected in HIF-1α-deficient cells might be a possible compensatory adaptation to the defect of glycolysis in the absence of HIF-1α. Interestingly, it was also noticed that HIF-1α-deficient cells used pyruvate more efficiently than wild-type cells.

However, direct evidence was still needed to implicate HIF-1α in the regulation of TCR-antigen activated T cells in vivo. This was investigated in mice with the Cre-lox-P-targeted deletion of the HIF-1α gene in a model of bacterial sepsis [28]. In these studies it was possible to overcome limitations in interpretations since HIF-1α has such a critical role in the development of cells of the immune system [29]. The deletion of the HIF-1α gene led to higher levels of pro-inflammatory cytokines, stronger anti-bacterial effects and much better survival of mice, which was partially explained by significantly increased NF-kB activation in TCR activated HIF-1α deficient T cells [28]. The important implication of these studies was that activated T cells can be recruited to more powerfully contribute to the anti-pathogen immune response by relief from inhibition by HIF-1α. These experiments uncovered the unappreciated reserve of anti-bacterial capacity of T cells and suggested novel therapeutic anti-pathogen and anti-tumor strategies based on targeted deletion or inhibition of HIF-1α in T cells [28].

These observations of immunosuppressive and T cell-autonomous functions of HIF-1α were further confirmed in studies of the genetic deletion of the HIF-1α isoform I.1 in T cells, which demonstrated enhanced antibacterial immunity and improved survival in a murine peritonitis model [30]. Subsequent studies of HIF-1α revealed that activated T cells preferentially express the novel alternative isoform I.3 of HIF-1α [31]. This novel mRNA isoform of human HIF-1α is upregulated in peripheral T lymphocytes after T-cell receptor stimulation and it encodes a protein that is 24 amino acids longer than the ubiquitous HIF-1α isoform. This mRNA isoform I.3 of HIF-1α is expressed in a tissue-specific manner with the highest expression found in peripheral blood leukocytes and the thymus.

Taken together, these mechanistically complementary observations in studies of i) bacterial superantigen T cell-mediated immunity in in vivo sepsis models [28]; ii) HIF-1α induced autoimmunity [2] and iii) anti-tumor immunity in cancer immunology models [8, 9, 32] provided strong genetic and pharmacological evidence for the immunosuppressive role of the hypoxia-HIF-1α and Adenosine-A2A Adenosine receptor axis. It was also demonstrated that the hypoxia-HIF-1α stage is indeed upstream and feeds into the Adenosine-A2A-Adenosinergic signaling downstream to inhibit the anti-tumor response [9].

An important advance in understanding of the immunosuppressive effects of Hypoxia-HIF-1α and Adenosine-A2AR signaling was the implication of the Hypoxia Response Element (HRE) and cAMP Response Element (CRE), which govern immunosuppressive transcription via HIF-1α (HRE) and cAMP (CRE) [7, 8]. The HRE directs the intermediate stages of the hypoxia-A2A-adenosinergic pathway by increasing adenosine-generating enzymes, with CD73 and CD39 attracting the most attention. CD39 and CD73 function in tandem to generate extracellular adenosine and therapeutic targeting of these ecto-enzymes has also reached the clinical stage. Important proof of principle and promise of the inhibition of adenosine-generating enzymes was provided in studies of the blockade of CD73 [17–22] as well as in studies of CD39 [23].

8.5 HIF-1α Prevents Tumor Recognition by the TCR on Tumor-Reactive T Cells

It was recently reported that Hypoxia-HIF-1α can prevent the recognition of tumor cells by down-regulating the expression of antigenic peptide-presenting MHC molecules [33]. This was explained by the observation that low oxygen tension (hypoxia) in the TME negatively regulates MHC expression in a HIF-dependent manner by downregulating antigen presenting proteins (e.g. TAP 1/2 and LMP7) that are required for the surface display of peptide-MHC complexes. This conclusion is supported by the decreased MHC class I levels in tumor cells with constitutive expression of HIF-1α due to genetic loss of von Hippel-Lindau (VHL) and increased MHC class I levels in tumor cells with genetically down-regulated expression of HIF-1α [33]. It was shown that tumor cells that were more hypoxic in vivo also had lower levels of MHC class I expression [8]. This is in agreement with the view that intratumoral hypoxia may serve as the trigger that lowers MHC class I levels, making tumor cells less recognizable and thereby promoting tumor evasion.

HIF-1α is at the "top of the pyramid" of other immunosuppressive molecules in hypoxic tissue microenvironments. Many data now support the view that the hypoxia-adenosinergic pathway has the potential to regulate all other known immunosuppressive molecules (e.g. IL-10, TGF-beta, COX-2) since the elimination of hypoxia also strongly decreases levels of these proteins and overall immunosuppression [7–9]. Since Hypoxia➔HIF-1α is upstream of CD73➔Adenosine➔A2AR, it was a very important finding that Adenosine➔A2AR signaling – and by implication the upstream Hypoxia➔HIF-1α signaling – trigger the generation of other immunosuppressors, including immunological negative regulators (e.g. CTLA4, PD1, LAG-3, $Foxp3^{+}LAG\text{-}3^{+}$ regulatory T cells) [34, 35].

8.6 New Immunotherapeutic Paradigm in Targeting Hypoxia and HIF-1α

It has been proposed that maximal tumor rejection during immunotherapy will occur only in combination with inhibitors of HIF-1α. While the tumor-promoting and protecting role of HIF-1α is well established [36, 37], it was the uncovering of the role of HIF-1α as a powerful orchestrator of individual immunosuppressive mechanisms [2, 8, 9, 28, 30] that postulated the need to combine HIF-1α inhibitors with immunotherapeutic protocols that induce sufficient numbers of anti-tumor T cells. In the absence of tumor-reactive T cells, the anti-tumor effects of eliminating HIF-α will likely not be observed since the reprogramming of an immunosuppressive TME into an immunopermissive TME will only lead to clinical benefits with existing anti-tumor immunity. This advance in understanding suggests that adoptive cell transfer might be the best clinical immunotherapy protocol to combine with inhibitors of HIF-1α to accomplish maximal tumor rejection [12].

8.7 Conceptually Novel Anti-A2-Adenosinergic Immunotherapeutic Motivation for Systemic Oxygenation

It is critical to address the acute medical need in preventing the inhibition of anti-tumor T and NK cells by eliminating the tumor hypoxia-driven immunosuppression. Extrapolating from the data reviewed above, this can be achieved by systemic oxygenation via hyperoxic breathing (60% oxygen). The choice of 60% oxygen was justified by its extensive clinical use with human patients as well as by our earlier studies where we established the anti-immunosuppressive effects in acute inflammation models [25]. However, this conceptually novel method of reprogramming the TME from immunosuppressive to immunopermissive by decreasing hypoxia and adenosine levels required additional investigation. Therefore, it was tested whether systemic oxygenation was sufficient to inhibit the tumor hypoxia-HIF-1α-driven processes, specifically the CD39/CD73-mediated accumulation of extracellular adenosine in tumors. Immunostaining of regions of low oxygen tension in the TME confirmed that breathing 60% oxygen does decrease hypoxia in cancerous tissues.

These data were confirmed by biochemical and proteomic studies of the oxygenation-induced effects, which compared whole cell proteomics analyses of the TME in mice breathing 21% (normoxic) or 60% oxygen. It was shown that 60% oxygen increases local oxygen tension, decreases hypoxia, and thereby destabilizes HIF-1α. The degradation of HIF-1α resulted in i) a marked decrease in protein expression of known HIF-1α downstream targets and ii) downregulation of many hypoxia-and HIF-1α-governed gene sets [8].

Interestingly, these assays also implicated other novel proteins which were not yet included among those that are dependent on hypoxia-HIF-1α signaling. As an example, Four and a Half LIM domain-1 protein (FHL-1), a known inhibitor of the HIF-1α-mediated pathway [26], was identified in proteomics screens of the TME as the most strongly upregulated in the physioxic (non-hypoxic) regions of tumors in mice breathing 60% oxygen. Since there are hypoxic markers, but no molecular markers of oxygenated TME, it was suggested to investigate the use of FHL-1 protein expression as an important identifier of oxygenated and immunopermissive TMEs [8]. One of the mechanisms by which FHL-1 might mediate the immunoenhancing effects of 60% oxygen is by preventing association of HIF-1α with its co-activator p300/CBP. In addition to FHL-1, the proteomics screen of the TME also identified Factor Inhibiting HIF-1 (FIH-1) and von Hippel-Lindau (VHL) as markers of those TME that have physioxic levels of oxygen [38].

Since systemic oxygenation was shown to decrease levels of Hypoxia➔HIF-1α, and the adenosine-generating enzyme CD73 is regulated by HIF-1α expression, it was proposed that levels of adenosine would also decrease with oxygen tension in the TME. Indeed, parallel studies were able to quantify levels of extracellular adenosine by HPLC-MS using samples withdrawn from equilibrium microdialysis probes placed in randomly selected areas in growing tumors. Mice breathing ambient 21% oxygen established the heterogeneity of hypoxia and extracellular adenosine levels in intra-tumoral microenvironments. In agreement with previous observations, the levels of extracellular adenosine in tumors were dramatically elevated compared with normal tissues [3, 39].

In more detailed studies, extracellular adenosine levels were measured in several areas of tumors in mice breathing 21% or 60% oxygen. It was demonstrated that while some TMEs were adenosine-rich, others were adenosine-poor, with levels of adenosine ranging from ~1400 nM to virtually undetectable (similar to the very low levels of adenosine in the normal tissue next to the tumor). Importantly, breathing 60% oxygen dramatically decreased levels of adenosine in the majority of tested TMEs. However, in a minority of probed areas of the TME, remaining levels of adenosine were reduced but still sufficiently high to activate the high affinity A2A adenosine receptor (but not the low affinity A2B receptor). These observations justify further research to clarify the ability of an A2AR antagonist to complement the effects of oxygen in relieving immunosuppression in the TME.

8.8 Multidirectional Weakening of Immunosuppression in Tumors by Supplemental Oxygen

Systemic oxygenation may be the most powerful blocker of Hypoxia-A2A-Adenosinergic signaling as well as other immunosuppressive mechanisms. Oxygenation not only blocks CD39/CD73-mediated accumulation of immunosuppressive extracellular adenosine in the TME, but also the expression of (i) CD39/CD73; (ii) A2A and A2B adenosine receptors; (iii) other anti-inflammatory molecules; and iv) pro-tumor angiogenesis molecules. Using systemic oxygenation as a tool to uncover the mechanism of immunosuppression in tumors revealed that oxygen-mediated decrease in extracellular adenosine in the TME is due to the decreased expression of the adenosine-generating ectoenzymes CD39 and CD73 on both tumor cells and on tumor-infiltrating CD8 and CD4 T cells [8].

Along with CD73/CD39 and A2A/A2B adenosine receptors, hyperoxic breathing also caused the downregulation of other immunosuppressive molecules such as prostaglandin E synthase-2 (COX-2). This provides additional independent evidence to support earlier reports that COX-2 is regulated by both hypoxia and A2A receptor-mediated induction of cAMP in tumors [40, 41]. The downregulation of both the levels of adenosine in the TME and A2A/A2B adenosine receptors by oxygenation is of additional mechanistic significance when considering the absence of A2A adenosine receptor reserve [42]. Indeed, it has been demonstrated that lower expression of A2AR will mediate smaller increases in intracellular cyclic AMP.

Thus, oxygenation powerfully weakens all known stages of the CD39/CD73-adenosine-A2AR/A2BR-cAMP pathway. Additionally, oxygenation also decreases the intensity of upstream hypoxia→ HIF-1α signaling by preventing the actions of HIF-1α-regulated downstream targets in the TME, such as angiogenesis factors like VEGF [8]. In studies of mice breathing 60% oxygen, the decreased levels of VEGF was consequential since it altered the neovascularization of tumors as reflected by a significant decrease in the expression of the intra-tumoral angiogenic marker (PECAM-1) [8]. In addition, oxygenation also prevented the escape of tumor cells from recognition by anti-tumor T cells. Oxygenation resulted in a higher expression of antigen-presenting MHC class I molecules on the surface of tumors, thereby increasing their recognition and susceptibility to cytotoxic T lymphocytes [8].

8.9 Conclusion

8.9.1 Future Translational Studies of the Therapeutic Potential Of Systemic Oxygenation

Taken together, these data show that oxygenation allows better recognition and lethal hit delivery to tumors by anti-tumor T cells. This, in turn, enhances tumor regression and improves survival of tumor-bearing mice.

The simplicity of using systemic oxygenation both as a therapeutic agent and as a valuable tool for mechanistic investigations of tumor biology and cancer immunology allowed for the discovery that oxygenation converts the TME from immunosuppressive to immunopermissive. Systemic oxygenation and/or oxygenation agents can be utilized to convert the hypoxic, adenosine-rich and tumor-protecting TME into a physioxic and extracellular adenosine-poor TME, thereby facilitating tumor regression.

The oxygenation-mediated inhibition of established tumor growth together with the genetic evidence for the anti-immunosuppressive mechanisms demonstrated the therapeutic potential of systemic oxygenation. We propose the exploration of therapies combining supplemental oxygen with existing immunotherapies of cancer. It is now important to test and confirm clinically the anti-tumor effects of systemic oxygenation in patients that are refractory to immunotherapies of cancer. Oxygenation and oxygenation agents may well be the drugs that address the well-recognized need to eliminate tumor hypoxia, an established poor prognosis factor. Recently, ongoing clinical trials by several pharmaceutical companies have demonstrated tumor-regression following blockade of the downstream A2A-adenosine receptor pathway [43]. This has provided proof of principle for the promise of targeting hypoxia-HIF-1α-mediated upstream events by oxygenation agents.

References

1. Ohta A, Sitkovsky M (2001) Role of G-protein-coupled adenosine receptors in downregulation of inflammation and protection from tissue damage. Nature 414(6866):916–920
2. Kojima H et al (2002) Abnormal B lymphocyte development and autoimmunity in hypoxia-inducible factor 1alpha -deficient chimeric mice. Proc Natl Acad Sci USA 99(4):2170–2174
3. Ohta A et al (2006) A2A adenosine receptor protects tumors from antitumor T cells. Proc Natl Acad Sci USA 103(35):13132–13137
4. Sitkovsky MV et al (2004) Physiological control of immune response and inflammatory tissue damage by hypoxia-inducible factors and adenosine A2A receptors. Annu Rev Immunol 22:657–682
5. Sitkovsky M, Lukashev D (2005) Regulation of immune cells by local-tissue oxygen tension: HIF1 alpha and adenosine receptors. Nat Rev Immunol 5(9):712–721
6. Sitkovsky MV et al (2008) Hypoxia-adenosinergic immunosuppression: tumor protection by T regulatory cells and cancerous tissue hypoxia. Clin Cancer Res 14(19):5947–5952
7. Sitkovsky MV (2009) T regulatory cells: hypoxia-adenosinergic suppression and re-direction of the immune response. Trends Immunol 30(3):102–108
8. Hatfield SM et al (2014) Systemic oxygenation weakens the hypoxia and hypoxia inducible factor 1alpha-dependent and extracellular adenosine-mediated tumor protection. J Mol Med (Berl) 92:1283–1292
9. Hatfield SM et al (2015) Immunological mechanisms of the antitumor effects of supplemental oxygenation. Sci Transl Med 7(277):277ra30
10. Kjaergaard J et al (2018) A2A adenosine receptor gene deletion or synthetic A2A antagonist liberate tumor-reactive CD8(+) T cells from tumor-induced immunosuppression. J Immunol 201(2):782–791
11. Eltzschig HK, Sitkovsky MV, Robson SC (2013) Purinergic signaling during inflammation. N Engl J Med 368(13):1260
12. Sitkovsky MV et al (2014) Hostile, hypoxia-A2-adenosinergic tumor biology as the next barrier to overcome for tumor immunologists. Cancer Immunol Res 2(7):598–605
13. Cronstein BN, Sitkovsky M (2017) Adenosine and adenosine receptors in the pathogenesis and treatment of rheumatic diseases. Nat Rev Rheumatol 13(1):41–51
14. Kojima H et al (2010) Differentiation stage-specific requirement in hypoxia-inducible factor-1alpha-regulated glycolytic pathway during murine B cell development in bone marrow. J Immunol 184(1):154–163
15. Abbott RK et al (2016) Germinal center hypoxia potentiates immunoglobulin class switch recombination. J Immunol 197(10):4014–4020
16. Abbott RK et al (2017) The GS protein-coupled A2a adenosine receptor controls T cell help in the germinal center. J Biol Chem 292(4):1211–1217
17. Stagg J et al (2011) CD73-deficient mice have increased antitumor immunity and are resistant to experimental metastasis. Cancer Res 71(8):2892–2900
18. Stagg J et al (2012) CD73-deficient mice are resistant to carcinogenesis. Cancer Res 72(9):2190–2196
19. Allard B et al (2013) Targeting CD73 enhances the antitumor activity of anti-PD-1 and anti-CTLA-4 mAbs. Clin Cancer Res 19(20):5626–5635
20. Loi S et al (2013) CD73 promotes anthracycline resistance and poor prognosis in triple negative breast cancer. Proc Natl Acad Sci USA 110(27):11091–11096
21. Jin D et al (2010) CD73 on tumor cells impairs antitumor T-cell responses: a novel mechanism of tumor-induced immune suppression. Cancer Res
22. Zhang B (2010) CD73: a novel target for cancer immunotherapy. Cancer Res 70(16):6407–6411
23. Deaglio S et al (2007) Adenosine generation catalyzed by CD39 and CD73 expressed on regulatory T cells mediates immune suppression. J Exp Med 204(6):1257–1265
24. Chouker A et al (2008) Critical role of hypoxia and A2A adenosine receptors in liver tissue-protecting physiological anti-inflammatory pathway. Mol Med 14(3–4):116–123
25. Thiel M et al (2005) Oxygenation inhibits the physiological tissue-protecting mechanism and thereby exacerbates acute inflammatory lung injury. PLoS Biol 3(6):e174

26. Semenza GL (2014) Hypoxia-inducible factor 1 and cardiovascular disease. Annu Rev Physiol 76:39–56
27. Synnestvedt K et al (2002) Ecto-5′-nucleotidase (CD73) regulation by hypoxia-inducible factor-1 mediates permeability changes in intestinal epithelia. J Clin Invest 110(7):993–1002
28. Thiel M et al (2007) Targeted deletion of HIF-1alpha gene in T cells prevents their inhibition in hypoxic inflamed tissues and improves septic mice survival. PLoS One 2(9):e853
29. Buras JA, Holzmann B, Sitkovsky M (2005) Animal models of sepsis: setting the stage. Nat Rev Drug Discov 4(10):854–865
30. Georgiev P et al (2013) Genetic deletion of the HIF-1alpha isoform I.1 in T cells enhances antibacterial immunity and improves survival in a murine peritonitis model. Eur J Immunol 43(3):655–666
31. Lukashev D, Sitkovsky M (2008) Preferential expression of the novel alternative isoform I.3 of hypoxia-inducible factor 1alpha in activated human T lymphocytes. Hum Immunol 69(7):421–425
32. Hatfield SM, Sitkovsky M (2015) Oxygenation to improve cancer vaccines, adoptive cell transfer and blockade of immunological negative regulators. Oncoimmunology 4(12):e1052934
33. Sethumadhavan S et al (2017) Hypoxia and hypoxia-inducible factor (HIF) downregulate antigen-presenting MHC class I molecules limiting tumor cell recognition by T cells. PLoS One 12(11):e0187314
34. Zarek PE et al (2008) A2A receptor signaling promotes peripheral tolerance by inducing T-cell anergy and the generation of adaptive regulatory T cells. Blood 111(1):251–259
35. Leone RD et al (2018) Inhibition of the adenosine A2a receptor modulates expression of T cell coinhibitory receptors and improves effector function for enhanced checkpoint blockade and ACT in murine cancer models. Cancer Immunol Immunother 67:1271–1284
36. Schito L, Semenza GL (2016) Hypoxia-inducible factors: master regulators of Cancer progression. Trends Cancer 2(12):758–770
37. Xiang L et al (2014) Ganetespib blocks HIF-1 activity and inhibits tumor growth, vascularization, stem cell maintenance, invasion, and metastasis in orthotopic mouse models of triple-negative breast cancer. J Mol Med (Berl) 92(2):151–164
38. Hubbi ME et al (2012) Four-and-a-half LIM domain proteins inhibit transactivation by hypoxia-inducible factor 1. J Biol Chem 287(9):6139–6149
39. Blay J, White TD, Hoskin DW (1997) The extracellular fluid of solid carcinomas contains immunosuppressive concentrations of adenosine. Cancer Res 57(13):2602–2605
40. Sittler T et al (2008) Concerted potent humoral immune responses to autoantigens are associated with tumor destruction and favorable clinical outcomes without autoimmunity. Clin Cancer Res 14(12):3896–3905
41. Takayama H, Sitkovsky MV (1988) Potential use of antagonists of cAMP-dependent protein kinase to block inhibition and modulate T-cell receptor-triggered activation of cytotoxic T-lymphocytes. J Pharm Sci 78:8–10
42. Armstrong JM et al (2001) Gene dose effect reveals no Gs-coupled A2A adenosine receptor reserve in murine T-lymphocytes: studies of cells from A2A-receptor-gene- deficient mice. Biochem J 354(Pt 1):123–130
43. Sheridan C (2018) Adorx dives into the adenosine antagonism pool with $10M series A. BioWorld 29(121):1–6

Hypoxia-Induced Resistance to Chemotherapy in Cancer

9

Lori M. Minassian, Tiziana Cotechini, Erin Huitema, and Charles H. Graham

Abstract

A major barrier to the successful management of cancer is the development of resistance to therapy. Chemotherapy resistance can either be an intrinsic property of malignant cells developed prior to therapy, or acquired following exposure to anti-cancer drugs. Given the impact of drug resistance to the overall poor survival of cancer patients, there is an urgent need to better understand the molecular pathways regulating this malignant phenotype. In this chapter we describe some of the molecular pathways that contribute to drug resistance in cancer, the role of a microenvironment deficient in oxygen (hypoxia) in malignant progression, and how hypoxia can be a significant factor in the development of drug resistance. We conclude by proposing potential therapeutic approaches that take advantage of a hypoxic microenvironment to chemosensitize therapy-resistant tumours.

Keywords

Hypoxia · Drug resistance · Chemotherapy · Tumor microenvironment · HIF-1 · Metastasis · Autophagy · Nitric oxide · Glyceryl trinitrate · PD-1/PD-L1

L. M. Minassian · T. Cotechini · E. Huitema
C. H. Graham (✉)
Department of Biomedical and Molecular Sciences, Queen's University, Kingston, ON, Canada
e-mail: grahamc@queensu.ca

9.1 Drug Resistance in Cancer

9.1.1 Intrinsic Drug Resistance

The development of drug resistance in cancer is complex and multifactorial; moreover, several mechanisms of drug resistance appear to be clinically relevant. Some of these mechanisms operate at the single cell level and include the overexpression of drug efflux proteins such as the multidrug resistance protein (MRP1; and related MRP2 and MRP3), as well as the P-glycoprotein (P-gp) efflux pump [2]; increased levels of detoxification and DNA repair enzymes such as glutathione-S-transferase (GST) and 06-alkylguanine DNA alkyltransferase [3, 4]; and mechanisms interfering with drug-induced apoptosis [1, 5]. Alternatively, drug resistance can occur at the multicellular level, where the tumour architecture plays an important role. In this case, cells can acquire resistance to several classes of drugs (multidrug resistance or MDR) via multiple mechanisms [6–9].

Cancer cell 'stemness' has emerged as a major contributor to drug resistance and recurrence. Cancer stem cells (CSCs) are capable of limited differentiation, self-renewal, and tumourigenicity, and exhibit enhanced proliferation [10] and survival [11] in response to chemotherapy. Importantly, resistance of these tumour initiating cells to both chemotherapy and radiation therapy [12] results in selective enrichment of heteroge-

D. M. Gilkes (ed.), *Hypoxia and Cancer Metastasis*, Advances in Experimental Medicine and Biology 1136,
https://doi.org/10.1007/978-3-030-12734-3_9

neous subpopulations [12, 13]. Acquisition of drug resistance in CSCs can arise as a result of multiple cell intrinsic mechanisms (Reviewed by Abdullah and Chow [14]). Most critically though, drug resistance in CSCs is dependent on maintenance of pluripotency, which contributes to tumour heterogeneity – a feature of MDR. Recent work has additionally highlighted the importance of autophagy in conservation of pluripotency [15] such that inhibition of autophagy sensitizes tumour cells to chemotherapy and thus represents a potential strategy to overcome drug resistance [16].

9.1.2 Role of the Tumour Microenvironment in Drug Resistance

Until relatively recently, tumour cell intrinsic pathways were the focus of mechanistic drug resistance studies. Accumulating evidence now shows that the tumour microenvironment (TME) plays a pivotal role in facilitating acquired drug resistance. Contributing to the TME is a network of fibroblasts, immune cells, host microbes, lymphatics and vasculature [17]. These cellular constituents reside within a complex stromal scaffolding made up of extracellular matrix (ECM) proteins often within an environment deficient in oxygen [18]. The TME contributes to drug resistance in a multifactorial manner. In the following paragraphs we describe some of the mechanisms by which elements of the TME can mediate drug resistance.

(a) Biomechanical and biophysical properties

Tissue architecture, including cellular organization, polarity [19] and deposition and composition of the ECM [20] can regulate apoptotic responses to chemotherapy. Relative to normal tissue architecture, the ECM of solid tumours is often rigid. The biomechanical properties of this stiffened ECM regulate and direct malignant cellular behaviours [21] including migration and invasion [22], dormancy, proliferation and chemosensitivity [20]. Increased tumour stiffness is predictive of neoadjuvant chemotherapy response in breast cancer [23] and is linked to chemoresistance in pancreatic cancer [24, 25]. High interstitial pressure resulting from a rigid ECM in solid tumours can also lead to drug resistance by preventing the transport of chemotherapeutic agents away from blood vessels (Reviewed by Munson and Shieh [26]). Poor delivery of molecules resulting from high interstitial pressure and inadequate blood perfusion can lead to a hypoxic environment and a deficit in nutrients such as glucose. Glucose deprivation was shown to induce resistance to doxorubicin and etoposide in Chinese hamster ovary cells, as well as human colon and ovarian cancer cell lines [27–29]. Similarly, changes in the pH of the tumour microenvironment resulting from increased anaerobic respiration and decreased removal of toxins cause alterations in cell membrane permeability, which in turn can limit cellular uptake of chemotherapeutic agents [30, 31].

Biomechanical properties of the ECM can also be altered by neoplastic progression. For example, tumour cell expression of the intracellular protein tyrosine kinase focal adhesion kinase (FAK) regulates local tissue fibrosis and promotes an immunosuppressive TME associated with therapy resistance in pancreatic ductal adenocarcinoma [25]. Indeed, stromal depletion has been utilized as an approach to enhance delivery of chemotherapeutic agents to desmoplastic tumours and has been successful at improving survival in mouse studies [32, 33].

(b) Host microbiome

The potential role of the microbiome in the modulation of therapy responses in cancer is an area of investigation that has received a great deal of attention in recent years [34], but is still a relatively new concept. It is becoming evident that gaining an understanding of pharmacomicrobiomics, the study of interactions between host microbes and drugs, is important to implementing effective cancer treatments [35]. While much of the work currently under way focuses on promoting a favourable intestinal microbiome for successful immune therapy [36], there is also evidence that intratumoural bacteria metabolize chemotherapeutic agents and thus contribute to chemoresistance [37].

(c) Immune microenvironment

Tumours are described as wounds that do not heal [38] and tumour-promoting inflammation is now widely accepted as a hallmark of cancer [39]. The tumour immune microenvironment (TiME) is dynamic and consists of innate and adaptive immune cells as well as humoral factors that are largely immunosuppressive [40]. Immune cells within the tumour microenvironment are functionally distinct from their counterpart immune cells of the adjacent normal stroma and are often described as being pro-tumourigenic. These tumour promoting immune cells within the TME negatively influence responses to radiotherapy and chemotherapy (Reviewed by Medler et al. [41]), in part by preventing tumour cell apoptosis [42]. Myeloid cells, including tumour associated macrophages (TAMs), neutrophils and myeloid-derived suppressor cells (MDSCs) are well-studied immune cells contributing to chemoresistance (Reviewed by Cotechini et al. [43]). However, cells of the adaptive immune system, including B cells and $CD4^+$ T cells, also contribute to chemoresistance and radioresistance in part by regulating the mobility and anti-tumour functions of cytotoxic $CD8^+$ T cells (CTL) [44–46].

An important signalling axis regulating immune cell activity and, in particular, $CD8^+$ cytotoxic T cell responses, is the immune checkpoint Programmed Death Receptor 1 (PD-1)/ Programmed Death Ligand 1 (PD-L1). PD-1 is a monomeric transmembrane receptor present on activated T cells, B cells, dendritic cells, NK cells and monocytes. Binding of PD-1 to its cognate ligands, PD-L1 or PD-L2, renders T cells hyporesponsive to antigen stimulation and manifests as inhibition of proliferation and dampened effector (cytotoxic) functions [47–49]. PD-L1 is expressed by many different cell types, including epithelial cells, B cells, T cells, monocytes and antigen presenting cells [50]. Importantly, tumour cells from various cancers including breast, colorectal, ovarian, bladder and lung cancers, as well as glioblastomas, lymphomas, melanomas and leukemias express PD-L1, and expression of this immune checkpoint is predictive of poor clinical prognosis [51–59].

There is recent evidence from our work [60] and that of others [61], that the PD-1/PD-L1 signalling axis is bi-directional and that reverse signalling endows tumour cells with enhanced resistance to conventional anti-cancer drugs. *In vitro* work from Azuma and colleagues revealed that PD-L1 overexpressing mouse mastocytoma (B7-H1/P815) cells are resistant to $PD-1^+$ CTL-mediated killing as well as to Fas and drug-mediated apoptosis [61]. Using *in vitro* and *in vivo* approaches, we recently discovered that PD-1/PD-L1 signalling endows human and mouse prostate and breast cancer cells with resistance to conventional chemotherapeutic agents likely via signalling through PI3K-AKT-mTOR and MEK-ERK pathways [60]. Importantly, our group has also shown that hypoxia induces PD-L1 expression in murine and human tumour cells leading to immune escape [158]. Hypoxia is a characteristic of solid tumours and occurs as a result of an imbalance between oxygen consumption and oxygen availability [62]. While a reduction in the local amount of oxygen can be initially detrimental to rapidly-proliferating cells, tumour cells adapt to hypoxia by activating oxygen sensitive transcription factors (described below). Tumour cells co-opt physiological adaptations to hypoxia in order to evade immune destruction and survive radiotherapy and chemotherapy.

9.2 Hypoxia and Malignant Progression

9.2.1 HIF-1: A Mediator of Hypoxia-Induced Malignant Phenotypes

The most well characterised transcription factor responsible for many cellular adaptations to hypoxia is the hypoxia-inducible transcription factor (HIF), a dimeric protein consisting of a constitutively active subunit (HIF-1β) as well as an oxygen-sensitive subunit (HIF-1α) [63]. Under well-oxygenated conditions, HIF-1α is unstable and rapidly degraded. HIF-1α is hydroxylated by the oxygen-dependent enzyme prolyl hydroxylase domain 2 (PHD2) and interacts with the von Hippel-Lindau tumour suppressor protein (pVHL). This interaction leads to the recruitment of E3 ubiquitin ligase that mediates the polyubiquitination of HIF-1α, which ultimately leads to the proteasomal degradation of HIF-1α [64, 65]. HIF-1α is also hydroxylated by factor inhibiting HIF-1 (FIH-1), which prevents binding of its coactivator p300/CBP and inhibits transcriptional activity [66, 67]. Due to the oxygen requirement for PHD2 activity, hypoxia prevents the hydroxylation of HIF-1α, thereby allowing it to bind to HIF-1β and mediate the transcription of hypoxia-inducible genes [63].

Many HIF-1 gene targets encode proteins involved in promoting tumour growth and malignant phenotypes such as angiogenesis, glucose metabolism, ECM remodelling, epithelial-to-mesenchymal transition, cell survival, and proliferation [68]. Glucose transporter 1 (*glut-1*) is a HIF-1 target gene and is involved in regulation of glucose uptake [69]. HIF-1 regulates angiogenesis by activating various genes, most notably vascular endothelial growth factor (*VEGF*), a master regulator of neo-vessel formation [70], as well as genes that mediate endothelial cell and pericyte proliferation, migration, adhesion, and maturation, vascular permeability and vasoactivity [71]. Despite activation of angiogenesis in response to hypoxia, blood vessels within the TME are tortuous and leaky and do not function in a normal capacity [72]. Hypoxia is also a central regulator of lymphatic vessel formation or lymphangiogenesis [73]. In addition to the presence of lymphatic vessels being associated with lymphogenous spread of disease, recent work examining lymphatic vessel density (LVD) in human melanoma revealed a positive correlation between LVD and the presence of immunosuppressive factors within the TME and tumour-draining lymph nodes [74]. Taken together, hypoxia enables tumour growth by promoting the classical hallmarks of cancer [39, 75].

9.2.2 Hypoxia and Radioresistance

As early as the 1950s, radiobiologists were aware that hypoxia within solid tumours reduces the efficacy of radiation therapy [76]. Gray and colleagues discovered that tumour cells were three times more sensitive to radiation under normoxic conditions compared to those in anoxia [76, 77]. Successful radiotherapy depends on the presence of relatively high levels of oxygen required for the generation of free radicals that cause irreversible DNA damage, and hence tumour cell death [78]. To overcome hypoxia-induced radio-resistance, studies have focused on developing therapeutics that function to increase oxygen delivery via improving blood flow, mimicking oxygen or targeting and destroying hypoxic cells [79]. Studies combining fractionated radiotherapy with oxygen mimetics such as 2-nitroimidazoles, or use of cytotoxic agents that specifically target hypoxic cells have shown increased tumour cell killing during radiotherapy [80]. However, despite decades of strong evidence revealing that modification of hypoxia is clinically efficacious in radiotherapy, it has yet to become a standard of care [81]. Similarly, it was discovered that many chemotherapeutic agents (*e.g.* carmustine and alkylating agents) display reduced cytotoxicity toward hypoxic tumour cells, as these drugs also require oxygen for maximal activity [82]. These early observations led to the development of novel chemotherapeutic bioreductive agents which are cytotoxically active only under limited levels of oxygen [82].

9.2.3 Hypoxia and Cancer Metastasis

Studies have demonstrated that hypoxia within the tumour mass is an independent marker of a poor prognosis for patients with various types of cancers such as carcinoma of the cervix [83], soft tissue sarcoma [84], carcinoma of the head and neck [85], cutaneous melanoma [86] and prostatic adenocarcinoma [87, 88]. In some of the above studies, disease-free survival for patients with tumours having median pO_2 values of less than 10 mmHg was found to be significantly lower than for patients with tumours having higher pO_2 values. Moreover, clinical studies now have provided evidence that low tumour oxygen levels are associated with increased tumour growth and metastasis [83, 84] and with biochemical relapse and recurrence of prostate cancer following radiotherapy [89].

Local tumour hypoxia is a serious impediment to the successful treatment of cancer in part as a result of hypoxia-mediated acquisition of malignant phenotypes that promote the spread of tumour cells. Experimental evidence in support of hypoxia having a direct stimulatory effect on metastasis was initially provided by the work of Hill and co-workers using various cell lines [90–93]. Their earlier studies demonstrated that exposure of mouse fibrosarcoma cells to hypoxia induces DNA over-replication and selects for tumour cell variants with increased metastatic potential [92]. More recently, we and others showed that hypoxia (both *in vitro* and *in vivo*) rapidly and transiently increases the invasiveness and metastatic potential of various tumour cell lines [90, 91, 93–99]. Our studies linked the hypoxia-mediated invasive ability of tumour cells to elevated expression of metastasis-associated molecules such as the urokinase plasminogen activator receptor (uPAR: a cell surface glycoprotein necessary for tumour cell invasion through the extracellular matrix) and the plasminogen activator inhibitor 1 as well as with decreased expression of tissue inhibitor of metalloproteinases 1 [95, 100, 101]. In support of these observations, Rofstad et al. reported that hypoxia promotes lymph node metastasis in human melanoma xenografts by up-regulating uPAR expression [102].

The above-mentioned hypoxia-associated tumour intrinsic mechanisms contributing to metastasis are often described as 'seed factors' [103] in relation to Stephen Paget's 'Seed and Soil' hypothesis [104]. It is becoming evident that the metastatic niche is an important factor to consider when discussing metastasis and metastatic potential since this fertile soil contributes to the metastatic microenvironment (MME). Similar to the TME, the MME is a hypoxic, immunosuppressive milieu consisting of dysregulated cellular and acellular components. Metastatic tumour cells often exhibit organotropism with respect to dissemination to secondary sites. For example, breast cancer frequently metastasizes to the lungs. There is evidence, using human breast cancer cells and metastatic murine models, that HIF-1 orchestrates metastatic programs driving lung-specific metastasis through various mechanisms [105]. In general, the MME favours seeding and outgrowth of disseminated metastatic tumour cells and thus contributes to malignancy. However, it is also important to consider the role of the pre-metastatic niche. This unique microenvironment is established prior to the dissemination of tumour cells and is primed by transformed cells within the primary neoplasm to enable colonization of metastatic cells. Indeed, recent work has shown that factors secreted by hypoxic tumour cells support the establishment of an immunosuppressive pre-metastatic niche [106, 107].

9.2.4 Role of Hypoxia in Autophagy

Autophagy – a cell-intrinsic process of 'self-eating' that maintains cellular homeostasis – is regulated by numerous stimuli and pathways, one of which is hypoxia [108]. Both HIF-1-dependent and HIF-1-independent mechanisms are known to control this process [109]. HIF-1-independent pathways tend to be activated in more severe hypoxic conditions, and work in concert with other cell stressors such as metabolic stress and nutrient starvation [109]. One important HIF-1α-independent pathway of hypoxia-induced autophagy is the unfolded protein response (UPR). Rouschop et al. demonstrated that hypoxia activates transcription of microtubule-associated protein 1 light chain 3β (*MAP1LC3B)* and autophagy-related gene 5 (*ATG5)* in multiple tumour cell lines via the UPR [110]. *MAP1LC3B*, encoding microtubule-associated protein light chain 3 (LC3B), and *ATG5* are important for autophagosome formation and thus play important roles in autophagic processes. Similar results were shown by Rzymski et al., who reported that autophagy-related gene 4 acts on *MAP1LC3B* transcription by directly binding to a cyclic AMP (cAMP) response element binding site in the promoter [111]. Both studies indicate that the increased transcriptional activation of *MAP1LC3B* leads to a replenishment of the LC3B pool, thus prolonging autophagy and allowing cells to survive through extended periods of hypoxia [110, 111]. There is also evidence that 5'-AMP-activated protein kinase (AMPK) regulates hypoxia-induced autophagy. In a study of androgen-dependent prostate cancer, it was discovered that hypoxia and androgen deprivation lead to activation of autophagy through AMPK and a mechanism partially mediated by Beclin-1 [112]. Similarly, Papandreou et al., showed that hypoxia increases autophagy in tumour cells via activation of AMPK, in a manner independent of HIF-1 and its target genes [113]. Recent studies have also shown an emerging role for micro-RNAs (miR) in regulating hypoxia-induced autophagy. For example, expression of miR-96, which can either promote or inhibit autophagy, is increased in response to hypoxia in prostate cancer cells [114].

Activation of autophagic processes provides a survival advantage to cancer cells subjected to hypoxic stress. HIF-1-dependent mechanisms of hypoxia-induced autophagy are thought to require the expression of Bcl-2/adenovirus E1B 19 kDa interacting protein 3 (BNIP3) as well as a similar protein, BNIP3L (also known as Nix) [115, 116]. It has been proposed that these molecules lead to autophagy by releasing Beclin-1 (a key mediator of autophagy) from Bcl-2/Beclin-1 or Bcl-XL/Beclin-1 complexes [117]. Tracy et al. showed that hypoxia-induced autophagy was dependent on HIF-1-mediated activation of BNIP3 [118]. HIF-1 has also been shown to induce autophagy via activation of miR-210, leading to a downregulation of Bcl-2 [119]. HIF-1 also stimulates autophagy in hypoxia via the p27-E2F1 pathway in esophageal cancer cells [120]. It is important to note that there is some controversy regarding whether hypoxia-induced autophagy is a cell survival or a cell death-inducing mechanism [121]. However, in the context of hypoxia-induced drug resistance in cancer cells, autophagy acts as a survival mechanism.

9.3 Mechanisms of Hypoxia-Induced Drug Resistance

As stated previously, many chemotherapeutic agents exhibit reduced cytotoxicity toward hypoxic tumour cells as such drugs often require oxygen for maximal activity [82]. Regardless of the oxygen requirement for anti-cancer drug activity, studies have also revealed that pre-incubation of certain human and non-human tumour cell lines under hypoxia alters their phenotype such that they transiently increase their resistance to drugs such as etoposide, and doxorubicin [4, 122–127]. Some explanations suggested for this form of resistance have included the upregulated expression of glucose- and oxygen-regulated proteins, DNA over-replication, cell cycle arrest, altered cellular metabolism, increased drug efflux pumps and greater genetic instability [82].

Various hypoxia-inducible genes with well-established roles in resistance to anticancer agents have been identified. For example, functional HIF-1 target hypoxia response elements (HREs) have been identified in genes encoding the multidrug resistance 1 protein (MDR1/ABCB1) and breast cancer resistance protein

(Bcrp/ABCG2), which are members of the ATP-binding cassette (ABC) transporter family that confer resistance through active efflux of a wide range of anti-cancer agents [128, 129]. As observed for most HRE-containing genes, increased HIF-1 binding activates the MDR1 or Bcrp gene promoters, resulting in increased expression of these drug transporters under hypoxic conditions [130].

Chemotherapeutic agents can trigger tumour cell death through the induction of pro-apoptotic pathways. However, it is important to recognise that tumour cells that have undergone drug-induced DNA damage can also be eliminated via other forms of programmed cell death such as autophagy, mitotic catastrophe, and necrosis [131]. Moreover, certain anti-cancer drugs are known to induce senescence in tumour cells [132–137], and drug-induced senescence and mitotic catastrophe may, in fact, be more prominent than apoptosis [132, 136]. Senescence is characterized by an irreversible arrest of the cell cycle and can be induced by various stresses including telomere dysfunction, oxidative damage, DNA damage, and aberrant expression of oncogenic proteins such as Ras [138]. Senescence is categorized as either replicative senescence, a physiological process triggered to limit the life span of non-malignant cells, or accelerated senescence, associated with a rapid onset of terminal proliferation arrest in response to cell damage such as drug- or radiation-induced DNA damage [138]. A study from our laboratory revealed that hypoxia-induced resistance to anti-cancer drugs is associated with decreased tumour cell senescence and that it requires HIF-1 activity [139]. While there is evidence that hypoxia can inhibit replicative senescence by increasing telomerase activity [140–143], it is doubtful that increased telomerase activity accounts for the hypoxia-mediated resistance to drug-induced senescence. It is rather likely that a lack of drug-induced senescence in hypoxic tumour cells is indirectly a result of hypoxia-triggered inhibition of DNA damage, as evidenced by another study from our group [144]. In that study we demonstrated that hypoxia prevents etoposide-induced DNA damage in cancer cells through a still to be characterised mechanism involving HIF-1α [144].

As mentioned previously, hypoxia is an important driver of autophagy. Although autophagy has been shown to have both pro-apoptotic and pro-survival roles in tumour cells, there is a link between hypoxia-induced autophagy and drug resistance [121]. As a protective response against chemotherapy- and radiotherapy-induced apoptosis, tumour cells undergo autophagic processes that degrade damaged cellular components [145]. For example, in response to cisplatin, glioma cell lines stimulate protective autophagic responses via up-regulation of AMPK and subsequent down-regulation of mammalian target of rapamycin (mTOR) [146]. It was subsequently revealed that hypoxia amplifies cisplatin-induced autophagy in a HIF-1-dependent manner and that inhibiting a crucial autophagy mediator, ATG5, restored sensitivity to cisplatin in lung cancer cells [147]. Similar results were observed in non-small cell lung cancer cells, where inhibition of LC3B restored cisplatin sensitivity under hypoxic conditions [145]. Hepatocellular carcinoma cells cultured in hypoxia also exhibited increased resistance to cisplatin, epirubicin, gemcitabine and mitomycin via hypoxia-mediated autophagic processes [148]. Furthermore, bladder cancer cells exposed to gemcitabine exhibit increased autophagy, which is augmented by hypoxia in a manner dependent on the HIF-1α/BNIP3/Beclin-1 signaling pathway [149]. The anti-cancer effects of other drugs, including paclitaxel, were also shown to be decreased under hypoxic conditions in a manner dependent on HIF-1 and autophagy [150, 151]. A study by Notte et al. showed that taxol induces the UPR in hypoxic breast cancer cells, and that upregulation of ATF4 leads to hypoxia-induced autophagy, as well as increased resistance to taxol [152].

While hypoxia-induced autophagy mediates resistance to conventional chemotherapeutic agents, there is also evidence that hypoxia-induced autophagy mediates resistance to anti-angiogenic agents in a HIF-1α/AMPK dependent manner [153], as well as resistance to ionizing radiation via the HIF-1α/miR-210/Bcl-2 pathway [119, 154].

In addition to the tumour-cell intrinsic mechanisms of hypoxia-mediated drug resistance described above, there are also tumour-cell extrinsic factors contributing to therapy resistance. The recent successes of immune therapy combined with the recognition that tumour-associated inflammation is a potentiator of malignant progression have led to recent exploration of links between hypoxia, the TiME and drug resistance. For example, hypoxic niches within the TME harbour CSCs [155], and thus contribute to cellular heterogeneity and drug resistance. In addition, hypoxia-induced release of macrophage chemoattractants results in recruitment of TAMs to the TME. These myeloid cells subsequently release factors that promote tumour cell survival and amplify resistance to therapy [156]. In addition, chemotherapy- and radiotherapy-induced immunogenic cell death, which is characterized by antigen-specific immune responses against dead-cell antigens, is hindered by the presence of TAMs [157]. Cytotoxic $CD8^+$ T cell responses against tumour antigens are dampened by checkpoint molecules expressed on immune cells and tumour cells within the TME [41]. Our group discovered that hypoxia is an important driver of PD-L1 expression in various human and mouse tumour cells. We found that HIF-1α-induced expression of PD-L1 results in resistance to CTL-mediated target cell lysis thus enabling immune escape [158]. Work by Noman and colleagues revealed that HIF-1α binds directly to the HRE in the proximal promoter of the PD-L1 gene and results in increased PD-L1 expression on various immune cells including macrophages, MDSCs and dendritic cells [159]. In addition, our more recent work describes a novel mechanism by which reverse signalling of PD-1/PD-L1 confers chemoresistance to tumour cells [60]. Thus, hypoxia within the TME drives an immunosuppressive phenotype, limits cytotoxicity, and promotes chemoresistance, which altogether potentiate malignancy and promote metastasis.

9.4 Therapeutic Opportunities

9.4.1 Nitric Oxide Mimetic Agents

There is evidence that cellular adaptations to hypoxia, such as the acquisition of malignant properties by tumour cells, are in part a consequence of a hypoxia-mediated inhibition of the nitric oxide/cyclic guanosine monophosphate (cGMP) signalling pathway [96, 97, 160–165]. The generation of nitric oxide results from the conversion of L-arginine into L-citrulline [166–168] (Fig. 9.1). This reaction is catalyzed by the enzyme nitric oxide synthase (NOS), of which there are three known isoforms: NOS-1, -2, and -3. Production of nitric oxide depends on the availability of several co-factors and co-substrates, including nicotinamide adenine dinucleotide phosphate (NADPH), flavin adenine dinucleotide (FAD), tetrahydrobiopterin, and oxygen [166, 169–171]. Consequently, the process of endogenous nitric oxide production is complex. Moreover, in the absence of oxygen, as it is the case in solid tumours, endogenous production of nitric oxide is limited [170, 171]. Under well-oxygenated conditions, nitric oxide generated by NOS binds to and activates soluble guanylyl cyclase (sGC), which in turn catalyzes the conversion of guanosine triphosphate (GTP) into cGMP. The latter is a known regulator of ion channel conductance, glycogenolysis, and apoptosis. It also causes smooth muscle relaxation and vasodilation. An important function of cGMP is the activation of protein kinase G (PKG), a serine/threonine-specific kinase that phosphorylates a number of biologically important targets.

Limited availability of cGMP under hypoxia leads to decreased activation of PKG and reduced protein phosphorylation, an important aspect of cellular adaptations to hypoxia. Our research has revealed that pharmacological inhibition of NOS, soluble guanylyl cyclase, or PKG in well oxygenated tumour cells results in the acquisition of phenotypes similar to those induced by hypoxia, such as increased invasive and metastatic ability [96, 97], as well as drug resistance [162, 172].

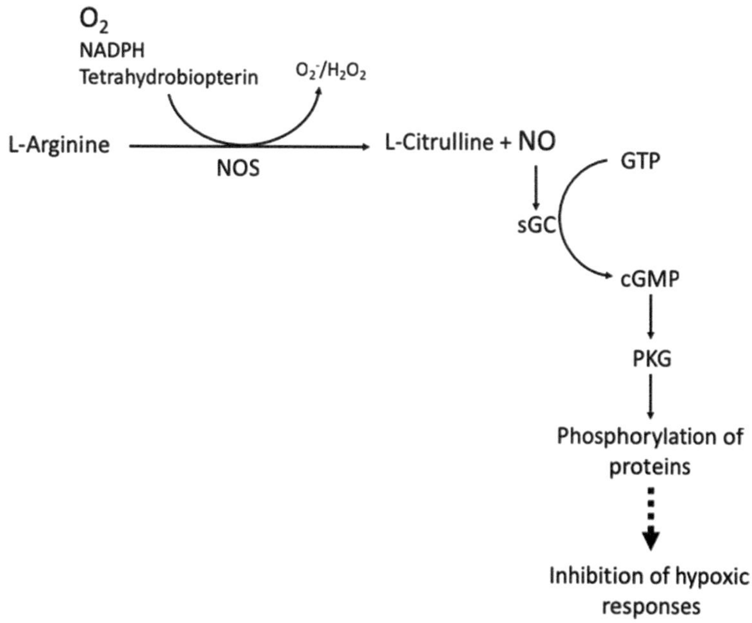

Fig. 9.1 Generation of nitric oxide (NO) and the NO/cGMP signalling pathway. NO is produced during the conversion of L-arginine into L-citrulline in an oxygen-dependent reaction catalyzed by nitric oxide synthase (NOS). Nitric oxide activates soluble guanylyl cyclase (sGC), which in turn leads to the activation of protein kinase G (PKG) and the phosphorylation of target proteins. Inhibition of this pathway due to hypoxia leads to adaptive responses

Furthermore, pharmacological activation of soluble guanylyl cyclase with various nitric oxide mimetic agents, such as glyceryl trinitrate (GTN; nitroglycerin), diethylenetriamine nitric oxide adduct (DETA-NO) and sodium nitroprusside, blocks the acquisition of malignant properties in tumour cells exposed to hypoxia [96, 97, 158, 162, 172]. A similar inhibition of hypoxia-induced acquisition of malignant phenotypes is achieved by direct activation of PKG using the non-hydrolysable cGMP analogue, 8-bromo-cGMP [96, 97, 172].

There is evidence that nitric oxide signalling interferes with tumour cell adaptations to hypoxia by inhibiting HIF-1α accumulation [173–175]. While high concentrations (>1 μM) of nitric oxide are capable of stabilizing HIF-1α during normoxic conditions, low concentrations of nitric oxide (<400 nM) have been reported to facilitate HIF-1α degradation thereby impairing HIF-1 signalling [176]. There is also evidence that under mildly hypoxic conditions inhibition of mitochondrial respiration by nitric oxide leads to a redistribution in intracellular oxygen and activation of the PHD enzymes responsible for HIF-1α degradation [174].

Thus, it appears that tumour cell adaptations to hypoxia are tightly regulated by nitric oxide and HIF-1 activity. These observations have led to the design of studies aimed at determining whether nitric oxide mimetic agents can delay disease progression or chemosensitize tumours in the clinical setting. We completed a phase II trial on prostate cancer patients with biochemical recurrence showing that continuous transdermal delivery of low doses (0.03 mg/h) of GTN may be effective at delaying disease progression [177]. This finding revealed that activation of nitric oxide signalling may have cancer inhibitory properties independent of potential chemosensitizing effects. Yasuda *et al.* reported improved response rates to vinorelbine plus cisplatin therapy in lung cancer patients treated with GTN for angina pectoris compared with patients without angina who did not use GTN [178]. This observation prompted subsequent studies to determine therapeutic benefits associated with clinical use of nitric oxide mimetics as adjuvants to chemotherapy. A Phase II trial involving patients with previously untreated stage IIIB/IV non-small-cell lung cancer revealed that, compared with patients treated with a placebo transdermal patch, transdermal delivery of GTN combined with vinorelbine and cisplatin was associated with significantly increased response rate and median time to progression [178]. A follow-up study revealed a lower incidence of cells immunoreactive for HIF-1α, P-gp, and vascular endothelial growth factor (VEGF), in lung adenocarcinomas from GTN treated patients relative to tumours from non-treated patients [179].

9.4.2 Checkpoint Inhibitors

Interfering with the PD-1/PD-L1 signalling axis using monoclonal antibodies has shown promising and unprecedented results for many types of cancers [180]. At the time of writing, there were two US Food and Drug Administration-approved anti-PD-1 therapies (Nivolumab and Pembrolizumab) and three anti-PD-L1 therapies (Atezolizumab, Durvalumab and Avelumab) for treatment of patients with melanoma, non-small cell lung cancer, metastatic urothelial bladder cancer, renal cell carcinoma, Hodgkin's lymphoma, advanced gastroesophageal cancer, metastatic colorectal cancer, hepatocellular carcinoma and Merkle cell carcinoma [181]. It is important to note, however, that most of these checkpoint inhibitors have not yet been approved for use as first-line therapy and, as such, patients will have received, or will concurrently be receiving, standard-of-care chemotherapy and radiotherapy. In addition, despite the successes of immune therapy, only a fraction of patients has shown durable responses. Therefore, targeting additional mechanisms of drug resistance may be important for achieving higher response rates in individuals receiving checkpoint blockade therapy. One such approach could involve simultaneous inhibition of HIF-1α in combination with PD-L1/PD-1 blockade. The findings from our own work and those of others discussed above call upon additional studies to elucidate the mechanism(s) behind the hypoxia-driven PD-L1 expression and its significance in cancer development. It is important to note that PD-L1 expression is known to be driven by several oncogenic pathways [50] of which hypoxia is an important regulator.

9.5 Conclusion

Here we have outlined tumour-cell intrinsic and extrinsic (microenvironmental) mechanisms by which hypoxia contributes to malignancy and drug resistance. It is clear that tumour hypoxia is an impediment to the successful management of cancer. An important challenge in developing successful therapeutic options to mitigate hypoxia-induced acquisition of malignant phenotypes is to identify therapies that selectively target hypoxic tumour cells and/or other cells in the tumour microenvironment that contribute to the acquisition of malignant phenotypes. Furthermore, identifying patients likely to respond to treatment and mechanisms of hypoxia-induced drug resistance is critical. Important also is the need to identify and to better understand what role, if any, hypoxia might have in patients that fail to respond to therapy. Thus, it is evident that more basic research is required to determine mechanisms by which hypoxia is associated with development of resistance to therapy. While relying on basic, pre-clinical research to inform and guide drug development and clinical trials is important, it is also worth noting that adopting a bedside-to-bench approach is an invaluable translational opportunity and will be beneficial in the design of strategies to overcome drug resistance.

References

1. Campos L et al (1993) High expression of bcl-2 protein in acute myeloid leukemia cells is associated with poor response to chemotherapy. Blood 81:3091–3096
2. Sugawara I, Akiyama S, Scheper RJ, Itoyama S (1997) Lung resistance protein (LRP) expression in human normal tissues in comparison with that of MDR1 and MRP. Cancer Lett 112:23–31. https://doi.org/10.1016/S0304-3835(96)04542-9
3. Britten RA, Green JA, Warenius HM (1992) Cellular glutathione (GSH) and glutathione S-transferase (GST) activity in human ovarian tumor biopsies fol-

lowing exposure to alkylating agents. Int J Radiat Oncol Biol Phys 24:527–531
4. Liang BC (1996) Effects of hypoxia on drug resistance phenotype and genotype in human glioma cell lines. J Neuro-Oncol 29:149–155
5. Bonetti A et al (1998) bcl-2 but not p53 expression is associated with resistance to chemotherapy in advanced breast cancer. Clin Cancer Res 4:2331–2336
6. St Croix B, Man S, Kerbel RS (1998) Reversal of intrinsic and acquired forms of drug resistance by hyaluronidase treatment of solid tumors. Cancer Lett 131:35–44
7. Graham CH et al (1994) Rapid acquisition of multicellular drug resistance after a single exposure of mammary tumor cells to antitumor alkylating agents. J Natl Cancer Inst 86:975–982
8. Kobayashi H et al (1993) Acquired multicellular-mediated resistance to alkylating agents in cancer. Proc Natl Acad Sci U S A 90:3294–3298
9. Croix BS et al (1996) Reversal by hyaluronidase of adhesion-dependent multicellular drug resistance in mammary carcinoma cells. J Natl Cancer Inst 88:1285–1296
10. Kurtova AV et al (2015) Blocking PGE2-induced tumour repopulation abrogates bladder cancer chemoresistance. Nature 517:209–213. https://doi.org/10.1038/nature14034
11. Prieto-Vila M, Takahashi RU, Usuba W, Kohama I, Ochiya T (2017) Drug resistance driven by Cancer stem cells and their niche. Int J Mol Sci 18:ARTN 2574. https://doi.org/10.3390/ijms18122574
12. Eyler CE, Rich JN (2008) Survival of the fittest: cancer stem cells in therapeutic resistance and angiogenesis. J Clin Oncol 26:2839–2845. https://doi.org/10.1200/JCO.2007.15.1829
13. Kreso A, Dick JE (2014) Evolution of the cancer stem cell model. Cell Stem Cell 14:275–291. https://doi.org/10.1016/j.stem.2014.02.006
14. Abdullah LN, Chow EK (2013) Mechanisms of chemoresistance in cancer stem cells. Clin Transl Med 2:3. https://doi.org/10.1186/2001-1326-2-3
15. Sharif T et al (2017) Autophagic homeostasis is required for the pluripotency of cancer stem cells. Autophagy 13:264–284. https://doi.org/10.1080/15548627.2016.1260808
16. Pagotto A et al (2017) Autophagy inhibition reduces chemoresistance and tumorigenic potential of human ovarian cancer stem cells. Cell Death Dis 8:e2943. https://doi.org/10.1038/cddis.2017.327
17. Balkwill FR, Capasso M, Hagemann T (2012) The tumor microenvironment at a glance. J Cell Sci 125:5591–5596. https://doi.org/10.1242/jcs.116392
18. Petrova V, Annicchiarico-Petruzzelli M, Melino G, Amelio I (2018) The hypoxic tumour microenvironment. Oncogene 7:10. https://doi.org/10.1038/s41389-017-0011-9
19. Weaver VM et al (2002) beta4 integrin-dependent formation of polarized three-dimensional architecture confers resistance to apoptosis in normal and malignant mammary epithelium. Cancer Cell 2:205–216
20. Schrader J et al (2011) Matrix stiffness modulates proliferation, chemotherapeutic response, and dormancy in hepatocellular carcinoma cells. Hepatology 53:1192–1205. https://doi.org/10.1002/hep.24108
21. Pickup MW, Mouw JK, Weaver VM (2014) The extracellular matrix modulates the hallmarks of cancer. EMBO Rep 15:1243–1253. https://doi.org/10.15252/embr.201439246
22. Paszek MJ et al (2005) Tensional homeostasis and the malignant phenotype. Cancer Cell 8:241–254. https://doi.org/10.1016/j.ccr.2005.08.010
23. Hayashi M et al (2012) Evaluation of tumor stiffness by elastography is predictive for pathologic complete response to neoadjuvant chemotherapy in patients with breast cancer. Ann Surg Oncol 19:3042–3049. https://doi.org/10.1245/s10434-012-2343-1
24. Rice AJ et al (2017) Matrix stiffness induces epithelial-mesenchymal transition and promotes chemoresistance in pancreatic cancer cells. Oncogene 6:e352. https://doi.org/10.1038/oncsis.2017.54
25. Jiang H et al (2016) Targeting focal adhesion kinase renders pancreatic cancers responsive to checkpoint immunotherapy. Nat Med 22:851–860. https://doi.org/10.1038/nm.4123
26. Munson JM, Shieh AC (2014) Interstitial fluid flow in cancer: implications for disease progression and treatment. Cancer Manag Res 6:317–328. https://doi.org/10.2147/Cmar.S65444
27. Ogiso Y, Tomida A, Lei S, Omura S, Tsuruo T (2000) Proteasome inhibition circumvents solid tumor resistance to topoisomerase II-directed drugs. Cancer Res 60:2429–2434
28. Shen J et al (1987) Coinduction of glucose-regulated proteins and doxorubicin resistance in Chinese hamster cells. Proc Natl Acad Sci U S A 84:3278–3282
29. Hughes CS, Shen JW, Subjeck JR (1989) Resistance to etoposide induced by three glucose-regulated stresses in Chinese hamster ovary cells. Cancer Res 49:4452–4454
30. Vaupel PW (1997) The influence of tumor blood flow and microenvironmental factors on the efficacy of radiation, drugs and localized hyperthermia. Klin Padiatr 209:243–249. https://doi.org/10.1055/s-2008-1043957
31. Vaupel PW, Frinak S, Bicher HI (1981) Heterogeneous oxygen partial pressure and pH distribution in C3H mouse mammary adenocarcinoma. Cancer Res 41:2008–2013
32. Olive KP et al (2009) Inhibition of Hedgehog signaling enhances delivery of chemotherapy in a mouse model of pancreatic cancer. Science 324:1457–1461. https://doi.org/10.1126/science.1171362
33. Whatcott CJ, Han H, Posner RG, Hostetter G, Von Hoff DD (2011) Targeting the tumor microenvironment in cancer: why hyaluronidase deserves a second look. Cancer Discov 1:291–296. https://doi.org/10.1158/2159-8290.CD-11-0136

34. Zitvogel L et al (2015) Cancer and the gut microbiota: an unexpected link. Sci Transl Med 7:271ps271. https://doi.org/10.1126/scitranslmed.3010473
35. Panebianco C, Andriulli A, Pazienza V (2018) Pharmacomicrobiomics: exploiting the drug-microbiota interactions in anticancer therapies. Microbiome 6:92. https://doi.org/10.1186/s40168-018-0483-7
36. Zitvogel L, Ma Y, Raoult D, Kroemer G, Gajewski TF (2018) The microbiome in cancer immunotherapy: diagnostic tools and therapeutic strategies. Science 359:1366–1370. https://doi.org/10.1126/science.aar6918
37. Geller LT, Straussman R (2018) Intratumoral bacteria may elicit chemoresistance by metabolizing anticancer agents. Mol Cell Oncol 5:e1405139. https://doi.org/10.1080/23723556.2017.1405139
38. Dvorak HF (1986) Tumors: wounds that do not heal. Similarities between tumor stroma generation and wound healing. N Engl J Med 315:1650–1659. https://doi.org/10.1056/NEJM198612253152606
39. Hanahan D, Weinberg RA (2011) Hallmarks of cancer: the next generation. Cell 144:646–674. doi:S0092-8674(11)00127-9 [pii] https://doi.org/10.1016/j.cell.2011.02.013
40. Binnewies M et al (2018) Understanding the tumor immune microenvironment (TIME) for effective therapy. Nat Med 24:541–550. https://doi.org/10.1038/s41591-018-0014-x
41. Medler TR, Cotechini T, Coussens LM (2015) Immune response to cancer therapy: mounting an effective antitumor response and mechanisms of resistance. Trends Cancer 1:66–75. https://doi.org/10.1016/j.trecan.2015.07.008
42. Zheng Y et al (2009) Macrophages are an abundant component of myeloma microenvironment and protect myeloma cells from chemotherapy drug-induced apoptosis. Blood 114:3625–3628. https://doi.org/10.1182/blood-2009-05-220285
43. Cotechini T, Medler TR, Coussens LM (2015) Myeloid cells as targets for therapy in solid tumors. Cancer J 21:343–350. https://doi.org/10.1097/PPO.0000000000000132
44. Affara NI et al (2014) B cells regulate macrophage phenotype and response to chemotherapy in squamous carcinomas. Cancer Cell 25:809–821. https://doi.org/10.1016/j.ccr.2014.04.026
45. DeNardo DG et al (2009) CD4(+) T cells regulate pulmonary metastasis of mammary carcinomas by enhancing protumor properties of macrophages. Cancer Cell 16:91–102. https://doi.org/10.1016/j.ccr.2009.06.018
46. Shiao SL et al (2015) TH2-polarized CD4(+) T cells and macrophages limit efficacy of radiotherapy. Cancer Immunol Res 3:518–525. https://doi.org/10.1158/2326-6066.CIR-14-0232
47. Parry RV et al (2005) CTLA-4 and PD-1 receptors inhibit T-cell activation by distinct mechanisms. Mol Cell Biol 25:9543–9553. doi:25/21/9543 [pii] https://doi.org/10.1128/MCB.25.21.9543-9553.2005
48. Dong H et al (2002) Tumor-associated B7-H1 promotes T-cell apoptosis: a potential mechanism of immune evasion. Nat Med 8:793–800. doi:nm730 [pii] https://doi.org/10.1038/nm730
49. Schwartz RH (2003) T cell anergy. Annu Rev Immunol 21:305–334. https://doi.org/10.1146/annurev.immunol.21.120601.141110
50. Chen J, Jiang CC, Jin L, Zhang XD (2015) Regulation of PD-L1: a novel role of pro-survival signalling in cancer. Ann Oncol 27:409–416. https://doi.org/10.1093/annonc/mdv615
51. Muenst S et al (2014) Expression of programmed death ligand 1 (PD-L1) is associated with poor prognosis in human breast cancer. Breast Cancer Res Treat 146:15–24. https://doi.org/10.1007/s10549-014-2988-5
52. Hamanishi J et al(2007) Programmed cell death 1 ligand 1 and tumor-infiltrating CD8+ T lymphocytes are prognostic factors of human ovarian cancer. Proc Natl Acad Sci U S A 104:3360–3365. doi:0611533104 [pii] https://doi.org/10.1073/pnas.0611533104
53. Inman BA et al (2007) PD-L1 (B7-H1) expression by urothelial carcinoma of the bladder and BCG-induced granulomata: associations with localized stage progression. Cancer 109:1499–1505. https://doi.org/10.1002/cncr.22588
54. Konishi J et al(2004) B7-H1 expression on non-small cell lung cancer cells and its relationship with tumor-infiltrating lymphocytes and their PD-1 expression. Clin Cancer Res 10:5094–5100. doi:10/15/5094 [pii] https://doi.org/10.1158/1078-0432.CCR-04-0428
55. Shi SJ et al (2013) B7-H1 expression is associated with poor prognosis in colorectal carcinoma and regulates the proliferation and invasion of HCT116 colorectal cancer cells. PLoS One 8:e76012. https://doi.org/10.1371/journal.pone.0076012
56. Nduom EK et al (2016) PD-L1 expression and prognostic impact in glioblastoma. Neuro-Oncology 18:195–205. https://doi.org/10.1093/neuonc/nov172
57. Madore J et al (2015) PD-L1 expression in melanoma shows marked heterogeneity within and between patients: implications for anti-PD-1/PD-L1 clinical trials. Pigment Cell Melanoma Res 28:245–253. https://doi.org/10.1111/pcmr.12340
58. Chen X et al (2008) Clinical significance of B7-H1 (PD-L1) expression in human acute leukemia. Cancer Biol Ther 7:622–627
59. Chen BJ et al (2013) PD-L1 expression is characteristic of a subset of aggressive B-cell lymphomas and virus-associated malignancies. Clin Cancer Res 19:3462–3473. https://doi.org/10.1158/1078-0432.CCR-13-0855
60. Black M et al (2016) Activation of the PD-1/PD-L1 immune checkpoint confers tumor cell chemoresistance associated with increased metastasis. Oncotarget 7:10557–10567. https://doi.org/10.18632/oncotarget.7235
61. Azuma T et al (2008) B7-H1 is a ubiquitous anti-apoptotic receptor on cancer cells. Blood 111:3635–

3643 doi:blood-2007-11-123141 [pii] https://doi.org/10.1182/blood-2007-11-123141
62. Harris AL (2002) Hypoxia--a key regulatory factor in tumour growth. Nat Rev Cancer 2:38–47
63. Semenza GL (2010) Defining the role of hypoxia-inducible factor 1 in cancer biology and therapeutics. Oncogene 29:625–634. https://doi.org/10.1038/onc.2009.441
64. Maxwell PH et al (1999) The tumour suppressor protein VHL targets hypoxia-inducible factors for oxygen-dependent proteolysis. Nature 399:271–275
65. Jaakkola P et al (2001) Targeting of HIF-alpha to the von Hippel-Lindau ubiquitylation complex by O2-regulated prolyl hydroxylation. Science 292:468–472. https://doi.org/10.1126/science.1059796
66. Lando D, Peet DJ, Whelan DA, Gorman JJ, Whitelaw ML (2002) Asparagine hydroxylation of the HIF transactivation domain a hypoxic switch. Science 295:858–861. https://doi.org/10.1126/science.1068592
67. Mahon PC, Hirota K, Semenza GL (2001) FIH-1: a novel protein that interacts with HIF-1alpha and VHL to mediate repression of HIF-1 transcriptional activity. Genes Dev 15:2675–2686. https://doi.org/10.1101/gad.924501
68. Semenza GL (2002) Involvement of hypoxia-inducible factor 1 in human cancer. Intern Med 41:79–83
69. Wood SM et al (1998) Selection and analysis of a mutant cell line defective in the hypoxia-inducible factor-1 alpha-subunit (HIF-1alpha). Characterization of hif-1alpha-dependent and -independent hypoxia-inducible gene expression. J Biol Chem 273:8360–8368
70. Forsythe JA et al (1996) Activation of vascular endothelial growth factor gene transcription by hypoxia-inducible factor 1. Mol Cell Biol 16:4604–4613
71. Krock BL, Skuli N, Simon MC (2011) Hypoxia-induced angiogenesis: good and evil. Genes Cancer 2:1117–1133. https://doi.org/10.1177/1947601911423654
72. Nagy JA, Chang SH, Dvorak AM, Dvorak HF (2009) Why are tumour blood vessels abnormal and why is it important to know? Br J Cancer 100:865–869. https://doi.org/10.1038/sj.bjc.6604929
73. Zampell JC et al (2012) HIF-1alpha coordinates lymphangiogenesis during wound healing and in response to inflammation. FASEB J 26:1027–1039. https://doi.org/10.1096/fj.11-195321
74. Bordry N et al (2018) Lymphatic vessel density is associated with CD8(+) T cell infiltration and immunosuppressive factors in human melanoma. Oncoimmunology 7:e1462878. https://doi.org/10.1080/2162402X.2018.1462878
75. Ruan K, Song G, Ouyang G (2009) Role of hypoxia in the hallmarks of human cancer. J Cell Biochem 107:1053–1062. https://doi.org/10.1002/jcb.22214
76. Gray LH, Conger AD, Ebert M, Hornsey S, Scott OC (1953) The concentration of oxygen dissolved in tissues at the time of irradiation as a factor in radiotherapy. Br J Radiol 26:638–648. https://doi.org/10.1259/0007-1285-26-312-638
77. Thomlinson RH, Gray LH (1955) The histological structure of some human lung cancers and the possible implications for radiotherapy. Br J Cancer 9:539–549
78. Fletcher GH (1984) Radiation and drug resistance of breast cancer. Am J Clin Oncol 7:617–624
79. Overgaard J (2007) Hypoxic radiosensitization: adored and ignored. J Clin Oncol 25:4066–4074. https://doi.org/10.1200/JCO.2007.12.7878
80. Bristow RG, Hill RP (2008) Hypoxia and metabolism. Hypoxia, DNA repair and genetic instability. Nat Rev Cancer 8:180–192. https://doi.org/10.1038/nrc2344
81. Horsman MR, Overgaard J (2016) The impact of hypoxia and its modification of the outcome of radiotherapy. J Radiat Res 57(Suppl 1):i90–i98. https://doi.org/10.1093/jrr/rrw007
82. Teicher BA (1994) Hypoxia and drug resistance. Cancer Metastasis Rev 13:139–168
83. Hockel M et al (1996) Association between tumor hypoxia and malignant progression in advanced cancer of the uterine cervix. Cancer Res 56:4509–4515
84. Brizel DM et al (1996) Tumor oxygenation predicts for the likelihood of distant metastases in human soft tissue sarcoma. Cancer Res 56:941–943
85. Brizel DM, Sibley GS, Prosnitz LR, Scher RL, Dewhirst MW (1997) Tumor hypoxia adversely affects the prognosis of carcinoma of the head and neck. Int J Radiat Oncol Biol Phys 38:285–289
86. Lartigau E et al (1997) Intratumoral oxygen tension in metastatic melanoma. Melanoma Res 7:400–406
87. Movsas B et al (2000) Increasing levels of hypoxia in prostate carcinoma correlate significantly with increasing clinical stage and patient age: an Eppendorf pO(2) study. Cancer 89:2018–2024
88. Movsas B et al (1999) Hypoxic regions exist in human prostate carcinoma. Urology 53:11–18
89. Milosevic M et al (2012) Tumor hypoxia predicts biochemical failure following radiotherapy for clinically localized prostate cancer. Clin Cancer Res 18:2108–2114. https://doi.org/10.1158/1078-0432.CCR-11-2711
90. Cairns RA, Kalliomaki T, Hill RP (2001) Acute (cyclic) hypoxia enhances spontaneous metastasis of KHT murine tumors. Cancer Res 61:8903–8908
91. Cairns RA, Hill RP (2004) Acute hypoxia enhances spontaneous lymph node metastasis in an orthotopic murine model of human cervical carcinoma. Cancer Res 64:2054–2061
92. Young SD, Marshall RS, Hill RP (1988) Hypoxia induces DNA overreplication and enhances metastatic potential of murine tumor cells. Proc Natl Acad Sci U S A 85:9533–9537
93. Zhang L, Hill RP (2004) Hypoxia enhances metastatic efficiency by up-regulating Mdm2 in KHT cells and increasing resistance to apoptosis. Cancer Res 64:4180–4189

94. Cuvier C, Jang A, Hill RP (1997) Exposure to hypoxia, glucose starvation and acidosis: effect on invasive capacity of murine tumor cells and correlation with cathepsin (L + B) secretion. Clin Exp Metastasis 15:19–25
95. Graham CH, Forsdike J, Fitzgerald CF, Macdonald-Goodfellow S (1999) Hypoxia-mediated stimulation of carcinoma cell invasiveness via upregulation of urokinase receptor expression. Int J Cancer 80:617–623
96. Postovit LM, Adams MA, Lash GE, Heaton JP, Graham CH (2002) Oxygen-mediated regulation of tumour cell invasiveness: involvement of a nitric oxide signalling pathway. J Biol Chem 277:35730–35737
97. Postovit LM, Adams MA, Lash GE, Heaton JP, Graham CH (2004) Nitric oxide-mediated regulation of hypoxia-induced B16F10 melanoma metastasis. Int J Cancer 108:47–53. https://doi.org/10.1002/ijc.11556
98. Graham CH, Fitzpatrick TE, McCrae KR (1998) Hypoxia stimulates urokinase receptor expression through a heme protein-dependent pathway. Blood 91:3300–3307
99. Graham CH, Postovit LM, Lash GE, Adams MA (2001) Ultra low levels of nitric oxide abrogate hypoxia-induced metastatic phenotypes: evidence that reduced nitric oxide mediates a key pathway in oxygen-regulated gene expression. Proc Am Assoc Cancer Res 42:794
100. Fitzpatrick TE, Graham CH (1998) Stimulation of plasminogen activator inhibitor-1 expression in immortalized human trophoblast cells cultured under low levels of oxygen. Exp Cell Res 245:155–162
101. Canning MT, Postovit LM, Clarke SH, Graham CH (2001) Oxygen-mediated regulation of gelatinase and tissue inhibitor of metalloproteinases-1 expression by invasive cells. Exp Cell Res 267:88–94
102. Rofstad EK et al (2002) Hypoxia promotes lymph node metastasis in human melanoma xenografts by up-regulating the urokinase-type plasminogen activator receptor. Cancer Res 62:1847–1853
103. Liu Q et al (2017) Factors involved in cancer metastasis: a better understanding to "seed and soil" hypothesis. Mol Cancer 16:176. https://doi.org/10.1186/s12943-017-0742-4
104. Paget S (1989) The distribution of secondary growths in cancer of the breast. 1889. Cancer Metastasis Rev 8:98–101
105. Rankin EB, Nam JM, Giaccia AJ (2016) Hypoxia: signaling the metastatic Cascade. Trends Cancer 2:295–304. https://doi.org/10.1016/j.trecan.2016.05.006
106. Sceneay J et al (2012) Primary tumor hypoxia recruits CD11b+/Ly6Cmed/Ly6G+ immune suppressor cells and compromises NK cell cytotoxicity in the premetastatic niche. Cancer Res 72:3906–3911. https://doi.org/10.1158/0008-5472.CAN-11-3873
107. Sceneay J, Parker BS, Smyth MJ, Moller A (2013) Hypoxia-driven immunosuppression contributes to the pre-metastatic niche. Oncoimmunology 2:e22355. https://doi.org/10.4161/onci.22355
108. Glick D, Barth S, Macleod KF (2010) Autophagy: cellular and molecular mechanisms. J Pathol 221:3–12. https://doi.org/10.1002/path.2697
109. Mazure NM, Pouyssegur J (2010) Hypoxia-induced autophagy: cell death or cell survival? Curr Opin Cell Biol 22:177–180. https://doi.org/10.1016/j.ceb.2009.11.015
110. Rouschop KM et al (2010) The unfolded protein response protects human tumor cells during hypoxia through regulation of the autophagy genes MAP1LC3B and ATG5. J Clin Invest 120:127–141. https://doi.org/10.1172/JCI40027
111. Rzymski T et al (2010) Regulation of autophagy by ATF4 in response to severe hypoxia. Oncogene 29:4424–4435. https://doi.org/10.1038/onc.2010.191
112. Chhipa RR, Wu Y, Ip C (2011) AMPK-mediated autophagy is a survival mechanism in androgen-dependent prostate cancer cells subjected to androgen deprivation and hypoxia. Cell Signal 23:1466–1472. https://doi.org/10.1016/j.cellsig.2011.04.008
113. Papandreou I, Lim AL, Laderoute K, Denko NC (2008) Hypoxia signals autophagy in tumor cells via AMPK activity, independent of HIF-1, BNIP3, and BNIP3L. Cell Death Differ 15:1572–1581. https://doi.org/10.1038/cdd.2008.84
114. Ma Y et al (2014) Biphasic regulation of autophagy by miR-96 in prostate cancer cells under hypoxia. Oncotarget 5:9169–9182. https://doi.org/10.18632/oncotarget.2396
115. Bellot G et al (2009) Hypoxia-induced autophagy is mediated through hypoxia-inducible factor induction of BNIP3 and BNIP3L via their BH3 domains. Mol Cell Biol 29:2570–2581. https://doi.org/10.1128/MCB.00166-09
116. Zhang H et al (2008) Mitochondrial autophagy is an HIF-1-dependent adaptive metabolic response to hypoxia. J Biol Chem 283:10892–10903
117. Mazure NM, Pouyssegur J (2009) Atypical BH3-domains of BNIP3 and BNIP3L lead to autophagy in hypoxia. Autophagy 5:868–869
118. Tracy K et al (2007) BNIP3 is an RB/E2F target gene required for hypoxia-induced autophagy. Mol Cell Biol 27:6229–6242. https://doi.org/10.1128/MCB.02246-06
119. Sun Y et al (2015) Hypoxia-induced autophagy reduces radiosensitivity by the HIF-1alpha/miR-210/Bcl-2 pathway in colon cancer cells. Int J Oncol 46:750–756. https://doi.org/10.3892/ijo.2014.2745
120. Wang P et al (2017) Hypoxia inducible factor-1alpha regulates autophagy via the p27-E2F1 signaling pathway. Mol Med Rep 16:2107–2112. https://doi.org/10.3892/mmr.2017.6794
121. Azad MB et al (2008) Hypoxia induces autophagic cell death in apoptosis-competent cells through a mechanism involving BNIP3. Autophagy 4:195–204
122. Kalra R, Jones AM, Kirk J, Adams GE, Stratford IJ (1993) The effect of hypoxia on acquired drug resis-

tance and response to epidermal growth factor in Chinese hamster lung fibroblasts and human breast-cancer cells in vitro. Int J Cancer 54:650–655
123. Rice GC, Ling V, Schimke RT (1987) Frequencies of independent and simultaneous selection of Chinese hamster cells for methotrexate and doxorubicin (adriamycin) resistance. Proc Natl Acad Sci U S A 84:9261–9264
124. Smith E, Stratford IJ, Adams GE (1980) Cytotoxicity of adriamycin on aerobic and hypoxic chinese hamster V79 cells in vitro. Br J Cancer 42:568–573
125. Wilson RE, Keng PC, Sutherland RM (1989) Drug resistance in Chinese hamster ovary cells during recovery from severe hypoxia. J Natl Cancer Inst 81:1235–1240
126. Luk CK, Veinot-Drebot L, Tjan E, Tannock IF (1990) Effect of transient hypoxia on sensitivity to doxorubicin in human and murine cell lines. J Natl Cancer Inst 82:684–692
127. Sanna K, Rofstad EK (1994) Hypoxia-induced resistance to doxorubicin and methotrexate in human melanoma cell lines in vitro. Int J Cancer 58:258–262
128. Comerford KM et al (2002) Hypoxia-inducible factor-1-dependent regulation of the multidrug resistance (MDR1) gene. Cancer Res 62:3387–3394
129. Krishnamurthy P et al (2004) The stem cell marker Bcrp/ABCG2 enhances hypoxic cell survival through interactions with heme. J Biol Chem 279:24218–24225. https://doi.org/10.1074/jbc.M313599200
130. Sullivan R, Graham CH (2008) Chemosensitization of cancer by nitric oxide. Curr Pharm Des 14:1113–1123
131. Ricci MS, Zong WX (2006) Chemotherapeutic approaches for targeting cell death pathways. Oncologist 11:342–357. https://doi.org/10.1634/theoncologist.11-4-342
132. Chang BD et al (1999) A senescence-like phenotype distinguishes tumor cells that undergo terminal proliferation arrest after exposure to anticancer agents. Cancer Res 59:3761–3767
133. Chang BD et al (2002) Molecular determinants of terminal growth arrest induced in tumor cells by a chemotherapeutic agent. Proc Natl Acad Sci U S A 99:389–394. https://doi.org/10.1073/pnas.012602599
134. Chang BD et al (1999) Role of p53 and p21waf1/cip1 in senescence-like terminal proliferation arrest induced in human tumor cells by chemotherapeutic drugs. Oncogene 18:4808–4818. https://doi.org/10.1038/sj.onc.1203078
135. Schwarze SR, Fu VX, Desotelle JA, Kenowski ML, Jarrard DF (2005) The identification of senescence-specific genes during the induction of senescence in prostate cancer cells. Neoplasia 7:816–823
136. te Poele RH, Okorokov AL, Jardine L, Cummings J, Joel SP (2002) DNA damage is able to induce senescence in tumor cells in vitro and in vivo. Cancer Res 62:1876–1883
137. Elmore LW et al (2002) Adriamycin-induced senescence in breast tumor cells involves functional p53 and telomere dysfunction. J Biol Chem 277:35509–35515. https://doi.org/10.1074/jbc.M205477200
138. Roninson IB, Broude EV, Chang BD (2001) If not apoptosis, then what? Treatment-induced senescence and mitotic catastrophe in tumor cells. Drug Resist Updat 4:303–313. https://doi.org/10.1054/drup.2001.0213
139. Sullivan R, Paré GC, Frederiksen, L. J., Semenza, G. L. & Graham, C. H. (2008) Hypoxia-induced resistance to anticancer drugs is associated with decreased senescence and requires hypoxia-inducible factor-1 activity. Mol Cancer Ther 7, 1961–1973. doi:7/7/1961 [pii] https://doi.org/10.1158/1535-7163.MCT-08-0198
140. Seimiya H et al (1999) Hypoxia up-regulates telomerase activity via mitogen-activated protein kinase signaling in human solid tumor cells. Biochem Biophys Res Commun 260:365–370. https://doi.org/10.1006/bbrc.1999.0910
141. Yatabe N et al (2004) HIF-1-mediated activation of telomerase in cervical cancer cells. Oncogene 23:3708–3715. https://doi.org/10.1038/sj.onc.1207460
142. Nishi H et al (2004) Hypoxia-inducible factor 1 mediates upregulation of telomerase (hTERT). Mol Cell Biol 24:6076–6083. https://doi.org/10.1128/MCB.24.13.6076-6083.2004
143. Anderson CJ, Hoare SF, Ashcroft M, Bilsland AE, Keith WN (2006) Hypoxic regulation of telomerase gene expression by transcriptional and post-transcriptional mechanisms. Oncogene 25:61–69. https://doi.org/10.1038/sj.onc.1209011
144. Sullivan R, Graham CH (2009) Hypoxia prevents etoposide-induced DNA damage in cancer cells through a mechanism involving hypoxia-inducible factor 1. Mol Cancer Ther 8:1702-1713. doi:1535-7163.MCT-08-1090 [pii] https://doi.org/10.1158/1535-7163.MCT-08-1090
145. Lee JG et al (2015) Autophagy contributes to the chemo-resistance of non-small cell lung cancer in hypoxic conditions. Respir Res 16:138. https://doi.org/10.1186/s12931-015-0285-4
146. Harhaji-Trajkovic L, Vilimanovich U, Kravic-Stevovic T, Bumbasirevic V, Trajkovic V (2009) AMPK-mediated autophagy inhibits apoptosis in cisplatin-treated tumour cells. J Cell Mol Med 13:3644–3654. https://doi.org/10.1111/j.1582-4934.2009.00663.x
147. Wu HM, Jiang ZF, Ding PS, Shao LJ, Liu RY (2015) Hypoxia-induced autophagy mediates cisplatin resistance in lung cancer cells. Sci Rep 5:12291. https://doi.org/10.1038/srep12291
148. Song J et al (2009) Hypoxia-induced autophagy contributes to the chemoresistance of hepatocellular carcinoma cells. Autophagy 5:1131–1144
149. Yang X et al (2018) Hypoxia-induced autophagy promotes gemcitabine resistance in human bladder cancer cells through hypoxia-inducible factor 1alpha activation. Int J Oncol 53:215–224. https://doi.org/10.3892/ijo.2018.4376

150. Liu XW et al (2010) HIF-1alpha-dependent autophagy protects HeLa cells from fenretinide (4-HPR)-induced apoptosis in hypoxia. Pharmacol Res 62:416–425. https://doi.org/10.1016/j.phrs.2010.07.002
151. Peng X et al (2014) Autophagy promotes paclitaxel resistance of cervical cancer cells: involvement of Warburg effect activated hypoxia-induced factor 1-alpha-mediated signaling. Cell Death Dis 5:e1367. https://doi.org/10.1038/cddis.2014.297
152. Notte A et al (2015) Taxol-induced unfolded protein response activation in breast cancer cells exposed to hypoxia: ATF4 activation regulates autophagy and inhibits apoptosis. Int J Biochem Cell Biol 62:1–14. https://doi.org/10.1016/j.biocel.2015.02.010
153. Hu YL et al (2012) Hypoxia-induced autophagy promotes tumor cell survival and adaptation to antiangiogenic treatment in glioblastoma. Cancer Res 72:1773–1783. https://doi.org/10.1158/0008-5472.CAN-11-3831
154. He WS, Dai XF, Jin M, Liu CW, Rent JH (2012) Hypoxia-induced autophagy confers resistance of breast cancer cells to ionizing radiation. Oncol Res 20:251–258
155. Li Z et al (2009) Hypoxia-inducible factors regulate tumorigenic capacity of glioma stem cells. Cancer Cell 15:501–513. https://doi.org/10.1016/j.ccr.2009.03.018
156. Henze AT, Mazzone M (2016) The impact of hypoxia on tumor-associated macrophages. J Clin Invest 126:3672–3679. https://doi.org/10.1172/JCI84427
157. Gebremeskel S, Johnston B (2015) Concepts and mechanisms underlying chemotherapy induced immunogenic cell death: impact on clinical studies and considerations for combined therapies. Oncotarget 6:41600–41619. https://doi.org/10.18632/oncotarget.6113
158. Barsoum IB, Smallwood CA, Siemens DR, Graham CH (2014) A mechanism of hypoxia-mediated escape from adaptive immunity in cancer cells. Cancer Res 74:665–674. https://doi.org/10.1158/0008-5472.CAN-13-0992
159. Noman MZ et al (2014) PD-L1 is a novel direct target of HIF-1alpha, and its blockade under hypoxia enhanced MDSC-mediated T cell activation. J Exp Med 211:781–790. https://doi.org/10.1084/jem.20131916
160. Bell EN et al (2007) Atrial natriuretic peptide attenuates hypoxia induced chemoresistance in prostate cancer cells. J Urol 177:751–756. doi:S0022-5347(06)02466-9 [pii] https://doi.org/10.1016/j.juro.2006.09.075
161. Frederiksen LJ et al (2003) Hypoxia induced resistance to doxorubicin in prostate cancer cells is inhibited by low concentrations of glyceryl trinitrate. J Urol 170:1003–1007
162. Matthews NE, Adams MA, Maxwell LR, Gofton TE, Graham CH (2001) Nitric oxide-mediated regulation of chemosensitivity in cancer cells. J Natl Cancer Inst 93:1879–1885
163. Muir CP, Adams MA, Graham CH (2006) Nitric oxide attenuates resistance to doxorubicin in three-dimensional aggregates of human breast carcinoma cells. Breast Cancer Res Treat 96:169–176
164. Postovit LM, Sullivan R, Adams MA, Graham CH (2005) Nitric oxide signalling and cellular adaptations to changes in oxygenation. Toxicology 208:235–248
165. Siemens DR et al (2008) Hypoxia increases tumor cell shedding of MHC class I chain-related molecule: role of nitric oxide. Cancer Res 68:4746–4753. doi:68/12/4746 [pii] https://doi.org/10.1158/0008-5472.CAN-08-0054
166. Alderton WK, Cooper CE, Knowles RG (2001) Nitric oxide synthases: structure, function and inhibition. Biochem J 357:593–615
167. Griffith OW, Stuehr DJ (1995) Nitric oxide synthases: properties and catalytic mechanism. Annu Rev Physiol 57:707–736
168. Hanafy KA, Krumenacker JS, Murad F (2001) NO, nitrotyrosine, and cyclic GMP in signal transduction. Med Sci Monit 7:801–819
169. Stuehr DJ (1999) Mammalian nitric oxide synthases. Biochim Biophys Acta 1411:217–230
170. Whorton AR, Simonds DB, Piantadosi CA (1997) Regulation of nitric oxide synthesis by oxygen in vascular endothelial cells. Am J Physiol 272:L1161–L1166
171. McCormick CC, Li WP, Calero M (2000) Oxygen tension limits nitric oxide synthesis by activated macrophages. Biochem J 350:709–716
172. Frederiksen LJ et al (2007) Chemosensitization of cancer in vitro and in vivo by nitric oxide signaling. Clin Cancer Res 13:2199–2206. https://doi.org/10.1158/1078-0432.CCR-06-1807
173. Brune B, Zhou J (2007) Nitric oxide and superoxide: interference with hypoxic signaling. Cardiovasc Res 75:275–282. https://doi.org/10.1016/j.cardiores.2007.03.005
174. Hagen T, Taylor CT, Lam F, Moncada S (2003) Redistribution of intracellular oxygen in hypoxia by nitric oxide: effect on HIF1alpha. Science 302:1975–1978
175. Barsoum IB et al (2011) Hypoxia induces escape from innate immunity in cancer cells via increased expression of ADAM10: role of nitric oxide. Cancer Res 71:7433–7441. https://doi.org/10.1158/0008-5472.CAN-11-2104
176. Mateo J, Garcia-Lecea M, Cadenas S, Hernandez C, Moncada S (2003) Regulation of hypoxia-inducible factor-1alpha by nitric oxide through mitochondria-dependent and -independent pathways. Biochem J 376:537–544. https://doi.org/10.1042/BJ20031155
177. Siemens DR, Heaton JP, Adams MA, Kawakami J, Graham CH (2009) Phase II study of nitric oxide donor for men with increasing prostate-specific antigen level after surgery or radiotherapy for prostate cancer. Urology 74:878–883. https://doi.org/10.1016/j.urology.2009.03.004

178. Yasuda H et al (2006) Randomized phase II trial comparing nitroglycerin plus vinorelbine and cisplatin with vinorelbine and cisplatin alone in previously untreated stage IIIB/IV non-small-cell lung cancer. J Clin Oncol 24:688–694. https://doi.org/10.1200/JCO.2005.04.0436
179. Yasuda H et al (2006) Nitroglycerin treatment may enhance chemosensitivity to docetaxel and carboplatin in patients with lung adenocarcinoma. Clin Cancer Res 12:6748–6757. https://doi.org/10.1158/1078-0432.CCR-06-1124
180. Swaika A, Hammond WA, Joseph RW (2015) Current state of anti-PD-L1 and anti-PD-1 agents in cancer therapy. Mol Immunol 67:4–17. https://doi.org/10.1016/j.molimm.2015.02.009
181. Gong J, Chehrazi-Raffle A, Reddi S, Salgia R (2018) Development of PD-1 and PD-L1 inhibitors as a form of cancer immunotherapy: a comprehensive review of registration trials and future considerations. J Immunother Cancer 6:8. https://doi.org/10.1186/s40425-018-0316-z

Therapeutic Strategies to Block the Hypoxic Response

10

Josh W. DiGiacomo and Daniele M. Gilkes

Abstract

Patients with the low levels of O_2 (hypoxia) in their primary tumors have a higher risk for metastasis and death, indicating a need to therapeutically inhibit the effectors of hypoxia. Many strategies have been developed and investigated to block the hypoxic response. For example, inhibitors of HIF-1 and HIF-2 function by altering the transcription, translation, dimerization, nuclear translocation, DNA-binding, or ubiquitination of the HIF proteins. Hypoxia-activated prodrugs inhibit the hypoxic response through hypoxia-mediated reduction of an inactive, or minimally active, chemical to a cytotoxic agent. Most hypoxia-activated prodrugs function by inducing DNA damage, but others with more novel functions, including prodrugs that release EGFR/HER2 inhibitors also exist. Despite the existence of many therapeutics to combat the hypoxic response, there has been very little success in late phase clinical trials, potentially due to a lack of biomarkers that can be used to stratify patients who would benefit from a hypoxia-targeted therapy.

Keywords

HIF targeting · Hypoxia targeting · Hypoxia-activated prodrugs

10.1 Introduction

Given that patients with poorly oxygenated primary tumors have a higher risk for metastasis and death [1, 2], there exists a need to therapeutically target the effectors of hypoxia. As a result, many therapies which target components of the hypoxia inducible factor (HIF) pathway have been developed over the past two decades [3, 4]. Many inhibitors are known to affect both HIF-1 and HIF-2, but specific HIF-1 and HIF-2 inhibitors exist as well. Furthermore, the HIF pathway can be enhanced or activated by oxygen-independent signaling mechanisms such as the protein kinase B (AKT) and mTOR pathways [5, 6], and the

J. W. DiGiacomo · D. M. Gilkes (✉)
Breast and Ovarian Cancer Program, Department of Oncology, The Johns Hopkins School of Medicine, Baltimore, MD, USA

Department of Chemical and Biomolecular Engineering, Johns Hopkins University, Baltimore, MD, USA
e-mail: dgilkes1@jhu.edu

D. M. Gilkes (ed.), *Hypoxia and Cancer Metastasis*, Advances in Experimental Medicine and Biology 1136,
https://doi.org/10.1007/978-3-030-12734-3_10

mutation of the von Hippel Lindau (VHL) tumor suppressor protein. This chapter will focus on inhibitors of HIFs as well as hypoxia-activated prodrugs (HAPs). HAPs are therapeutics designed to become activated under hypoxia and are proving to be another class of drugs that can reduce or inhibit the detrimental effects of hypoxia. These therapies not only reduce tumor progression directly through targeting hypoxic cancer cells or components of the HIF pathway, but also may play an important role in preventing inflammation in the tumor microenvironment as well [7]. Despite the wide variety of inhibitors available, many are still in the early stages of development [8]. This chapter will serve as an introduction to the current progress in the development of therapeutic strategies to block the hypoxic response. We will review the mechanisms, results, and status of agents targeting various components of the HIF pathway. Refer to Fig. 10.1 for a summary schematic demonstrating how the discussed HIF inhibitors and HAPs interfere with the hypoxic response. See Table 10.1 for a review of their status in clinical development.

10.2 HIF-Directed Therapies

HIFs themselves are the most widely targeted proteins in therapies aimed at inhibiting the hypoxic response. Chemical and nucleic-acid-based forms of HIF-targeting agents affect HIF functionality and expression at multiple levels (Fig. 10.1).

10.2.1 Inhibitors of HIF Transcriptional Activity

3-(5′-hydroxymethyl-2′-furyl)-1-benzylindazole (YC-1) is a small molecule inhibitor that inter-

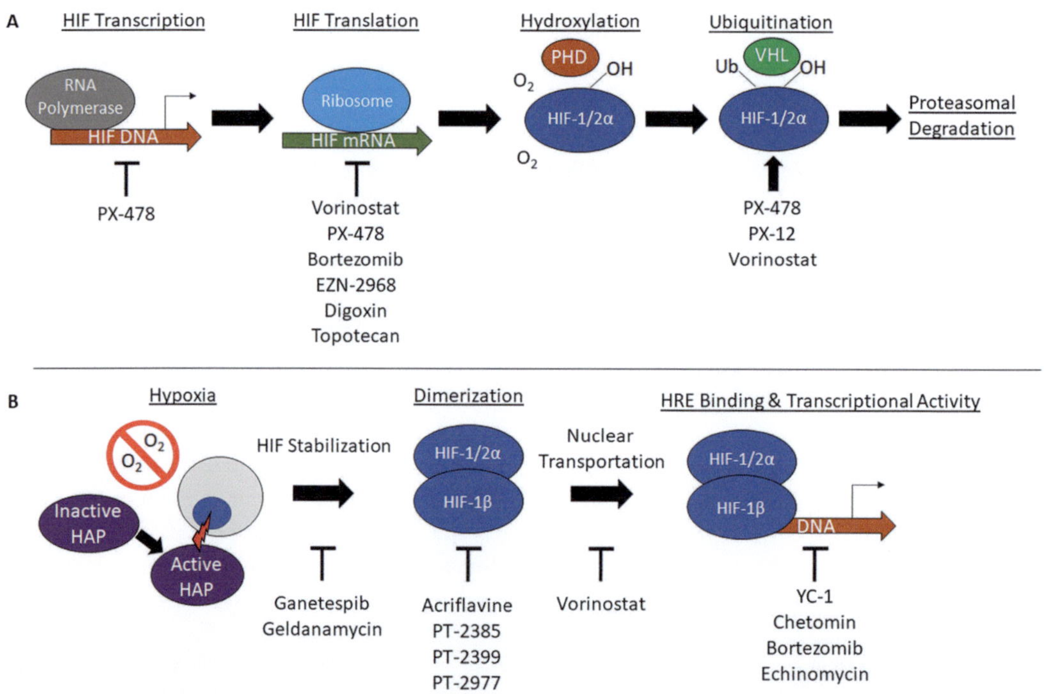

Fig. 10.1 Therapeutic inhibition of the hypoxic response. (**a**) Schematic of HIF-1α and/or HIF-2α mRNA and protein regulation under normal oxygen levels. (**b**) Schematic of the HIF-1 and/or HIF-2 protein regulation under hypoxia. The unique way in which each discussed drug affects HIF regulation and the hypoxic response is also displayed. Abbreviations: HIF - Hypoxia-inducible factor, PHD - Prolyl hydroxylase, VHL - von Hippel-Lindau, HAP - Hypoxia Activated Prodrug, HRE - Hypoxia response element

Table 10.1 Summary of therapeutic strategies to block the hypoxic response

Name	Target	Clinical Status	References
YC-1	HIF-1 HIF-2	No current clinical trials registered.	–
Chetomin	HIF-1 HIF-2	No current clinical trials registered.	–
Vorinostat	HIF-1 HIF-2	FDA-approved for the treatment of cutaneous T cell lymphoma. No ongoing or completed trials specifically investigating the role of inhibiting HIFs on drug efficacy. Several ongoing and completed trials.	[65, 66] NCT01720602 NCT01059552 NCT00268385
Digoxin	HIF-1 HIF-2	Active phase II trial examining the effects of digoxin on HIF-1 in breast cancer tissues.	NCT01763931
Acriflavine	HIF-1 HIF-2	No current registered clinical trials for cancer.	–
PX-12	HIF-1	Phase II study for advanced pancreatic cancer terminated due to lack of significant antitumor activity and low baseline Trx-1 levels. Phase Ib trial for advanced gastrointestinal cancers suggests further development as an intravenous infusion is not feasible. No currently ongoing trials.	[49, 50]
Bortezomib	HIF-1	FDA-approved for the treatment of multiple myeloma that has progressed on prior therapy. A trial studying combination of TH-302 with bortezomib has yet to post results.	[52] NCT01522872
Echinomycin	HIF-1	Many phase II trials demonstrated toxicities along with minimal or no antitumor effects. No currently ongoing trials.	[20–22]
EZN-2968	HIF-1	Completed phase I study for advanced solid tumors or lymphoma and phase Ib proof-of-mechanism, open label study for hepatocellular carcinoma with no results posted. No currently ongoing trials.	NCT00466483 NCT02564614
Ganetespib	HIF-1	Failure to meet endpoints in a phase III study for non-small cell lung cancer and two-phase II studies for metastatic prostate and breast cancer. Phase I combination trial with a secondary outcome investigating the modulation of HIF-1 was terminated due to supplier's suspension of further clinical development.	[40, 41] NCT01798485 NCT02192541
Geldanamycin	HIF-1	Tested in several trials, but none investigating its role as a mediator of HIF. Its derivative, 17-AAG, was investigated in a completed phase II trial of kidney cancer with a secondary endpoint of HIF modulation in resected tumor tissue.	NCT00088374
PX-478	HIF-1	Phase I dose-escalation trial for advanced solid tumors demonstrated acceptable tolerability and HIF-α inhibition proportional to dosage. No ongoing trials.	[31]
Topotecan	HIF-1 HIF-2?	Phase I trial for advanced solid tumors demonstrated capability to reduce expression of HIF-1α. Another phase I trial with a secondary outcome involving hypoxia-induced plasma proteins completed with no results posted. Currently being used in other clinical trials outside the scope of the hypoxic response.	[62] NCT00388089
PT-2385	HIF-2	Phase I dose-escalation trial demonstrated good tolerance and no dose-limiting toxicity. Phase II trials for recurrent glioblastoma and clear cell renal cell carcinoma are ongoing.	[74] NCT03216499 NCT03108066
PT-2399	HIF-2	No current registered clinical trials.	–
PT-2977	HIF-2	Ongoing phase I multiple-dose, dose-escalation and expansion trial for advanced solid tumors and phase II open-label study for clear cell renal cell carcinoma. Early phase I results show acceptable tolerance with signs of clinical activity.	[76] NCT02974738 NCT03401788

(continued)

Table 10.1 (continued)

Name	Target	Clinical Status	References
TPZ	Hypoxic cells	Several phase III trials in which combination with other therapies didn't show additional benefit. Ongoing early phase trials for hepatocellular carcinoma. A phase I trial investigating for cervical cancer with a secondary outcome measurement of hypoxia has completed with results yet to be posted.	[83–85] NCT02174549 NCT03145558 NCT03259867 NCT00098995
SN300000	Hypoxic cells	No current registered clinical trials.	–
TH-302	Hypoxic cells	Failure to meet endpoints in two phase III studies of combination with other therapies for soft-tissue sarcoma and pancreatic ductal adenocarcinoma. Ongoing and recruiting for early phase trials.	[96, 97] NCT02402062 NCT03098160 NCT01497444 NCT02342379
TH-4000	Hypoxic cells	Two phase II trials using TH-4000 as a monotherapy were terminated early due to lack of efficacy. No ongoing trials.	NCT02454842 NCT02449681
EO9	Hypoxic cells	Two phase III trials completed without reaching a primary endpoint of significant difference in 2-year tumor recurrence. Three phase III trials were terminated for business or sponsor decisions. One phase III trial remains active and another is recruiting.	NCT00598806 NCT00461591 NCT01410565 NCT01469221 NCT01475266 NCT02563561 NCT03224182
PR-104	Hypoxic cells	Several early phase trials indicated poor tolerance and dose-limiting toxicities. No currently registered trials.	[109, 110] NCT00862082
AQ4N	Hypoxic cells	Three early phase trials for glioblastoma, non-Hodgkin's lymphoma and leukemia, and solid malignancies and non-Hodgkin's lymphoma currently of unknown status.	NCT00394628 NCT00109356 NCT00090727

feres with the transcriptional activity of HIFs, potentially indirectly by preventing the binding of coactivator, p300, to HIF-1α [9]. HIF-2 protein levels and functional activities are affected by YC-1 as well. YC-1 is effective in suppressing vascular endothelial growth factor (VEGF) expression and tumor development in mouse models of breast cancer [10]. YC-1 treatment also reduces gastric carcinoma cell line invasion in Matrigel invasion assays and reduces metastasis in murine xenograft models of hepatoma and non-small cell lung cancer [11]. Interestingly, YC-1 treatment after radiotherapy suppresses upregulation of HIF-1 activity and delays tumor growth in murine xenografts of human cervical cancer, however YC-1 treatment before radiation leads to an increase in hypoxia which diminishes the radiotherapy's beneficial effects [12]. Most recently, low-dose YC-1 in combination with glucose and insulin has been shown to induce apoptosis in hypoxic gastric carcinoma cells by inhibiting anaerobic glycolysis [13]. The combination also inhibits tumor growth in xenograft models. There are currently no registered clinical trials investigating the use of YC-1 treatment in humans.

Chetomin is a small molecule inhibitor which disrupts the structure of the CH1 domain of p300 and therefore suppresses its ability to interact with HIFs [14]. Chetomin treatment reduces VEGF and carbonic anhydrase IX transcription under hypoxia and radiosensitizes hypoxic fibrosarcoma cells *in vitro* [15]. Chetomin reduces the hypoxia-induced radioresistance of malignant glioma cell lines as well [16]. Binding of p300 to both HIF-1α and HIF-2α is affected by chetomin treatment, and in xenograft models chetomin reduces both tumor growth and erythropoietin enhancer-controlled luciferase activity [14]. Despite preclinical results, a clinical trial for chetomin has yet to be registered on clinicaltrials.gov.

Echinomycin is a small-molecule inhibitor which intercalates DNA and potently inhibits HIF-1 activity by blocking the binding of HIF-1α/HIF-1β dimer to hypoxia response element

(HRE) sequences in DNA [17]. It has been noted that although echinomycin reduces HIF-1 activity under hypoxia in liver cancer cell lines, the effect is nonspecific and affects other transcription factors as well [18]. Additionally, a small increase in HIF-1 activity under normal oxygen concentrations with low concentrations of echinomycin is observed, suggesting a dual-effect. Effects of echinomycin on HIF-2 have not been described. A derivative of echinomycin, YK-2000, also demonstrates apoptotic activity in colon cancer cells preclinically [19]. Echinomycin was clinically studied in the 1990s, with many trials launched in multiple cancers [20–22]. Unfortunately, toxicities, most commonly nausea and vomiting, along with minimal or no antitumor effects were observed. Currently, echinomycin is not being actively investigated in a registered clinical trials.

Acriflavine is another small molecule that functions by inhibiting the dimerization of both HIF-1α and HIF-2α with HIF-1β, and has been shown to reduce tumor growth and vascularization in prostate cancer xenografts [23]. Acriflavine blocks lysyl oxidase (LOX) and lysyl oxidase-like protein 1 (LOXL) expression, collagen cross-linking, bone-marrow derived cell recruitment, and lung metastasis in orthotopic breast cancer models [24]. In addition, acriflavine inhibits hypoxia-mediated, coculture-induced signaling between mesenchymal stem cells and breast cancer cells [25]. Similar to the other inhibitors described in this section, no clinical trials for acriflavine have been registered on clincaltrials.gov.

10.2.2 Inhibitors of the Transcription & Translation of HIFs

S-2-amino-3-[4-N,N,-bis(2-chloroethyl)amino] phenyl propionic acid N-oxide dihydrochloride (PX-478) is an experimental inhibitor which affects HIF-1α at the mRNA and protein level by reducing HIF transcription, translation, and deubiquitination [26]. PX-478 enhances radiosensitivity of prostate carcinoma cells, although the effectiveness varies between cell lines [27]. The combination of gemcitabine with PX-478 reduces tumor growth in pancreatic ductal adenocarcinoma murine models [28]. The number of infiltrating T-lymphocytes present increases with combination therapy, suggesting PX-478 mediates gemcitabine-induced immunogenic cell death. Antitumor activity of PX-478 is also observed in esophageal squamous cell cancer, as the drug inhibits proliferation and epithelial-to-mesenchymal transition *in vitro* as well as murine xenograft tumor growth *in vivo* [29]. PX-478 treatment of mice with colon cancer xenografts does not affect HIF-2α levels in tumor macrophages [30]. Despite positive preclinical results, only one clinical trial involving PX-478 has been reported to date. The Phase I dose-escalation for advanced solid tumors demonstrated acceptable tolerance for PX-478, and inhibition of HIF-1α in peripheral blood mononuclear cells was proportional to drug dosage [31].

EZN-2968, also known as RO7070179, is a locked nucleic acid RNA antagonist designed to bind to HIF-1α mRNA and prevent protein translation [32]. EZN-2968 reduces HIF-1 mRNA and protein levels under both normal oxygen and hypoxic conditions in human prostate and glioblastoma cells *in vitro*, as well as inhibits tumor growth in murine prostate cancer xenografts. EZN-2968 harbors a 3-base pair mismatch in the nucleotide sequence of HIF-2α and therefore downregulates HIF-2α mRNA minimally both *in vitro* and in mouse models. A 2014 pilot trial of EZN-2968 for patients with refractory solid tumors demonstrated reduced HIF-1α mRNA levels in post-treatment tumor biopsies compared to pre-treatment biopsies in 4 of 6 patients [33]. Two patients from this study also presented reduced HIF-1 target gene mRNA expression. However, this trial closed prematurely due to suspension of clinical development by the pharmaceutical sponsor. Although there are not active clinical trials for EZN-2968, a Phase I study in adult patients with advanced solid tumors or lymphoma (NCT00466483) and a Phase 1b, proof-of-mechanism, open label study in adult patients with hepatocellular carcinoma (NCT02564614) have been completed with results yet to be published.

Digoxin is a cardiac glycoside that has FDA-approval for the treatment of heart disease, but also inhibits the translation of both HIF-1α and HIF-2α and decreases the growth of tumor xenografts [34]. Digoxin treatment inhibits lung metastasis of breast cancer cells in orthotopic models, in part by blocking the expression of angiopoietin-like 4 and L1 cell adhesion molecule, both of which are regulated by HIF-1 [35]. HIF-dependent transcriptional activation of LOX and LOXL expression, collagen cross-linking, bone-marrow derived cell recruitment, and lung metastasis are also blocked by digoxin treatment in an orthotopic breast cancer model [24]. Several clinical trials utilizing digoxin alone or in combination for the treatment of several cancers such as head and neck cancer (NCT02906800) and Kaposi's sarcoma (NCT02212639) are ongoing or recruiting patients. One active Phase II trial (NCT01763931) will examine the effect of digoxin on HIF-1 expression after 2 weeks of daily oral digoxin therapy, compared to no study drug, as measured by immunohistochemistry (IHC), in surgically resected breast cancer tissue.

10.2.3 HIF Protein Stability & Degradation

STA-9090 (ganetespib) is a small molecule inhibitor of the HIF-chaperone, heat shock protein 90 (HSP90) [36]. HSP90 stablizes HIF-1α by preventing its degradation [37]. Ganetespib reduces HIF-1α, but not HIF-2α, protein levels and HIF-1 transcriptional activity, as well as inhibits tumor growth, vascularization, stem cell maintenance, invasion, and metastasis in orthotopic murine models of triple-negative breast cancer [38]. Ganetespib reduces HIF-mediated angiogenesis in colorectal cancer models as shown through reduced Matrigel plug vascularization and inhibits tumor xenograft growth [39]. Because of preclinical success, ganetespib, alone or in combination with other therapies, has been tested in several clinical trials. However, a recent Phase III study combining ganetespib with docetaxel for advanced non-small cell lung cancer was terminated early after a first interim analysis due to futility (NCT01798485). A Phase II study for ganetespib in patients with docetaxel-pretreated, metastatic, castrate-resistant prostate cancer was terminated at interim analysis as no patient attained 6-month progression free survival [40]. A Phase II trial for metastatic breast cancer patients also failed to reach specified criteria for overall response rate, but did demonstrate clinical activity in human epidermal growth factor receptor 2 (HER2)-positive and triple-negative patients [41]. A Phase I trial combining the use of ganetespib with ziv-aflibercept for the treatment of refractory gastrointestinal carcinomas, non-squamous non-small cell lung carcinomas, urothelial carcinomas, and sarcomas included a secondary outcome measurement to investigate the level of HIF-1 (NCT02192541) after treatment but was terminated early due to the drug supplier's suspension of further clinical development of ganetespib. No currently registered clinical trials utilizing ganetespib are investigating the effects of the drug on HIF-1 as an endpoint.

Geldanamycin is an HSP90 inhibitor that reduces HIF-1α protein levels and transcriptional activity in cancer cells *in vitro* [37, 42]. In addition to promoting HIF-1 protein degradation and VEGF mRNA reduction, geldanamycin treatment reduces the ability of conditioned media from hypoxic prostate cancer cells to stimulate invasion *in vitro* [43]. Additionally, low-dose geldanamycin treatment decreases *in vitro* bladder cancer cell invasion stimulated by treatment with hepatocyte growth factor, cobalt chloride, and phenanthroline [44]. The effects of geldanamycin on HIF-2 have not been reported. Although geldanamycin has been investigated in clinical trials, the ability for the drug to modulate HIF levels *in vivo* has not been reported. Tanespimycin (17-AAG), a derivative of geldanamycin, has been studied in a completed Phase II trial (NCT00088374) in patients with kidney tumors characterized as having VHL mutations. HIF-modulation was considered as a secondary outcome in this trial, but results have not been reported.

10.2.4 Indirect HIF Inhibitors

1-methylpropyl 2-imidazolyl disulfide (PX-12) is an inhibitor of thioredoxin-1 (Trx-1), a small redox protein known to increase HIF-1α protein levels in cancer [45]. One study suggests PX-12 treatment reduces HIF-1α levels, through a mechanism independent of Trx-1 inhibition, by inducing spermidine/spermine N(1)-acetyl transferase 1 expression which leads to enhanced HIF-1 ubiquitination and therefore degradation [46]. PX-12 decreases the hypoxic induction of HIF-1α and VEGF expression *in vitro* in human breast and colon cancer cell lines [47]. Breast cancer murine xenografts also exhibit lower levels of HIF-1α and VEGF protein in PX-12 treated mice. Additionally, PX-12 treatment lowers VEGF levels in cancer patient plasma [48]. Whether or how PX-12 might affect HIF-2 has not yet been described. A randomized Phase II study of 3-hour intravenous PX-12 treatment daily for 5 days on 21-day intervals in patients with advanced pancreatic cancer was terminated early due to a lack of significant antitumor activity and low baseline Trx-1 levels [49]. A Phase IB trial utilizing a 24-h, continuous, intravenous infusion of PX-12 once every 7 or 14 days for advanced gastrointestinal cancers demonstrated no significant changes in VEGF, which is induced by Trx-1 through HIF-1α, after treatment [50]. There are no current clinical trials registered for the treatment of cancer with PX-12.

Bortezomib is a proteasome inhibitor that interferes with HIF-1α activity and exhibits antitumor effects [51]. Bortezomib is Food & Drug Administration (FDA)-approved for the treatment of multiple myeloma that has progressed on prior therapy [52]. Several mechanisms of action have been proposed for HIF regulation by bortezomib, including affecting the HIF-1α C-terminal activation domain, reducing HIF-1α protein translation through eIF2α phosphorylation, and inhibiting protein expression and nuclear accumulation through inhibition of the phosphatidylinositol-4,5-bisphosphate 3-kinase/protein kinase B (PI3K/Akt) and mitogen-activated protein kinase pathways [51, 53, 54]. Bortezomib reduces HIF-1, but not HIF-2-mediated transcription *in vitro* [55]. HIF-1α expression enhances the bortezomib-induced apoptosis of of human umbilical vein endothelial cells [56]. Bortezomib is cytotoxic against hypoxic tumor cells *in vitro* by inducing endoplasmic reticulum stress [57]. There are nearly 200 active or recruiting clinical trials utilizing bortezomib alone or in combination with other anticancer drugs. One trial (NCT01522872) is studying the effects of the HAP TH-302 (evofosfamide) in combination with bortezomib and other agents in relapsed/refractory multiple myeloma, although the current status of the trial is unknown. As a secondary outcome measurement, the trial will investigate the relationship between hypoxia in the bone marrow and response to the drug regimens using hypoxia-markers such as pimonidazole. Evofosfamide will be discussed in greater detail in Sect. 10.4.

Topotecan is a topoisomerase I inhibitor used as an anticancer agent which demonstrates the ability to inhibit the translation of HIF-1α [58]. HIF-2α protein levels are also strongly affected by topotecan in neuroblastoma cells [59]; however, endothelial cells demonstrate minimal HIF-2α reduction upon treatment [60]. Daily administration of topotecan in glioblastoma xenografts inhibits HIF-1α protein expression and significantly reduces tumor growth and HIF-1 target gene expression in tumor tissue [61]. Topotecan blocks hypoxia-induced VEGF expression in human neuroblastoma cells [59]. A Phase I pilot trial of oral topotecan demonstrated that topotecan decreases HIF-1α expression *in vivo* [62]. In the study, 4 of 16 patients that had detectable HIF-1α nuclear staining prior to treatment had undetectable HIF-1α after treatment. Four patients had decreased levels of VEGF and GLUT-1 mRNA when comparing pre- and post-treatment tumor biopsies. Another Phase I study investigating both topotecan and bortezomib in advanced solid tumors (NCT00388089) with a secondary outcome plasma measurement to detect hypoxia-inducible proteins has been completed with results yet to be posted. Outside the scope of hypoxia, topotecan is being utilized in many other clinical trials for the treatment of cancer.

8-(hydroxyamino)-8-oxo-N-phenyl-octanamide (vorinostat) is a histone deacetylase inhibitor which nonspecifically decreases HIF-1α protein levels. Vorinostat modulates HIF-1α at several levels, including translational repression, reduced nuclear translocation, and enhanced ubiquitination [63, 64]. Vorinostat treatment also reduces HIF-2 protein levels [64]. Currently, vorinostat is FDA-approved for the treatment of cutaneous T cell lymphoma [65]. A recent Phase I/II trial for metastatic clear-cell renal cell carcinoma (ccRCC) patients previously treated with chemotherapy utilized vorinostat with a VEGF inhibitor, bevacizumab, and noted the combination is relatively well tolerated and results in an objective response rate of 18%, suggesting clinical activity [66]. Although no currently listed trials are evaluating the role of vorinostat on HIF levels after treatment, many clinical trials for vorinostat alone or in combination with other anti-cancer agents are ongoing or recruiting patients for multiple types of malignancies including metastatic breast cancer (NCT01720602), non-small cell lung cancer (NCT01059552), and malignant gliomas (NCT00268385).

10.2.5 Emerging HIF Inhibitors

Even though many established HIF inhibitors exist, investigation and development of new therapeutics is ongoing. Lee et al. synthesized and evaluated a variety of derivatives of the HIF-1α inhibitor, oltipraz, to optimize its antagonist potential [67]. The group ultimately established a derivative that reduces insulin-induced levels of HIF-1α mRNA and protein levels in a human colon cancer cell line to a greater extent than YC-1 treated controls. mRNA transcripts of HIF target genes, VEGF and glucose transporter 1 (GLUT-1), are also reduced by treatment with the derivative.

Koivunen et al. determined that 3-oxoglutarate, an analog of 2-oxoglutarate which acts as a co-factor for prolyl hydroxylases (PHD), reduces HIF-1α levels under normal oxygen concentrations in several cancer cell lines independent of PHD [68]. Combination of 3-oxoglutarate with the anticancer drug, vincristine, induces apoptosis *in vitro* in several cancer cell lines and reduces tumor growth in murine colon cancer xenografts better than vincristine alone. Wang et al. has developed a novel chalcone derivative which reduces HIF-1α mRNA and protein expression in a hepatoma cell line [69]. This derivative is also able to inhibit the VEGF-induced migration of endothelial and hepatoma cells *in vitro* and reduces tumor growth and angiogenesis in murine xenografts.

10.3 HIF-2 Specific Therapeutics

ccRCC is the most common form of kidney cancer. Inactivation of the VHL tumor suppressor and consequently the deregulation of HIF-2α is a common event [70], although VHL is known to regulate HIF-1α as well. HIF-2α may also play a pro-tumorigenic role in neuroblastoma as well [71]. As such, several HIF-2α specific inhibitors are currently being developed and investigated for clinical efficacy.

Developed by Peloton Therapeutics, PT-2385 is a small-molecule antagonist that functions by allosterically blocking the dimerization of HIF-2α, but not HIF-1α, with HIF-1β, thus specifically preventing formation of a functional HIF-2 protein [72]. Recent preclinical testing in ccRCC cell lines and murine xenografts demonstrate reduced expression of HIF-2α target genes with PT-2385 treatment. Additionally, PT-2385 treatment causes tumor regression in mice. In advanced HCC, sorafenib is used as a standard treatment and has been shown to be more potent in the presence of the androgen receptor (AR); however, sorafenib has a side effect of inducing HIF-2α, which transcriptionally suppresses AR [73]. Therefore, PT-2385 can enhance sorafenib efficacy by preventing AR inhibition. This has been demonstrated in hepatocellular carcinoma (HCC) cell lines both *in vitro* and in xenograft models. A Phase I dose-escalation trial of PT-2385 in patients with previously treated advanced ccRCC demonstrated that PT-2385 is well-tolerated as no dose-limiting toxicity was observed at any given dosage [74]. In this study, 52% of patients demonstrated stable disease as best response, with 2% and 12% achieving a complete or partial response, respectively. In addition, all doses of PT-2385 treatment resulted in decreases in plasma erythropoietin reductions, indicating biologic activity as erythropoietin is regulated by HIF-2. Currently, PT-2385 is being studied in a Phase II trial for patients with recurrent glioblastoma (NCT03216499) and an open-label Phase II trial of VHL disease-associated ccRCC (NCT03108066).

The HIF-2 inhibitor PT-2977, also developed by Peloton Therapeutics, demonstrates increased potency in preclinical tumor models of ccRCC compared to PT-2385 [75]. A potential interaction between HIF-1 and PT-2977 has not yet been demonstrated. Two clinical trials utilizing PT-2977 are already underway. The first trial (NCT02974738) is a Phase I multiple-dose, dose-escalation and expansion trial in patients with advanced solid tumors. Early results from the trial demonstrate that the inhibitor is well tolerated with signs of clinical activity in the treatment of ccRCC [76]. A second trial utilizing PT-2977 (NCT03401788) is a Phase II open-label study for the treatment of patients with a germline VHL alteration.

Another product of Peloton Therapeutics, and close analogue of PT-2385, PT-2399 also inhibits HIF-2 by preventing the dimerization of HIF-2α with HIF-1β by disrupting necessary hydrogen bonds and hydrophobic interactions [77]. Cho et al. has shown that PT-2399 causes tumor regression in mouse models of primary and metastatic VHL-defective ccRCC [78]. Xenograft models in this study demonstrated decreased circulating tumor-derived VEGF and decreased angiogenesis. ccRCC cell lines have variable sensitivity to PT-2399, and more sensitive ccRCC lines have higher HIF-2α levels than insensitive lines and exhibit greater suppression of HIF target genes. The variability in cell lines likely reflect differences in dependence on HIF-2α rather than drug efficacy. Chen et al. found that PT-2399 suppresses tumorigenesis in 10 out of 18 ccRCC cell lines, and that some VHL-mutant ccRCCs are resistant despite dissociation of the HIF-2 dimer, suggesting variability in dependence on HIF-2 as Cho et al. described [79]. Chen et al. also noted that PT-2399 treatment leads to resistance and found two mutations in HIF-2α and HIF-1β which prevent PT-2399-induced dissociation. This study demonstrates that PT-2399 specifically disassembles HIF-2, but not HIF-1, dimers and affects HIF-2, but not HIF-1, transcriptional target regulation. Despite these preclinical studies of PT-2399, no clinical trials utilizing the drug have been registered. The recent developments and trials involving these specific HIF-2 inhibitors have provided promising results for the blockade of the hypoxic response through HIF-2; however, the development of therapeutic strategies to block the hypoxic response through methods other than targeting HIFs are also being developed and investigated.

10.4 Hypoxia Activated Prodrugs (HAPs)

As mentioned above, HAPs are initially inactive drugs that become chemically reduced, and activated, under low oxygen conditions. In the context of cancer, under low oxygen HAPs are reduced to an activated, cytotoxic, DNA-damaging agent which inhibits the ability of nearby hypoxic cells to proliferate [7]. HAPs can be divided into five classes of reducible chemical compounds: nitro groups, quinones, aromatic N-oxides, aliphatic N-oxides, and transition metals [80]. The compounds are generally reduced by one-electron or two-electron oxidoreductases which function as catalysts for the reaction. One-electron oxidoreductases generate an oxygen-sensitive radical intermediate. This reversible reaction allows the HAP to return to an inactivated state, mitigating the occurrence of HAP activation in more-oxygenated regions and preventing damage to non-hypoxic tissues. Two-electron oxidoreductases do not generate intermediates, causing an irreversible reduction and increasing the likelihood of oxygen-independent or off-target activation. As such, two strategies in the design and consideration of potential HAPs exist: activation at intermediate hypoxia and activation at severe hypoxia [81]. The oxygen state required for activation can be quantified by a HAP's quantitative oxygen dependence (K_{O_2}), with lower K_{O_2} (~0.1 μM O_2 in cell culture medium) indicating activation in severe hypoxia and higher K_{O_2} (1–10 μM O_2) for intermediate hypoxia. As expected, a HAP with a lower K_{O_2} is more likely to accurately target tumor hypoxia instead of off-target activation in healthy, physiologically low oxygen regions; however, HAPs with higher K_{O_2} will affect more tumor cells that experience at least some degree of hypoxia. To offset this effect, low K_{O_2} HAPs are generally desired to produce activated forms which can diffuse to nearby regions of higher oxygenation and affect neighboring cells as well, a phenomenon known as the bystander effect. This section will introduce and summarize several cancer-targeting HAPs that have been developed, investigated preclinically, and studied in clinical trials.

Tirapazamine (TPZ) is one of the earliest and most studied HAPs. An aromatic N-oxide, TPZ functions, as most HAPs do, by reducing under hypoxic conditions to an agent which can cause DNA damage through single and double-stranded breaks. TPZ was studied most extensively in the 1980s and 1990s, and demonstrated preclinical anti-cancer activity when combined with both radiation and chemotherapy, most notably cisplatin and carboplatin [82]. As such, TPZ moved quickly to clinical trials, where many successful early phase trials established that TPZ alone and in combination with cisplatin and other chemotherapies is generally well-tolerated and warranted further study. However, in the 2000s several Phase III trials of TPZ in combination with other anti-cancer agents failed to demonstrate additional benefit [83], although one of these trials was later criticized for lack of stratification based on patients with hypoxic tumors as well as issues with radiation delivery [84]. In addition, a more recent Phase III trial of cisplatin and irradiation versus cisplatin, TPZ, and irradiation in cervical carcinoma demonstrated that although the combination of TPZ and cisplatin is tolerable at the starting dose, it is not more effective than cisplatin alone in promoting progression free or overall survival [85]. Despite setbacks, research with TPZ continues. The addition of TPZ enhances the efficacy of irinotecan in hepatocellular carcinoma both *in vitro* and in liver cancer xenograft models. This may be related to a correlation between high levels of HIF-1α and topoisomerase I inhibitor resistance [86]. Three clinical trials (one Phase I, two Phase II) utilizing TPZ in the treatment of hepatocellular carcinoma are also currently recruiting patients (NCT02174549, NCT03145558, NCT03259867). One Phase I trial investigating TPZ, cisplatin, and radiation in cervical cancer (NCT00098995) with a secondary outcome measurement of hypoxia by ^{18}F-azomycinarabinoside PET scan has completed with results yet to be posted.

SN30000 is an analog of TPZ which was developed to improve TPZ half-life and diffusion in tumors [84]. SN30000 is reduced selectively under hypoxia *in vitro* by colon cancer cells with an oxygen dependence similar to TPZ [87]. Diffusion of SN30000 across multicellular layers of both colon and cervical cancer cell lines is faster than TPZ. Human tumor murine xenografts also demonstrate increased efficacy of SN30000 with radiation compared to TPZ with radiation. The diflavin one-electron reductases POR, MTRR, NOS2A, and NDOR1 have been identified as reducing agents of SN30000 [88]. Preclinical pharmacokinetic research of SN30000 in NIH-III mice demonstrate that the drug confers acute toxicity, induces kidney toxicity comparable to another well-known HAP (TH-302), and can actually be reduced in normal tissues despite displaying cytotoxic selectivity for tumor tissue [89]. The drug has yet to be tested clinically.

TH-302 is another well studied and clinically active HAP. In human breast and prostate cancer murine xenografts, TH-302 induces maximum toxicity in hypoxic and near-hypoxic regions, but also demonstrates activity in vascularized tumor tissue [90]. This study also indicated that TH-302 enhances the efficacy of both doxorubicin and docetaxel. The combination of TH-302 and paclitaxel is more effective than paclitaxel alone in reducing tumor growth in both the mammary gland and bone in osteolytic breast cancer murine models [91]. In human lung carcinoma murine xenografts, TH-302 has both a more favorable safety profile and, at equal hematoxicity levels, superior antitumor activity than the FDA-approved agent ifosfamide [92]. Additionally, in human neuroblastoma and rhabdomyosarcoma murine xenografts, the combination of TH-302 with topotecan enhances tumor response and increases animal survival compared to monotherapy [93]. Despite much preclinical and early phase clinical trial success [94, 95], TH-302 recently endured failures in two Phase III studies which combined TH-302 with doxorubicin to treat soft-tissue sarcoma (NCT01440088, did not improve overall survival) and with gemcitabine in the treatment of pancreatic ductal adenocarcinoma (NCT01746979, difference in overall survival time was not statistically significant) [96, 97]. These failures suggest that improved identification of patients which would benefit most from HAP therapies is warranted and requires additional investigation. Despite these setbacks, early phase trials utilizing drug regimens including TH-302 are ongoing and recruiting patients (NCT02402062, NCT03098160, NCT01497444, NCT02342379).

Tarloxotinib bromide (TH-4000) functions differently than most HAPs, as its reduced form releases an irreversible epidermal growth factor receptor (EGFR)/HER2 inhibitor rather than a DNA damaging agent [98]. As such, TH-4000 has been applied to non-small cell lung cancer (NSCLC), in which HIF-dependent mechanisms upregulate wild-type EGFR and EGFR tyrosine kinase inhibitor resistance is observed [99]. In preclinical models, TH-4000 demonstrates more antiproliferative potential than the EGFR inhibitor, erlotinib, against human cancer cell lines *in vitro* as well as the ability to silence wild type EGFR signaling in human xenograft models [98]. Another study determined that TH-4000 alone or in combination with erlotinib is more effective in achieving tumor regression than erlotinib alone in NSCLC mice models [99]. However, TH-4000 has not yet achieved clinical success. A Phase II study of EGFR mutant NSCLC (NCT02454842) and a Phase II study on recurrent or metastatic squamous cell carcinoma of the head, neck, or skin (NCT02449681) using TH-4000 as a monotherapy were both recently terminated early due to a lack of efficacy. As such, there are no other currently recruiting or ongoing clinical trials for TH-4000.

Apaziquone (EO9) is another HAP which was studied preclinically in the 1990s and early 2000s, with many results demonstrating preferential cytotoxicity towards hypoxic cells along with antitumor activity or enhanced efficacy in combination with radiation in animal models [100, 101]. Early clinical trials of EO9 failed to show activity with intravenous delivery, and it was determined that EO9 has poor avascular tissue penetration and undergoes rapid removal from the body [102]. To address this issue, EO9 was tested in clinical trials of bladder cancer with intravesical delivery directly into the bladder. This method of drug delivery is well tolerated and demonstrated positive results in early phase trials [103, 104]. However, EO9 has undergone several Phase III trials without success. Two trials (NCT00598806, NCT00461591) completed without reaching a primary endpoint of significant difference in 2-year tumor recurrence [105]. Three more Phase III trials were later terminated due to business reasons (NCT01410565, NCT01469221) or sponsor decisions (NCT01475266). Despite these setbacks, one Phase III trial remains active (NCT02563561) and another is currently recruiting patients (NCT03224182).

PR-104 is another HAP which shows more activity against both hypoxic and non-hypoxic xenografts of multiple cancer types than TPZ, and demonstrates efficacy both as a monotherapy and in combination with chemotherapy and radiotherapy [106]. PR-104 prolongs survival and decreases leukemia burden in mice bearing primary acute lymphoblastic leukemia xenografts [107]. However, PR-104 has experienced several problems in early phase clinical trials. A randomized Phase I/II open-label trial (NCT00862082) of PR-104 in combination with sorafenib for the treatment of HCC was terminated due to poor tolerance. Although a Phase I trial for solid tumors with triweekly intravenous infusion demonstrated tolerability [108], another later Phase I trial for solid tumors with weekly infusion demonstrated thrombocytopenia and neutropenia as dose-limiting toxicities [109]. As thrombocytopenia occurred with a delayed-onset, PR-104 was only recommended for a short course of treatment. Additionally, a Phase Ib study on PR-104 in combination with gemcitabine or docetaxel in patients with advanced solid tumors concluded that the combinations cause dose-limiting and severe myelotoxicity, although prophylactic granulocyte colony stimulating factor treatment allows PR-104 escalation with docetaxel [110]. Clinical trials to test PR-104 are not currently active or recruiting.

Banoxatrone (AQ4N) is a unique HAP which functions by reducing to AQ4, a potent topoisomerase II inhibitor, under hypoxic conditions [111]. AQ4N in its non-reduced form demonstrates relatively low cytotoxic activity, and preclinical studies have demonstrated that the reduced form may enhance the antitumor efficacy in combination with radiation and chemotherapy. A study on the treatment of pancreatic mouse model xenografts with AQ4N demonstrated reduced tumor growth and progression at a level comparable to gemcitabine treatment [112]. This study also determined that AQ4N is rapidly cleared from circulation but also quickly accumulates in tumor tissues at high levels. AQ4N is efficiently reduced and activated by CYP2S1, an extrahepatic P450 enzyme expressed at higher levels in hypoxic cells [113]. Three early phase trials for the study of AQ4N treatment in glioblastoma (NCT00394628), non-Hodgkin's lymphoma and leukemia (NCT00109356), and solid malignancies and non-Hodgkin's lymphoma (NCT00090727) were initiated in the mid-2000s but results were never reported. No other registered clinical trials for AQ4N are currently active or recruiting patients.

Even with extensive development and clinical trials for established HAPs, newer HAPs continue to be designed and investigated. TAT-ODD-procaspase-3 (TOP3) is a recently developed fusion protein utilizing a human immunodeficiency virus-derived TAT domain to enhance cell penetration and an oxygen-dependent degradation domain from HIF-1α to increase hypoxia specificity [114]. TOP3 combination with gemcitabine and TS-1 improves efficacy under hypoxia *in vitro* and enhances survival of mice in an orthotopic pancreatic cancer xenograft model.

Zhang et al. has created a catalase-loaded cisplatin-prodrug-constructed liposome which relieves hypoxia through catalase degradation of hydrogen peroxide to water and oxygen and simultaneously delivers chemotherapy through cisplatin [115]. Combination treatment of the liposomes with radiation therapy results in reduced tumor growth in murine models with 4T1 tumors.

OCT1002 is a HAP which operates through a two-electron reductase and reduces tumor growth and lung metastases in combination with bicalutamide in prostate cancer xenograft models [116]. A further study of OCT1002 determined it also enhances the efficacy of abiraterone, cabazitaxel, and docetaxel *in vivo* [117]. As the aforementioned three HAPs are relatively new in development, none have reached the stage of clinical trials yet.

The plethora of emerging and established HAPs provide hope that targeting hypoxia can become a clinically achievable therapeutic strategy. However, the variability of tumor hypoxia between patients and a lack of predictive biomarkers for hypoxia-targeted therapies create concern over the use of HAPs and where they could be most effectively applied in the clinical setting [118]. Despite this, the preclinical trend of both HAPs and HIF-directed therapies improving the efficacies of common chemotherapies and radiotherapy cannot be understated, and further investigation of these strategies is warranted to assist clinical development.

10.5 Summary & Conclusions

This chapter briefly reviewed several established and emerging therapies for blocking tumor response to hypoxia. Molecularly, most hypoxia-targeting therapies are designed to block HIF function. This effect is achieved by inhibiting HIF transcription, translation, dimerization, nuclear transport, DNA-binding, and even degradation. Specific HIF-2 inhibitors largely function through inhibition of dimerization of the HIF-2α and HIF-1β subunits. HAPs are another method for blocking the hypoxic response. HAPs are chemicals which exist in an inactive or minimally active form in higher oxygen conditions but are reduced to cytotoxic agents under hypoxia. Most HAPs function by inducing DNA damage in their reduced form such that the cell can no longer divide, thereby targeting cells that experience hypoxia as well as their neighbors if the drug can diffuse locally. Some HAPs have more novel abilities as well, including but not limited to reduction to a topoisomerase II inhibitor or reduction to an EGFR/HER2 inhibitor. Although not discussed in this chapter, the hypoxic response has been shown to be indirectly blocked through targeting the PI3K/AKT pathway upstream of HIF, which may also serve as a potential therapeutic strategy [119, 120]. Preclinical research into these therapeutic strategies shows that they can often enhance the efficacy of pre-existing chemotherapeutic agents or radiotherapy, demonstrating the real potential for blocking the hypoxic response in the clinical setting. Despite many existing hypoxia-targeting strategies and preclinical data, there has been very little late phase clinical trial success. This is potentially due to the lack of biomarkers for patients who would benefit from hypoxia-directed therapies as well as large variability in tumor hypoxia between patients. Further research into this area is warranted for a future where hypoxic status and treatment can be integrated into cancer treatment.

References

1. Hockel M et al (1996) Association between tumor hypoxia and malignant progression in advanced cancer of the uterine cervix. Cancer Res 56(19):4509–4515
2. Hockel M et al (1993) Intratumoral pO2 predicts survival in advanced cancer of the uterine cervix. Radiother Oncol 26(1):45–50
3. Semenza GL (2003) Targeting HIF-1 for cancer therapy. Nat Rev Cancer 3(10):721–732
4. Hu Y, Liu J, Huang H (2013) Recent agents targeting HIF-1alpha for cancer therapy. J Cell Biochem 114(3):498–509
5. Stiehl DP et al (2002) Normoxic induction of the hypoxia-inducible factor 1alpha by insulin and interleukin-1beta involves the phosphatidylinositol 3-kinase pathway. FEBS Lett 512(1–3):157–162
6. Hudson CC et al (2002) Regulation of hypoxia-inducible factor 1alpha expression and function by the mammalian target of rapamycin. Mol Cell Biol 22(20):7004–7014
7. DiGiacomo JW, Gilkes DM (2018) Tumor hypoxia as an enhancer of inflammation-mediated metastasis: emerging therapeutic strategies. Target Oncol 13(2):157–173
8. Ban HS et al (2016) Hypoxia-inducible factor (HIF) inhibitors: a patent survey (2011–2015). Expert Opin Ther Pat 26(3):309–322
9. Li SH et al (2008) A novel mode of action of YC-1 in HIF inhibition: stimulation of FIH-dependent p300 dissociation from HIF-1{alpha}. Mol Cancer Ther 7(12):3729–3738
10. Carroll CE et al (2013) The anticancer agent YC-1 suppresses progestin-stimulated VEGF in breast cancer cells and arrests breast tumor development. Int J Oncol 42(1):179–187
11. Shin DH et al (2007) Preclinical evaluation of YC-1, a HIF inhibitor, for the prevention of tumor spreading. Cancer Lett 255(1):107–116
12. Harada H et al (2009) Treatment regimen determines whether an HIF-1 inhibitor enhances or inhibits the effect of radiation therapy. Br J Cancer 100(5):747–757
13. Wakiyama K et al (2017) Low-dose YC-1 combined with glucose and insulin selectively induces apoptosis in hypoxic gastric carcinoma cells by inhibiting anaerobic glycolysis. Sci Rep 7(1):12653
14. Kung AL et al (2004) Small molecule blockade of transcriptional coactivation of the hypoxia-inducible factor pathway. Cancer Cell 6(1):33–43
15. Staab A et al (2007) Effects of HIF-1 inhibition by chetomin on hypoxia-related transcription and radiosensitivity in HT 1080 human fibrosarcoma cells. BMC Cancer 7:213
16. Kessler J et al (2010) HIF-1alpha inhibition by siRNA or chetomin in human malignant glioma cells: effects on hypoxic radioresistance and monitoring via CA9 expression. BMC Cancer 10:605
17. Kong D et al (2005) Echinomycin, a small-molecule inhibitor of hypoxia-inducible factor-1 DNA-binding activity. Cancer Res 65(19):9047–9055
18. Vlaminck B et al (2007) Dual effect of echinomycin on hypoxia-inducible factor-1 activity under normoxic and hypoxic conditions. FEBS J 274(21):5533–5542
19. Park JY et al (2004) Echinomycin and a novel analogue induce apoptosis of HT-29 cells via the activation of MAP kinases pathway. Pharmacol Res 50(2):201–207

20. Marshall ME et al (1993) Phase II trial of echinomycin for the treatment of advanced renal cell carcinoma. A Southwest Oncology Group study. Investig New Drugs 11(2–3):207–209
21. Chang AY et al (1998) A randomized phase II trial of echinomycin, trimetrexate, and cisplatin plus etoposide in patients with metastatic nonsmall cell lung carcinoma: an Eastern Cooperative Oncology Group Study (E1587). Cancer 82(2):292–300
22. Gradishar WJ et al (1995) A phase II clinical trial of echinomycin in metastatic soft tissue sarcoma. An illinois Cancer Center Study. Investig New Drugs 13(2):171–174
23. Lee K et al (2009) Acriflavine inhibits HIF-1 dimerization, tumor growth, and vascularization. Proc Natl Acad Sci U S A 106(42):17910–17915
24. Wong CC et al (2012) Inhibitors of hypoxia-inducible factor 1 block breast cancer metastatic niche formation and lung metastasis. J Mol Med (Berl) 90(7):803–815
25. Chaturvedi P et al (2013) Hypoxia-inducible factor-dependent breast cancer-mesenchymal stem cell bidirectional signaling promotes metastasis. J Clin Invest 123(1):189–205
26. Koh MY et al (2008) Molecular mechanisms for the activity of PX-478, an antitumor inhibitor of the hypoxia-inducible factor-1alpha. Mol Cancer Ther 7(1):90–100
27. Palayoor ST et al (2008) PX-478, an inhibitor of hypoxia-inducible factor-1alpha, enhances radiosensitivity of prostate carcinoma cells. Int J Cancer 123(10):2430–2437
28. Zhao T et al (2015) Inhibition of HIF-1alpha by PX-478 enhances the anti-tumor effect of gemcitabine by inducing immunogenic cell death in pancreatic ductal adenocarcinoma. Oncotarget 6(4):2250–2262
29. Zhu Y et al (2017) Inhibition of HIF-1alpha by PX-478 suppresses tumor growth of esophageal squamous cell cancer in vitro and in vivo. Am J Cancer Res 7(5):1198–1212
30. Welsh S et al (2004) Antitumor activity and pharmacodynamic properties of PX-478, an inhibitor of hypoxia-inducible factor-1alpha. Mol Cancer Ther 3(3):233–244
31. Tibes R et al (2010) Results from a phase I, dose-esclation study of PX-478, an orally available inhibitor of HIF-1alpha. J Clin Oncol 28(15(Suppl.)):3076
32. Greenberger LM et al (2008) A RNA antagonist of hypoxia-inducible factor-1alpha, EZN-2968, inhibits tumor cell growth. Mol Cancer Ther 7(11):3598–3608
33. Jeong W et al (2014) Pilot trial of EZN-2968, an antisense oligonucleotide inhibitor of hypoxia-inducible factor-1 alpha (HIF-1alpha), in patients with refractory solid tumors. Cancer Chemother Pharmacol 73(2):343–348
34. Zhang H et al (2008) Digoxin and other cardiac glycosides inhibit HIF-1alpha synthesis and block tumor growth. Proc Natl Acad Sci U S A 105(50):19579–19586
35. Zhang H et al (2012) HIF-1-dependent expression of angiopoietin-like 4 and L1CAM mediates vascular metastasis of hypoxic breast cancer cells to the lungs. Oncogene 31(14):1757–1770
36. Jhaveri K, Modi S (2015) Ganetespib: research and clinical development. Onco Targets Ther 8:1849–1858
37. Isaacs JS et al (2002) Hsp90 regulates a von Hippel Lindau-independent hypoxia-inducible factor-1 alpha-degradative pathway. J Biol Chem 277(33):29936–29944
38. Xiang L et al (2014) Ganetespib blocks HIF-1 activity and inhibits tumor growth, vascularization, stem cell maintenance, invasion, and metastasis in orthotopic mouse models of triple-negative breast cancer. J Mol Med (Berl) 92(2):151–164
39. Nagaraju GP et al (2013) Antiangiogenic effects of ganetespib in colorectal cancer mediated through inhibition of HIF-1alpha and STAT-3. Angiogenesis 16(4):903–917
40. Thakur MK et al (2016) A phase II trial of ganetespib, a heat shock protein 90 (Hsp90) inhibitor, in patients with docetaxel-pretreated metastatic castrate-resistant prostate cancer (CRPC)-a prostate cancer clinical trials consortium (PCCTC) study. Investig New Drugs 34(1):112–118
41. Jhaveri K et al (2014) A phase II open-label study of ganetespib, a novel heat shock protein 90 inhibitor for patients with metastatic breast cancer. Clin Breast Cancer 14(3):154–160
42. Mabjeesh NJ et al (2002) Geldanamycin induces degradation of hypoxia-inducible factor 1alpha protein via the proteosome pathway in prostate cancer cells. Cancer Res 62(9):2478–2482
43. Alqawi O, Moghaddas M, Singh G (2006) Effects of geldanamycin on HIF-1alpha mediated angiogenesis and invasion in prostate cancer cells. Prostate Cancer Prostatic Dis 9(2):126–135
44. Koga F, Tsutsumi S, Neckers LM (2007) Low dose geldanamycin inhibits hepatocyte growth factor and hypoxia-stimulated invasion of cancer cells. Cell Cycle 6(11):1393–1402
45. Welsh SJ et al (2002) The redox protein thioredoxin-1 (Trx-1) increases hypoxia-inducible factor 1alpha protein expression: Trx-1 overexpression results in increased vascular endothelial growth factor production and enhanced tumor angiogenesis. Cancer Res 62(17):5089–5095
46. Kim YH et al (2011) Antitumor agent PX-12 inhibits HIF-1alpha protein levels through an Nrf2/PMF-1-mediated increase in spermidine/spermine acetyl transferase. Cancer Chemother Pharmacol 68(2):405–413
47. Welsh SJ et al (2003) The thioredoxin redox inhibitors 1-methylpropyl 2-imidazolyl disulfide and pleurotin inhibit hypoxia-induced factor 1alpha and vascular endothelial growth factor formation. Mol Cancer Ther 2(3):235–243

48. Baker AF et al (2006) The antitumor thioredoxin-1 inhibitor PX-12 (1-methylpropyl 2-imidazolyl disulfide) decreases thioredoxin-1 and VEGF levels in cancer patient plasma. J Lab Clin Med 147(2):83–90
49. Ramanathan RK et al (2011) A randomized phase II study of PX-12, an inhibitor of thioredoxin in patients with advanced cancer of the pancreas following progression after a gemcitabine-containing combination. Cancer Chemother Pharmacol 67(3):503–509
50. Baker AF et al (2013) A phase IB trial of 24-hour intravenous PX-12, a thioredoxin-1 inhibitor, in patients with advanced gastrointestinal cancers. Investig New Drugs 31(3):631–641
51. Befani CD et al (2012) Bortezomib represses HIF-1alpha protein expression and nuclear accumulation by inhibiting both PI3K/Akt/TOR and MAPK pathways in prostate cancer cells. J Mol Med (Berl) 90(1):45–54
52. Kane RC et al (2003) Velcade: U.S. FDA approval for the treatment of multiple myeloma progressing on prior therapy. Oncologist 8(6):508–513
53. Kaluz S, Kaluzova M, Stanbridge EJ (2006) Proteasomal inhibition attenuates transcriptional activity of hypoxia-inducible factor 1 (HIF-1) via specific effect on the HIF-1alpha C-terminal activation domain. Mol Cell Biol 26(15):5895–5907
54. Zhu K et al (2009) Control of HIF-1alpha expression by eIF2 alpha phosphorylation-mediated translational repression. Cancer Res 69(5):1836–1843
55. Abd-Aziz N, Stanbridge EJ, Shafee N (2015) Bortezomib attenuates HIF-1- but not HIF-2-mediated transcriptional activation. Oncol Lett 10(4):2192–2196
56. Veschini L et al (2007) Hypoxia-inducible transcription factor-1 alpha determines sensitivity of endothelial cells to the proteosome inhibitor bortezomib. Blood 109(6):2565–2570
57. Fels DR et al (2008) Preferential cytotoxicity of bortezomib toward hypoxic tumor cells via overactivation of endoplasmic reticulum stress pathways. Cancer Res 68(22):9323–9330
58. Rapisarda A et al (2004) Topoisomerase I-mediated inhibition of hypoxia-inducible factor 1: mechanism and therapeutic implications. Cancer Res 64(4):1475–1482
59. Puppo M et al (2008) Topotecan inhibits vascular endothelial growth factor production and angiogenic activity induced by hypoxia in human neuroblastoma by targeting hypoxia-inducible factor-1alpha and -2alpha. Mol Cancer Ther 7(7):1974–1984
60. Calvani M et al (2006) Hypoxic induction of an HIF-1alpha-dependent bFGF autocrine loop drives angiogenesis in human endothelial cells. Blood 107(7):2705–2712
61. Rapisarda A et al (2004) Schedule-dependent inhibition of hypoxia-inducible factor-1alpha protein accumulation, angiogenesis, and tumor growth by topotecan in U251-HRE glioblastoma xenografts. Cancer Res 64(19):6845–6848
62. Kummar S et al (2011) Multihistology, target-driven pilot trial of oral topotecan as an inhibitor of hypoxia-inducible factor-1alpha in advanced solid tumors. Clin Cancer Res 17(15):5123–5131
63. Hutt DM et al (2014) The histone deacetylase inhibitor, Vorinostat, represses hypoxia inducible factor 1 alpha expression through translational inhibition. PLoS One 9(8):e106224
64. Zhang C et al (2017) Vorinostat suppresses hypoxia signaling by modulating nuclear translocation of hypoxia inducible factor 1 alpha. Oncotarget 8(34):56110–56125
65. Mann BS et al (2007) FDA approval summary: vorinostat for treatment of advanced primary cutaneous T-cell lymphoma. Oncologist 12(10):1247–1252
66. Pili R et al (2017) Combination of the histone deacetylase inhibitor vorinostat with bevacizumab in patients with clear-cell renal cell carcinoma: a multicentre, single-arm phase I/II clinical trial. Br J Cancer 116(7):874–883
67. Lee YH et al (2016) Synthesis and biological evaluation of 1,2-dithiol-3-thiones and pyrrolo[1,2-a]pyrazines as novel hypoxia inducible factor-1 (HIF-1) inhibitor. Bioorg Med Chem 24(12):2843–2851
68. Koivunen P et al (2016) The 2-oxoglutarate analog 3-oxoglutarate decreases normoxic hypoxia-inducible factor-1alpha in cancer cells, induces cell death, and reduces tumor xenograft growth. Hypoxia (Auckl) 4:15–27
69. Wang L et al (2015) Novel chalcone derivatives as hypoxia-inducible factor (HIF)-1 inhibitor: synthesis, anti-invasive and anti-angiogenic properties. Eur J Med Chem 89:88–97
70. Shen C, Kaelin WG Jr (2013) The VHL/HIF axis in clear cell renal carcinoma. Semin Cancer Biol 23(1):18–25
71. Holmquist-Mengelbier L et al (2006) Recruitment of HIF-1alpha and HIF-2alpha to common target genes is differentially regulated in neuroblastoma: HIF-2alpha promotes an aggressive phenotype. Cancer Cell 10(5):413–423
72. Wallace EM et al (2016) A small-molecule antagonist of HIF2alpha is efficacious in preclinical models of renal cell carcinoma. Cancer Res 76(18):5491–5500
73. Xu J et al (2017) Increasing AR by HIF-2alpha inhibitor (PT-2385) overcomes the side-effects of sorafenib by suppressing hepatocellular carcinoma invasion via alteration of pSTAT3, pAKT and pERK signals. Cell Death Dis 8(10):e3095
74. Courtney KD et al (2018) Phase I dose-escalation trial of PT2385, a first-in-class hypoxia-inducible factor-2alpha antagonist in patients with previously treated advanced clear cell renal cell carcinoma. J Clin Oncol 36(9):867–874
75. Wong TW et al (2018) PT2977, a novel HIF-2a antagonist, has potent antitumor activity and remodels the immunosuppressive tumor microenvironment in clear cell renal cell cancer. Mol Targets Cancer Ther 17(1(Suppl.))

76. Papadopoulos KP et al (2018) A first-in-human phase 1 dose-escalation trial of the oral HIF-2a inhibitor PT2977 in patients with advanced solid tumors. J Clin Oncol 36(15(Suppl.)):2508
77. Sun DR et al (2018) Exploring the inhibition mechanism on HIF-2 by inhibitor PT2399 and 0X3 using molecular dynamics simulations. J Mol Recognit 31:e2730
78. Cho H et al (2016) On-target efficacy of a HIF-2alpha antagonist in preclinical kidney cancer models. Nature 539(7627):107–111
79. Chen W et al (2016) Targeting renal cell carcinoma with a HIF-2 antagonist. Nature 539(7627):112–117
80. Guise CP et al (2014) Bioreductive prodrugs as cancer therapeutics: targeting tumor hypoxia. Chin J Cancer 33(2):80–86
81. Wilson WR, Hay MP (2011) Targeting hypoxia in cancer therapy. Nat Rev Cancer 11(6):393–410
82. Brown JM, Wang LH (1998) Tirapazamine: laboratory data relevant to clinical activity. Anticancer Drug Des 13(6):529–539
83. Reddy SB, Williamson SK (2009) Tirapazamine: a novel agent targeting hypoxic tumor cells. Expert Opin Investig Drugs 18(1):77–87
84. Mistry IN et al (2017) Clinical advances of hypoxia-activated prodrugs in combination with radiation therapy. Int J Radiat Oncol Biol Phys 98(5):1183–1196
85. DiSilvestro PA et al (2014) Phase III randomized trial of weekly cisplatin and irradiation versus cisplatin and tirapazamine and irradiation in stages IB2, IIA, IIB, IIIB, and IVA cervical carcinoma limited to the pelvis: a Gynecologic Oncology Group Study. J Clin Oncol 32(5):458–464
86. Cai TY et al (2014) Tirapazamine sensitizes hepatocellular carcinoma cells to topoisomerase I inhibitors via cooperative modulation of hypoxia-inducible factor-1alpha. Mol Cancer Ther 13(3):630–642
87. Hicks KO et al (2010) Pharmacokinetic/pharmacodynamic modeling identifies SN30000 and SN29751 as tirapazamine analogues with improved tissue penetration and hypoxic cell killing in tumors. Clin Cancer Res 16(20):4946–4957
88. Wang J et al (2014) Identification of one-electron reductases that activate both the hypoxia prodrug SN30000 and diagnostic probe EF5. Biochem Pharmacol 91(4):436–446
89. Gu Y et al (2017) Reductive metabolism influences the toxicity and pharmacokinetics of the hypoxia-targeted benzotriazine di-oxide anticancer agent SN30000 in mice. Front Pharmacol 8:531
90. Saggar JK, Tannock IF (2014) Activity of the hypoxia-activated pro-drug TH-302 in hypoxic and perivascular regions of solid tumors and its potential to enhance therapeutic effects of chemotherapy. Int J Cancer 134(11):2726–2734
91. Liapis V et al (2016) Anticancer efficacy of the hypoxia-activated prodrug evofosfamide (TH-302) in osteolytic breast cancer murine models. Cancer Med 5(3):534–545
92. Sun JD et al (2016) Comparison of hypoxia-activated prodrug evofosfamide (TH-302) and ifosfamide in preclinical non-small cell lung cancer models. Cancer Biol Ther 17(4):371–380
93. Zhang L et al (2016) Combined antitumor therapy with metronomic topotecan and hypoxia-activated prodrug, evofosfamide, in neuroblastoma and rhabdomyosarcoma preclinical models. Clin Cancer Res 22(11):2697–2708
94. Ganjoo KN et al (2011) A phase I study of the safety and pharmacokinetics of the hypoxia-activated prodrug TH-302 in combination with doxorubicin in patients with advanced soft tissue sarcoma. Oncology 80(1–2):50–56
95. Chawla SP et al (2014) Phase II study of the safety and antitumor activity of the hypoxia-activated prodrug TH-302 in combination with doxorubicin in patients with advanced soft tissue sarcoma. J Clin Oncol 32(29):3299–3306
96. Tap WD et al (2017) Doxorubicin plus evofosfamide versus doxorubicin alone in locally advanced, unresectable or metastatic soft-tissue sarcoma (TH CR-406/SARC021): an international, multicentre, open-label, randomised phase 3 trial. Lancet Oncol 18(8):1089–1103
97. Cutsem EV et al (2016) MAESTRO: A randomized, double-blind phase III study of evofosfamide (Evo) in combination with gemcitabine (Gem) in previously untreated patients (pts) with metastatic or locally advanced unresectable pancreatic ductal adenocarcinoma (PDAC). J Clin Oncol 34(15 (Suppl.)):4007
98. Silva S et al (2015) Preclinical efficacy of tarloxotinib bromide (TH-4000), a hypoxia-activated EGFR/HER2 inhibitor: rationale for clinical evaluation in EGFR mutant, T790M-negative NSCLC following progression on EGFR-TKI therapy. Mol Cancer Ther 14(12 (Suppl.))
99. Patterson AV et al (2015) TH-4000, a hypoxia-activated EGFR/Her2 inhibitor to treat EGFR-TKI resistant T790M-negative NSCLC. J Clin Oncol 33(15 (Suppl.))
100. Hendriks HR et al (1993) EO9: a novel bioreductive alkylating indoloquinone with preferential solid tumour activity and lack of bone marrow toxicity in preclinical models. Eur J Cancer 29A(6):897–906
101. Adams GE et al (1992) Bioreductive drugs as post-irradiation sensitizers: comparison of dual function agents with SR 4233 and the mitomycin C analogue EO9. Int J Radiat Oncol Biol Phys 22(4):717–720
102. Phillips RM, Hendriks HR, Peters GJ (2013) EO9 (Apaziquone): from the clinic to the laboratory and back again. Br J Pharmacol 168(1):11–18
103. van der Heijden AG et al (2006) Phase II marker lesion study with intravesical instillation of apaziquone for superficial bladder cancer: toxicity and marker response. J Urol 176(4 Pt 1):1349–1353. discussion 1353
104. Hendricksen K et al (2009) Two-year follow-up of the phase II marker lesion study of intravesical

apaziquone for patients with non-muscle invasive bladder cancer. World J Urol 27(3):337–342
105. Phillips RM (2016) Targeting the hypoxic fraction of tumours using hypoxia-activated prodrugs. Cancer Chemother Pharmacol 77(3):441–457
106. Patterson AV et al (2007) Mechanism of action and preclinical antitumor activity of the novel hypoxia-activated DNA cross-linking agent PR-104. Clin Cancer Res 13(13):3922–3932
107. Benito J et al (2011) Pronounced hypoxia in models of murine and human leukemia: high efficacy of hypoxia-activated prodrug PR-104. PLoS One 6(8):e23108
108. Jameson MB et al (2010) A phase I trial of PR-104, a nitrogen mustard prodrug activated by both hypoxia and aldo-keto reductase 1C3, in patients with solid tumors. Cancer Chemother Pharmacol 65(4):791–801
109. McKeage MJ et al (2011) A phase I trial of PR-104, a pre-prodrug of the bioreductive prodrug PR-104A, given weekly to solid tumour patients. BMC Cancer 11:432
110. McKeage MJ et al (2012) PR-104 a bioreductive pre-prodrug combined with gemcitabine or docetaxel in a phase Ib study of patients with advanced solid tumours. BMC Cancer 12:496
111. Patterson LH (2002) Bioreductively activated antitumor N-oxides: the case of AQ4N, a unique approach to hypoxia-activated cancer chemotherapy. Drug Metab Rev 34(3):581–592
112. Lalani AS et al (2007) Selective tumor targeting by the hypoxia-activated prodrug AQ4N blocks tumor growth and metastasis in preclinical models of pancreatic cancer. Clin Cancer Res 13(7):2216–2225
113. Nishida CR, Lee M, de Montellano PR (2010) Efficient hypoxic activation of the anticancer agent AQ4N by CYP2S1 and CYP2W1. Mol Pharmacol 78(3):497–502
114. Hoang NT et al (2016) Hypoxia-inducible factor-targeting prodrug TOP3 combined with gemcitabine or TS-1 improves pancreatic cancer survival in an orthotopic model. Cancer Sci 107(8):1151–1158
115. Zhang R et al (2017) Catalase-loaded cisplatin-prodrug-constructed liposomes to overcome tumor hypoxia for enhanced chemo-radiotherapy of cancer. Biomaterials 138:13–21
116. Nesbitt H et al (2017) Targeting hypoxic prostate tumors using the novel hypoxia-activated prodrug OCT1002 inhibits expression of genes associated with malignant progression. Clin Cancer Res 23(7):1797–1808
117. Nesbitt H et al (2017) The unidirectional hypoxia-activated prodrug OCT1002 inhibits growth and vascular development in castrate-resistant prostate tumors. Prostate 77(15):1539–1547
118. Hunter FW, Wouters BG, Wilson WR (2016) Hypoxia-activated prodrugs: paths forward in the era of personalised medicine. Br J Cancer 114(10):1071–1077
119. Zhou J et al (2004) PI3K/Akt is required for heat shock proteins to protect hypoxia-inducible factor 1alpha from pVHL-independent degradation. J Biol Chem 279(14):13506–13513
120. Zhong H et al (2000) Modulation of hypoxia-inducible factor 1alpha expression by the epidermal growth factor/phosphatidylinositol 3-kinase/PTEN/AKT/FRAP pathway in human prostate cancer cells: implications for tumor angiogenesis and therapeutics. Cancer Res 60(6):1541–1545

Index

D. M. Gilkes (ed.), *Hypoxia and Cancer Metastasis*, Advances in Experimental Medicine and Biology 1136,
https://doi.org/10.1007/978-3-030-12734-3

MIX
Papier aus verantwortungsvollen Quellen
Paper from responsible sources
FSC® C105338

If you have any concerns about our products,
you can contact us on
ProductSafety@springernature.com

In case Publisher is established outside the EU,
the EU authorized representative is:
Springer Nature Customer Service Center GmbH
Europaplatz 3, 69115 Heidelberg, Germany

Printed by Libri Plureos GmbH
in Hamburg, Germany